International Workshop on Discovery Physics at the LHC (Kruger2018)

Journal of Physics: Conference Series
Volume 1271

Kruger, South Africa
3 – 7 December 2018

ISBN: 978-1-5108-9188-3
ISSN: 1742-6588

Printed from e-media with permission by:

Curran Associates, Inc.
57 Morehouse Lane
Red Hook, NY 12571

Some format issues inherent in the e-media version may also appear in this print version.

Institute of Physics
Dirac House, Temple Back
Bristol BS1 6BE UK

Phone: 44 1 17 929 7481
Fax: 44 1 17 920 0979

techtracking@iop.org

Additional copies of this publication are available from:

Curran Associates, Inc.
57 Morehouse Lane
Red Hook, NY 12571 USA
Phone: 845-758-0400
Fax: 845-758-2633
Email: curran@proceedings.com
Web: www.proceedings.com

Table of contents

Volume 1271

International Workshop on Discovery Physics at the LHC (Kruger2018)

3–7 December 2018, Kruger, South Africa

Accepted papers received: 5 June 2019
Published online: 26 July 2019

Preface

Papers

ALICE

Belle II

Theory

IOP Conf. Series: Journal of Physics: Conf. Series **1271** (2019) 011001 doi:10.1088/1742-6596/1271/1/011001

Foreword Proceedings Kruger2018

The fifth biennial conference on "Discovery Physics at the LHC " was held December 3-7 2018 at the Casa do Sol Hotel, Hazyview in South Africa. In total there were about 100 participants coming from Canada, France, Germany, Italy, Korea, Norway, Poland, Russia, Switzerland, UK and USA. The latest results from the Large Hadron Collider, the Relativistic Heavy Ion Collider as well the latest theoretical insights were presented. With the majestic Kruger national Park in the background this led to a very stimulating conference with many exchanges taking place. The proceedings reflect the high level of the conference.

The financial contributions from iThemba LABS, the Department of Science and Technology, the SA-CERN programme, the UCT-CERN Research Centre, the Universities of Cape Town and of the Witwatersrand are gratefully acknowledged.

Local Organizing Committee:

Faiçal Azaiez
Zinhle Buthelezi
Jean Cleymans
Simon H. Connell
Alan S. Cornell (Chair)
Thomas Dietel
Siegfied Förtsch
Naomi Haasbroek (Conference Manager)
William A. Horowitz
Deepak Kar
Betty Kibirige
Bruce Mellado
Sahal Yacoob

IOP Conf. Series: Journal of Physics: Conf. Series **1271** (2019) 011002 doi:10.1088/1742-6596/1271/1/011002

Peer review statement

All papers published in this volume of *Journal of Physics: Conference Series* have been peer reviewed through processes administered by the proceedings Editors. Reviews were conducted by expert referees to the professional and scientific standards expected of a proceedings journal published by IOP Publishing.

IOP Conf. Series: Journal of Physics: Conf. Series **1271** (2019) 012001 doi:10.1088/1742-6596/1271/1/012001

Highlights of the production of (anti-)(hyper-)nuclei and exotica with ALICE at the LHC

Benjamin Dönigus (for the ALICE Collaboration)

Institut für Kernphysik, Goethe-Universität Frankfurt, Max-von-Laue-Str. 1, 60438 Frankfurt, Germany

E-mail: benjamin.doenigus@cern.ch

Abstract. Measurements of the production of light (anti-)nuclei and (anti-)hypernuclei have been performed in different colliding systems at the LHC, namely pp, p–Pb and Pb–Pb. The results of the production in Pb–Pb collisions of light (anti-)(hyper-)nuclei follows the expectation of the thermal model, whereas the production of nuclei in pp and p–Pb collisions shows a closer agreement with the expectations from coalescence models. Both models can give predictions for the production yields of hypothetical states such as bound states of two Λ hyperons or of a Λ and a neutron, which are expected to decay weakly. These states are searched for with the ALICE apparatus. Since no signal is found, upper limits, which are significantly below the model expectations, are set. Furthermore, the most recent measurement of the lifetime of the hypertriton determined in Pb–Pb collisions at $\sqrt{s_{\rm NN}} = 5.02$ TeV at the LHC gives a value exactly between the world average and the free Λ lifetime. In addition, the new result is in agreement with other recent measurements in heavy-ion collisions.

1. Introduction

The ALICE (A Large Ion Collider Experiment) Collaboration has, in addition to the usual heavy-ion physics research, developed a program to investigate many different kinds of (multi-)baryon states predicted by QCD and QCD inspired models [1]. One way of addressing this research topic is to start by measuring rarely observed (anti-)nuclei and (anti-)hypernuclei, which are measured copiously in Pb–Pb collisions at the LHC due to the large data sets. With these data the model predictions, e.g. the thermal (statistical) model and coalescence approaches, can be tested. Ultimately, the aim is to understand the production mechanisms at play in these collisions, resulting in large production yields for (anti-)nuclei and (anti-)hypernuclei. Furthermore, one can then use these models to predict the production yields of exotic bound states, still to be discovered or excluded, by setting strong upper limits, compared to these models.

The main assumption of the thermal (statistical) model used in the following is a grand-canonical ensemble, which is characterized by three parameters, temperature, (baryon) chemical potential and volume. In this approach all production yields of light flavored hadrons (containing only u, d and s quarks) are described by these three parameters through the grand-canonical partition function [2,3]. The thermal model at the LHC has basically only one key parameter: the chemical freeze-out temperature T_{ch}. This statement is valid because the baryo-chemical potential μ_B is close to zero, and therefore it is often treated as being zero in thermal model fits. The nuclei themselves are subject to a strong sensitivity on T_{ch}, because of their large mass and the exponential behaviour of the expected particle yields $\mathrm{d}N/\mathrm{d}y$ on the particle mass m

($\mathrm{d}N/\mathrm{d}y \propto \exp(-m/T_{ch})$). The interesting fact to mention here is that the binding energies of these objects are very small compared to the chemical freeze-out temperature T_{ch} [2,3,4].

In the coalescence approach, the constituents of a (hyper-)nucleus, produced from the fireball, are expected to come close together in phase space and finally form the (hyper-)nucleus. In this model the produced nuclei could break apart and be created again by final-state coalescence.

More details and the current status of these two different production models, connected to the here presented data, are given in [5].

The measured cross-section in heavy-ion collisions is often split into so called centrality intervals or classes. These are extracted from Monte-Carlo simulations to model the fraction of the cross-section as a percentage of the total cross-section based on detectors in the forward direction. In nucleus-nucleus collisions, these intervals are interpreted as a proxy for the degree of centrality of the collision. This interpretation is no longer true for pp or p–Pb collisions, where the cross-section is divided in multiplicity classes.

2. (Anti-)nuclei

The measurement of (anti-)nuclei is performed by the ALICE experiment by using the Time Projection Chamber (TPC) and the Time-Of-Flight detector (TOF). The TOF is used in addition whenever the TPC is not sufficient to separate the different nuclei from other hadrons using the specific energy-loss $\mathrm{d}E/\mathrm{d}x$. One way to combining the information of both detectors is to measure the mass of the investigated particle. This is accomplished by using the velocity $\beta = \mathrm{v}/c$ given by TOF and the measured momentum (mainly determined by the curvature of the track in the TPC), for a pre-selection of particles which are based on the TPC $\mathrm{d}E/\mathrm{d}x$ measurement. A result of the measured $(m/z)^2$ for $^3\overline{\mathrm{He}}$ and $^4\overline{\mathrm{He}}$ in Pb–Pb collisions at $\sqrt{s_{\mathrm{NN}}} = 5.02$ TeV is shown in Fig. 1. This figure shows that the data sample of 2015 contains 16 clearly identifed anti-alphas, by using the TPC and TOF information.

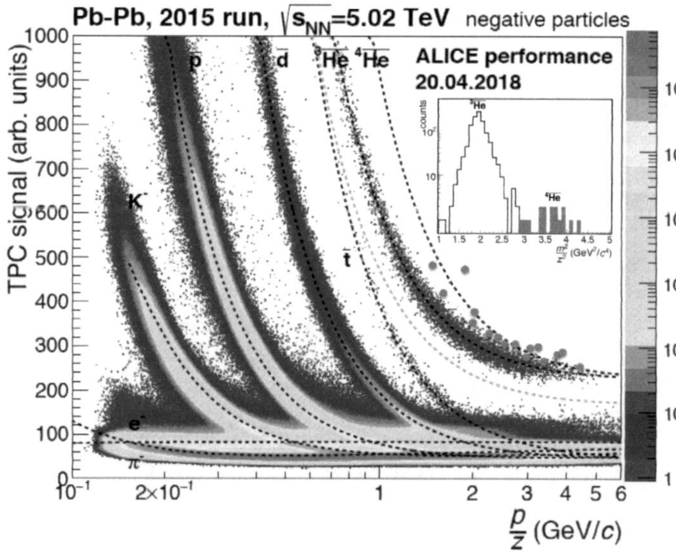

Figure 1. $\mathrm{d}E/\mathrm{d}x$ for negative tracks, measured in the ALICE TPC, of events containing at least one particle with $|Z| > 1$ (selected by requiring only events where one particle is found above the grey dashed line between $\bar{\mathrm{t}}$ and $^3\overline{\mathrm{He}}$). All light nuclei are clearly visible and shown together with the parameterization of the Bethe-Bloch expectation. The inset also shows the mass obtained by combining the measured time-of-flight and rigidity p/z, for a 3σ selection of alphas in the TPC.

Furthermore, spectra are extracted by dividing the signal in p_T slices and correcting them for acceptance and efficiency. The result for deuterons in pp collisions at $\sqrt{s} = 7$ TeV and for five different centrality bins in Pb–Pb at $\sqrt{s_{NN}} = 2.76$ TeV is displayed in Fig. 2, together with the ratios of the anti-nuclei to nuclei in the same centrality classes. In addition, this ratio is also shown for ^{3}He. All ratios are compatible with unity as a function of transverse momentum as expected from thermal and coalescence models. Thus, nuclei and anti-nuclei are produced with equal probability at the LHC [4].

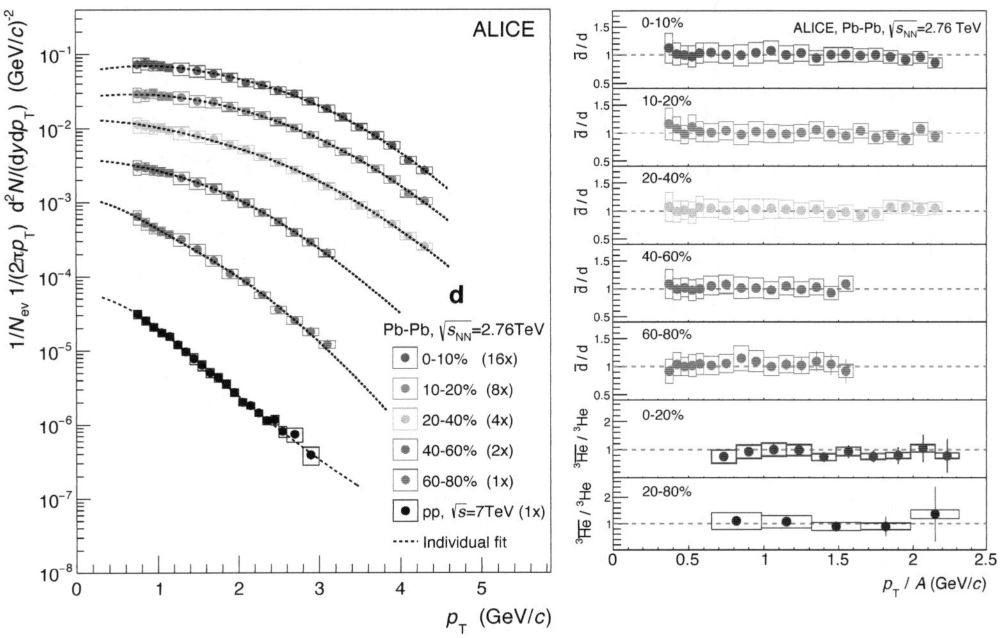

Figure 2. Invariant transverse momentum spectra for deuterons in pp at $\sqrt{s} = 7$ TeV and 10 centrality classes in Pb–Pb at $\sqrt{s_{NN}} = 2.76$ TeV (left panel) and ratio between anti-nuclei and nuclei ($\bar{d}$/d and $^3\overline{He}/^3$He) for the different centrality classes in the right panel. Figures are taken from [8].

The spectra in Pb–Pb show a clear hardening (higher mean p_T) with increasing centrality and they are fitted with blast-wave distributions [6], mainly to extrapolate to the unmeasured region of the spectra in order to extract the integrated production yield. The blast-wave function is used to take into account the radial push, the so called radial flow, all particles are exposed to in a heavy-ion collision during the evolution of the fireball.

The development of the radial flow from minimum bias pp events towards the Pb–Pb centrality intervals is even better visible in the data at higher energies, namely the pp collisions at $\sqrt{s} = 13$ TeV and Pb–Pb at $\sqrt{s_{NN}} = 5.02$ TeV, representing the current top energies in these colliding systems at the LHC. The aforementioned effect is displayed in Fig. 3 (left), together with the pp data split in multiplicity intervals, where a similar hardening can be seen.

The ratios of the production yields of deuterons to protons in pp, p–Pb and Pb–Pb collisions as a function of the mean multiplicity $\langle dN_{ch}/d\eta \rangle$ are depicted in Fig. 4. These show an increase going from pp multiplicities over p-Pb multiplicities until a saturation seems to be reached in peripheral Pb–Pb collisions. The increase with multiplicity is predicted by coalescence models whereas the basically constant behaviour at higher multiplicities is expected from the thermal

IOP Conf. Series: Journal of Physics: Conf. Series **1271** (2019) 012001 doi:10.1088/1742-6596/1271/1/012001

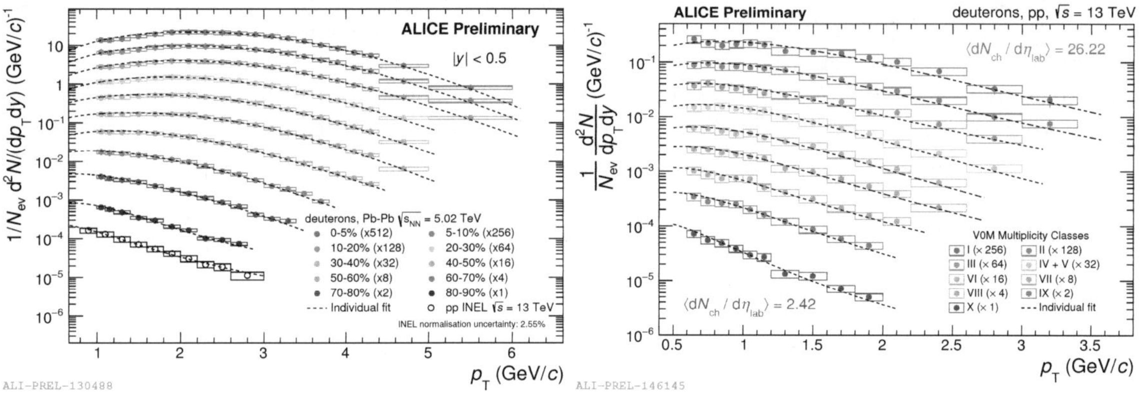

Figure 3. Transverse momentum spectra for deuterons in pp at $\sqrt{s} = 13$ TeV and five centrality classes in Pb–Pb at $\sqrt{s_{\rm NN}} = 5.02$ TeV (left panel) and transverse momentum spectra for deuterons in pp at $\sqrt{s} = 13$ TeV split in 9 different multiplicity classes (right panel). The dashed lines indicate individual fits to the data, for pp using a Levy-Tsallis function [9,10,11] and in Pb–Pb using a blast-wave function.

model.

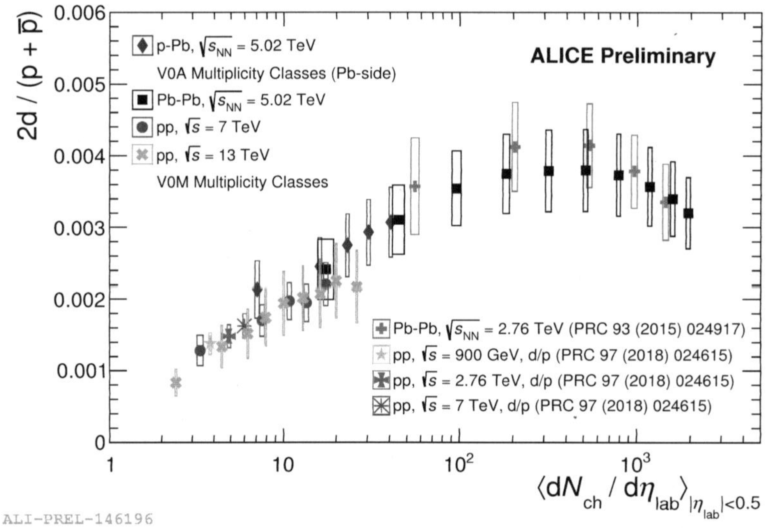

Figure 4. Ratios of the production yields of deuterons to protons $(2\text{d}/(\text{p}+\bar{\text{p}}))$ as a function of the mean multiplicity $\langle \text{d}N_{\rm ch}/\text{d}\eta \rangle$ measured in pp, p–Pb and Pb–Pb collisions at different energies.

Figure 5 shows the production yield of nuclei in Pb–Pb and p–Pb collisions as a function of their mass together with exponential fits. As mentioned above an exponential decrease with mass is expected from the thermal model. The penalty factor for adding a baryon to the system is about 300 in Pb–Pb and around 600 in p-Pb, which would mean that the suppression in the yields of d to ^{3}He is 300 in Pb–Pb and 600 in p–Pb.

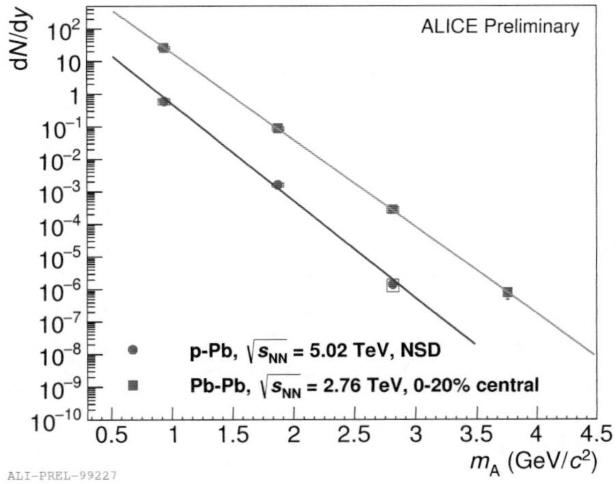

Figure 5. Production yields of different nuclei as a function of their mass. The lines indicate exponential fits through the data points at the two energies.

Figure 6 displays the production yields of different light flavour particle species compared to three different thermal model fits from three different model implementations. These model fits agree very well with each other. From these fits a temperature T_{ch} of about 153 MeV can be extracted. All nuclei and hypernuclei production yields are rather well described with this model. The biggest tension in this fit is visible for the K*, which is a strongly-decaying resonance with a lifetime shorter than the lifetime of the fireball. This result will be discussed in more detail in Sec. 5.

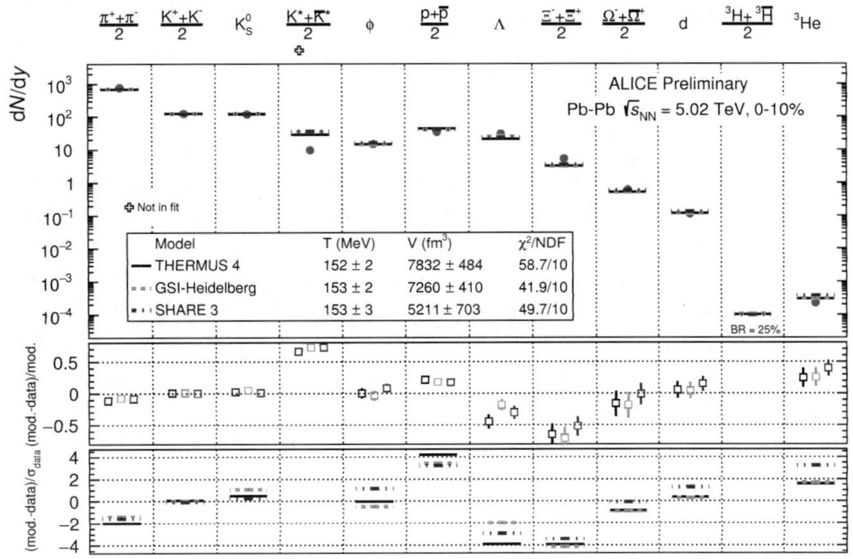

Figure 6. Production yields of different light flavour particle species compared to three different thermal model implementations.

3. (Anti-)hypernuclei

The hypertriton is the lightest hypernucleus, consisting of a proton, a neutron and a Λ hyperon. It decays weakly, for instance by its two-body decay mode into a ^{3}He and a π^-, with a lifetime close to the one of the free Λ hyperon. The separation energy of the Λ is only 130 keV, which implies, from simple quantum mechanical calculations, that the hypertriton has a size of about 10 fm [5]. This already indicates that the lifetime should be very close to the one of the free Λ. A similar prediction is given by a more sophisticated model involving also three-body forces [7]. In addition, this model gives a branching ratio of the hypertriton decay into the ^{3}He+π^- of about 25%. The branching ratios (B.R.) of the hypertriton decay modes are not well known because many channels involve neutral particles such as neutrons and π^0 that are not easily accessible by production experiments. The most recent calculation includes also the modelling of the π final-state interaction, which leads to a value very close to the current world average (about 81% of the free Λ hyperon lifetime) [12]. First results on the hypertriton by the ALICE Collaboration were published in [13].

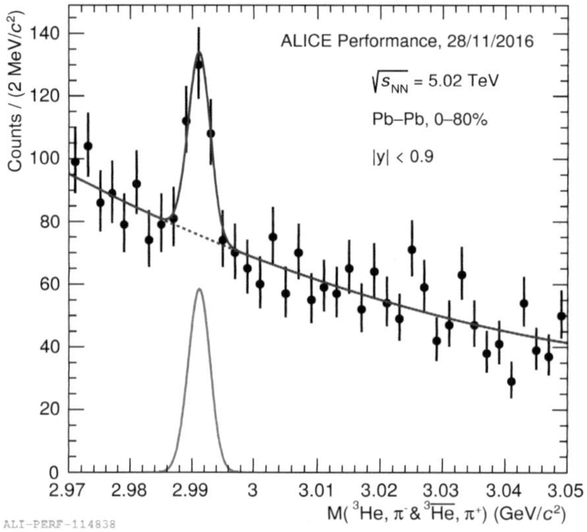

Figure 7. Invariant mass of the hypertriton, reconstructed by using a secondary decay topology of a ^{3}He and π^- together with their charge conjugate from the 2015 data sample.

The reconstructed invariant mass of the hypertriton in 0-80% central Pb–Pb collisions at $\sqrt{s_{\rm NN}}$ = 5.02 TeV is shown in Fig. 7. The extracted signal is fitted by a Gaussian and the extracted mass is $m = (2.991 \pm 0.001)$ GeV/c^2 and the width $\sigma = (1.7 \pm 0.3)$ MeV/c^2. From such fits to invariant mass distributions, differential analyses in centrality and $p_{\rm T}$ are performed.

The measured transverse momentum spectrum of the hypertriton (red points) and anti-hypertriton (blue points) in 10-40% central Pb–Pb events at $\sqrt{s_{\rm NN}}$ = 5.02 TeV are shown in Fig. 8, together with blast-wave functions based on the measurement of light nuclei. The fact that these spectra can be parameterised by blast-wave functions indicates that the hypertriton experiences the same flow and the same evolution as all other particles produced in the hot fireball.

The measured dN/dy×B.R. of the hypertriton in 0-10% central Pb–Pb collisions at $\sqrt{s_{\rm NN}}$ = 5.02 TeV as a function of the branching ratio is depicted in Fig. 9, together with model calculations describing its production. The thermal model, which describes well the production of light hadrons and nuclei, is in agreement with the measurement around the 25% branching ratio expected from the previously mentioned calculation.

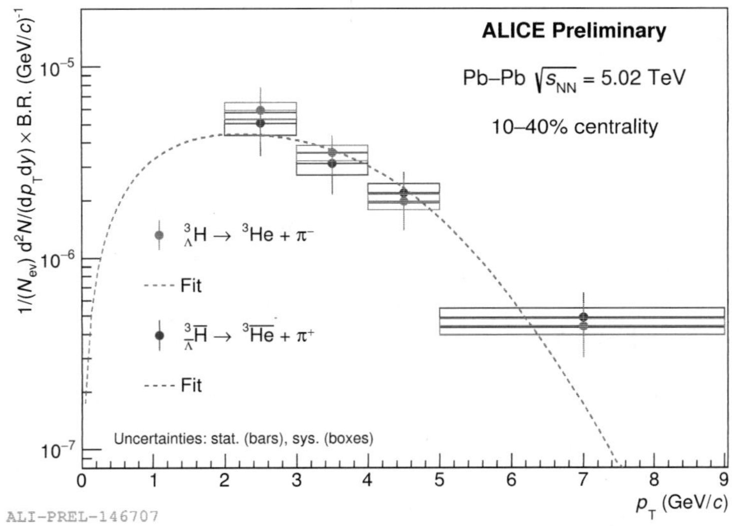

Figure 8. Transverse momentum spectrum of the hypertriton in 10-40% central Pb–Pb events at $\sqrt{s_{NN}} = 5.02$ TeV. The dashed lines are blast-wave functions based on the measurement of light nuclei.

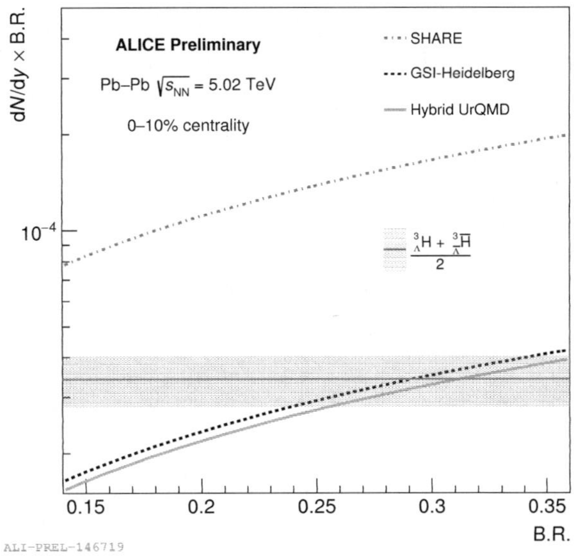

Figure 9. p_T-integrated rapidity density times branching ratio as a function of branching ratio (dN/dy×B.R. vs B.R.). The red horizontal line is the measured dN/dy×B.R. and the band around it represents the quadratic sum of statistical and systematic uncertainties. The other lines indicate different theoretical expectations.

The extracted lifetime is displayed in Fig. 10 and compared with previous measurements. An average lifetime calculated following the PDG description, which deviates from the free Λ

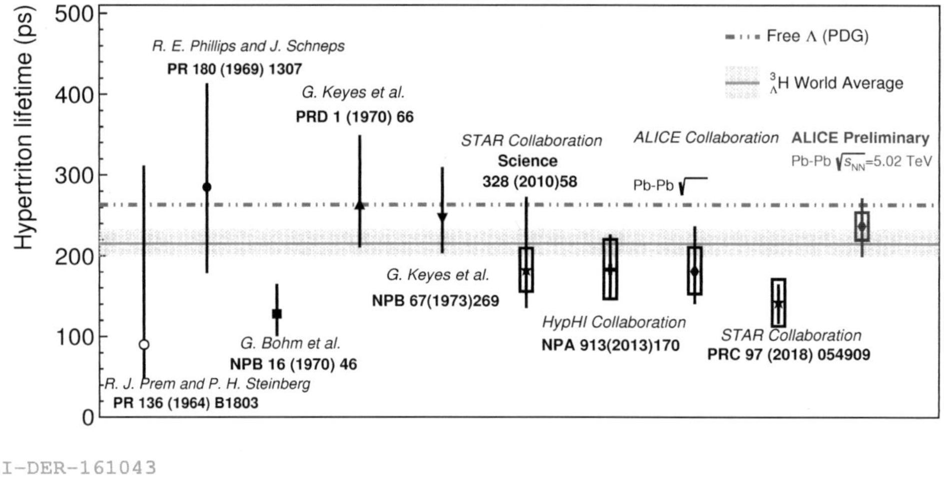

Figure 10. Measured hypertriton lifetime (blue symbol) compared to previously published results (in black). The yellow band represents the world average of the hypertriton lifetime measurements. The dashed line indicates the lifetime of the Λ hyperon as reported by the Particle Data Group.

hyperon lifetime by more than 2σ, is also shown. The measurements before 2010 were done with visualization techniques (emulsions and bubble chambers) whereas the latest ones are all from heavy-ion collisions. The fact that the lifetimes extracted from heavy-ion collisions are lower than the lifetime of the free Λ hyperon leads to the suggestion, in the hyper-nuclei community, that the measurements in heavy-ion collisions might be biased because of the the rougher collision environment, which could influence the observed lifetime of the hypertriton. In particular, the most recent value from the STAR Collaboration [14] is more than 3σ below the free Λ lifetime. Therefore measurements of the lifetime in elementary collisions are desirable.

An unbinned fit was used as cross-check of the result from the exponential fit in ct bins. This unbinned fit is shown in Fig. 11. With this method sideband regions are used for fitting the background and the signal region for extracting the lifetime of the hypertriton.

4. Exotica searches

Since the production yields of composite objects (such as light (anti-)(hyper-)nuclei) are rather well described by both production models they can be used to predict the production of exotic objects such as bound states among hyperons or hyperons and nucleons. Of particular interest is the so-called H-dibaryon, a hexaquark state composed of $uuddss$ ($\Lambda\Lambda$) predicted by R. Jaffe in a bag model calculation in 1977 [15]. It is investigated in the weak decay mode $\Lambda + \mathrm{p} + \pi^-$. Another investigated bound state is a possible bound state of Λn that would decay into d $+\pi^-$. Since no signal was observed, upper limits have been extracted for a wide phase space of lifetime and branching ratio.

The upper limits of the investigated bound states are depicted in Fig. 12 as a function of the branching ratio assuming the same lifetime of the free Λ hyperon. All production models are far away from the extracted upper limits in reasonable regions of the branching ratio. Only if the branching ratios were very small, the model predictions would reach down to the upper-limit value.

The dependency of the upper limit on the lifetime is displayed in Fig. 13 and shows that only

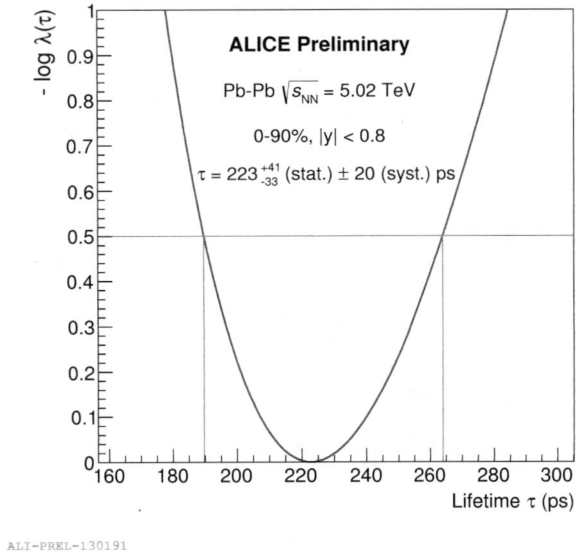

Figure 11. Negative logarithmic likelihood of the measured hypertriton lifetime from the unbinned fit method.

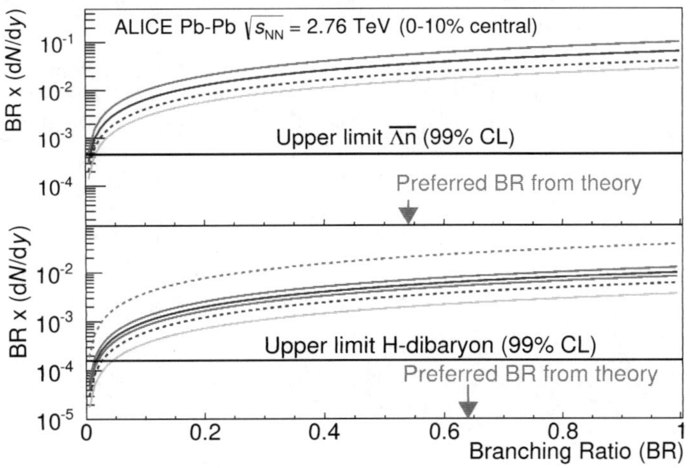

Figure 12. Extracted upper limit dN/dy as a function of the branching ratio compared with different models. For details see [16].

at very long lifetimes (about 3 m) the upper limit of the H-dibaryon comes close to the thermal model expectation. For the Λn bound state the upper limits are for all configurations under study below the model expectation. More details are given in [16].

5. Resonances

In Fig. 6 one can see a deviation for the $K^*(892)$ when the measured production yield is compared to the thermal model expectation. On the other hand, the measured $\phi(1020)$ production yield is well compatible with the thermal model. This can be understood from the different lifetimes

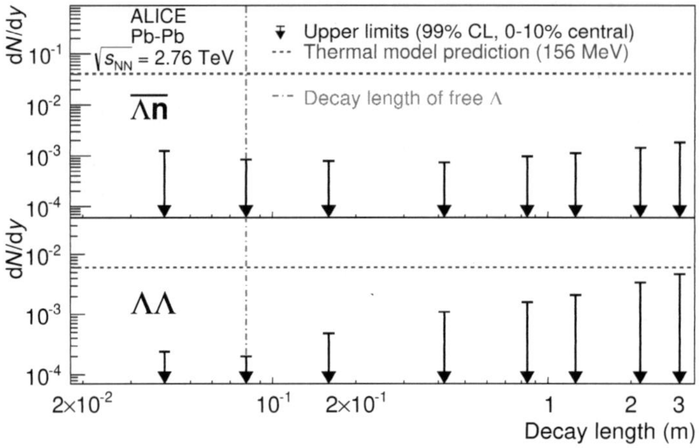

Figure 13. Upper limit $\mathrm{d}N/\mathrm{d}y$ as a function of the decay length compared with the the thermal model. For details see [16].

of the two resonances, namely 46.1 fm/c for the $\phi(1020)$ and 4.2 fm/c for the $\mathrm{K}^*(892)$. These are respectively longer and shorter than the lifetime of the fireball which is about 10 fm/c for central heavy-ion collisions at the LHC. Consequently some of the $\mathrm{K}^*(892)$ are decaying already inside the expanding fireball and interact with it, so that these $\mathrm{K}^*(892)$ can not be reconstructed anymore. This can be seen in Fig. 14 from the ratio of $\mathrm{K}^*(892)$ to K^-, which decreases as the number of produced charged particles (here as $\langle \mathrm{d}N_{\mathrm{ch}}/\mathrm{d}\eta\rangle^{1/3}$) increases (going from pp multiplicity towards central Pb–Pb events). At the same time the ratio of $\phi(1020)$ to K^- stays rather constant. One can see that at low multiplicities the measured values agree with the thermal model.

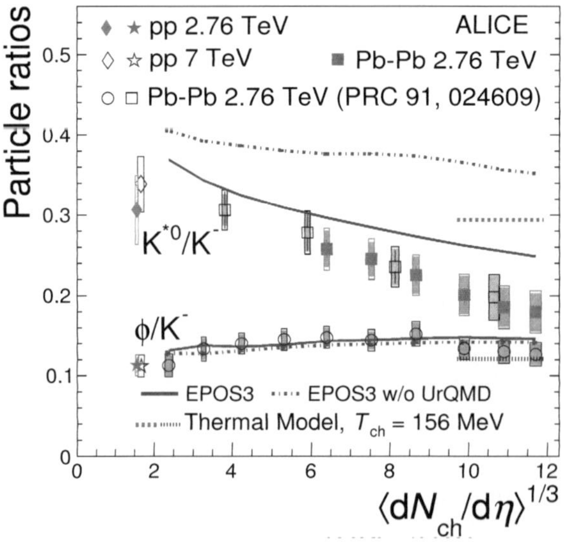

Figure 14. Ratio of $\mathrm{K}^*(892)$ to K^- and $\phi(1020)$ to K^- as function of $\langle \mathrm{d}N_{\mathrm{ch}}/\mathrm{d}\eta\rangle^{1/3}$ compared to different models. For details see [17].

6. Summary and outlook

The results from the production of (anti-)nuclei (in pp, p–Pb and Pb–Pb collisions) and (anti-)hypernuclei (in Pb–Pb collisions) are described rather well by the thermal and the coalescence models, whereas some aspects in pp and p–Pb are in closer agreement to coalescence than thermal models. Both models can make predictions for the production of exotic objects, such as the H-dibaryon and a Λn bound state. Since in the search of these exotica no signals were observed, upper limits were set which are significantly below the expectations of both models. A comparison of the thermal model involving all discussed states is displayed in Fig. 15. It shows that the thermal model describes the yields of the known states very well. Even the hypertriton, which has a Λ separation energy of only 130 keV, is in good agreement. Instead, the upper limits of the exotic states are more than a factor of 25 below the expectation. This makes the existence of these particles in the investigated phase space rather unlikely.

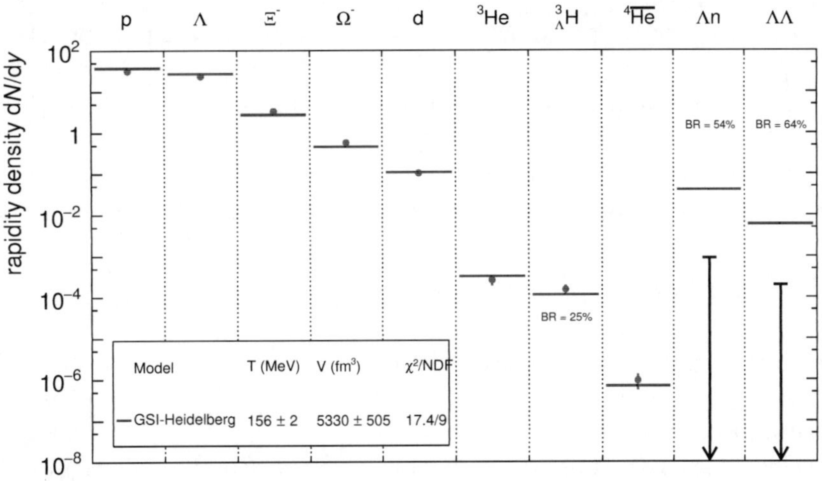

Figure 15. Comparison of production yields dN/dy of baryons, light (hyper-)nuclei and exotica with thermal model predictions. Hypertriton was corrected by 25% for the branching ratio, Λn by 54% and $\Lambda\Lambda$ by 64% as shown in [18].

Further, the most recent measurement of the lifetime of the hypertriton, determined in Pb–Pb collisions at $\sqrt{s_{\rm NN}} = 5.02$ TeV at the LHC, gives a value around the expectations of the most recent calculation [12] and it is consistent with the free Λ hyperon lifetime and the world average. Thus it is in agreement with other recent measurements in heavy-ion collisions and from emulsions. The repeat of this measurement in more elementary collisions is one of the aims of run 2, which started in 2015 and will deliver significantly larger data sets at higher collision energies as before ($\sqrt{s} = 13$ TeV in pp).

After 2018 in the so-called long shutdown 2 (LS2) two main detectors of the ALICE setup will be upgraded. The current Inner Tracker, consisting of six layers of silicon detectors based on three different technologies, will be replaced by a new tracker made of seven layers of monolithic silicon pixel sensors that will significantly increase the (secondary) vertex resolution [19]. The multi-wire proportional chambers of the TPC will be replaced by four layers of GEM detectors that will allow for continuous readout at rates of 50 kHz [20]. These upgrades will allow the ALICE detector to cope with the higher luminosity expected in run 3 (starting in 2021) and run 4 (starting in 2026). In runs 3 and 4, a data sample of 10^{10} central Pb–Pb events corresponding to 10 nb^{-1} integrated luminosity is expected to be collected. This sample will allow to study the production of the particles presented here with higher precision and will allow ALICE to

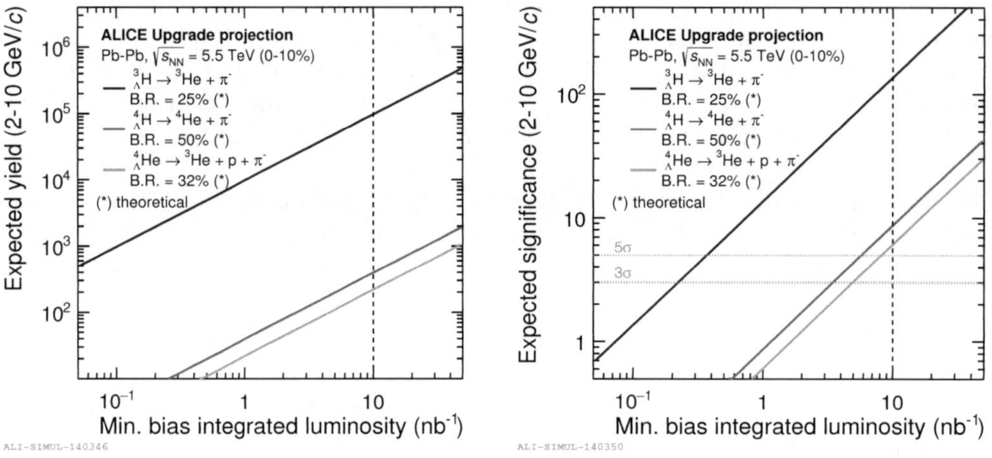

Figure 16. Expected yields of hyper-nuclei ($^3_\Lambda$H, $^4_\Lambda$H and $^4_\Lambda$He) in 0-10% central Pb–Pb collisions in runs 3 and 4 as a function of the integrated minimum bias luminosity (left panel). Expected significance of hyper-nuclei ($^3_\Lambda$H, $^4_\Lambda$H and $^4_\Lambda$He) in 0-10% central Pb-Pb collisions in runs 3 and 4 as a function of the integrated minimum bias luminosity (right panel). The vertical dashed line represents the target luminosity of 10 nb^{-1}. The horizontal dashed lines at 3σ and 5σ help to guide the eye.

study also hyper-nuclei with A=4 as indicated in Fig. 16.

7. Acknowledgments

The author acknowledges the support from the German Bundesministerium für Bildung, Wissenschaft, Forschung und Technologie (BMBF) though the project grant 05P2015 - ALICE at High Rate (BMBF-FSP202) (Förderkennzeichen 05P15RFCA1).

8. References

[1] Abelev B et al. (ALICE Collaboration), 2014 *J. Phys. G: Nucl. Part. Phys.* **41** 087001
[2] Andronic A et al., 2018 *Nature* **561** 321
[3] Andronic A et al., 2011 *Phys. Lett. B* **697** 203
[4] Cleymans J et al., 2011 *Phys. Rev. C* **84** 054916
[5] Braun-Munzinger P, Dönigus B, 2019 *Nucl. Phys. A* **987** 144
[6] Schnedermann E et al., 1993 *Phys. Rev. C* **48** 2462
[7] Kamada H et al., 1998 *Phys. Rev. C* **57** 1595
[8] Adam J et al. (ALICE Collaboration), 2016 *Phys. Rev. C* **93** 02491
[9] Tsallis C, 1988 *J. Statist. Phys.* **52** 479
[10] Abelev B.I. et al. (STAR Collaboration), 2007 *Phys. Rev. C* **75** 064901
[11] Cleymans J, Worku D, 2012 *J. Phys. G: Nucl. Part. Phys.* **39** 025006
[12] Gal A, Garzilaco H, 2019 *Phys. Lett. B* **791** 48
[13] Adam J et al. (ALICE Collaboration), 2016 *Phys. Lett. B* **754** 360
[14] Adamczyk L et al. (STAR Collaboration), 2018 *Phys. Rev. C* **97** 054909
[15] Jaffe R L, 1977 *Phys. Rev. Lett.* **38** 195 and 617
[16] Adam J et al. (ALICE Collaboration), 2016 *Phys. Lett. B* **752** 267
[17] Adam J et al. (ALICE Collaboration), 2017 *Phys. Rev. C* **95** 064606
[18] Braun-Munzinger P, Dönigus B, Löher N, 2015 *CERN Courier* **September**
[19] Abelev B et al. (ALICE Collaboration), 2014 *J. Phys. G* **41** 087002
[20] Adam J et al. (ALICE Collaboration), 2014 *CERN-LHCC-2013-020, ALICE-TDR-016*

Collectivity in small and large collision systems at the LHC with ALICE

Katarína Křížková Gajdošová (for the ALICE Collaboration)

Faculty of Nuclear Sciences and Physical Engineering, Czech Technical University in Prague, Břehová 7, 115 19 Prague, Czech Republic

Abstract. Interactions of protons (pp) or of a proton with a lead nucleus (p–Pb) were not expected to form a deconfined hot and dense matter, called the Quark-Gluon Plasma (QGP), which is instead produced in heavy-ion collisions. However, high-multiplicity events in small collision systems exhibit signs of collectivity, which are understood as a signature of the QGP emergence in heavy-ion collisions. An excellent tool to probe the presence of collectivity is the anisotropic flow measured with two- and multi-particle cumulants. In these proceedings, we present the first measurement of flow coefficients and their magnitude correlations using symmetric cumulants for charged particles in pp collisions at centre-of-mass energies of $\sqrt{s} = 13$ TeV, p–Pb at $\sqrt{s_{\rm NN}} = 5.02$ TeV, Xe–Xe at $\sqrt{s_{\rm NN}} = 5.44$ TeV and Pb–Pb collisions at $\sqrt{s_{\rm NN}} = 5.02$ TeV, collected during the LHC Run 2 programme. In addition, the flow coefficients of identified particles in p–Pb collisions are shown. Such a broad spectrum of colliding systems with different energies and a wide range of multiplicity allows for a detailed investigation of their collision dynamics. Non-flow effects, which are azimuthal correlations not originating from a common symmetry plane, are suppressed with the pseudorapidity separation and the subtraction method. These results provide an important insight into the nature of collective phenomena in different collision systems.

1. Introduction

Heavy-ion collisions at ultrarelativistic energies are used to create a state of strongly-interacting matter called the Quark–Gluon Plasma (QGP), where quarks and gluons are in a deconfined state. In a hydrodynamic picture, the collective evolution of this medium translates the initial spatial anisotropies in the overlap region of the colliding heavy nuclei into an anisotropy of final-state particles. To characterise the anisotropy, the azimuthal distribution of emitted particles can be decomposed into a Fourier expansion relative to a common symmetry plane Ψ_n with anisotropic flow coefficients $v_n = \langle \cos[n(\varphi - \Psi_n)]\rangle$ [1]:

$$\frac{dN}{d\varphi} \propto 1 + 2\sum_{n=1}^{\infty} v_n \cos[n(\varphi - \Psi_n)]. \tag{1}$$

The flow harmonics v_n thus quantify the preferred direction of emitted particles, and represent a collective response of the QGP to the initial spatial anisotropies, which makes them one of the most suitable probes to study the properties of this medium. An extensive set of measurements of anisotropic flow performed at the Large Hadron Collider (LHC), together with quantitative model comparisons have improved our knowledge about the QGP [2, 3, 4, 5, 6, 7, 8]. It is now well established that this medium behaves like a nearly perfect, collectively expanding liquid;

this behaviour manifests itself in the form of long-range multi-particle correlations between the final-state particles.

Small collision systems, such as pp or p–Pb collisions, were originally considered to lack the conditions necessary to create a hot and dense medium such as the QGP. However, recent measurements of anisotropic flow revealed features, similar to those observed in heavy-ion collisions, that are believed to indicate the presence of the QGP. In particular, measurements of two-particle correlations as a function of the pseudorapidity difference $\Delta\eta$ and the azimuthal angle difference $\Delta\varphi$ revealed a near side "ridge" structure in high-multiplicity pp and p–Pb collisions [9]. A negative sign of the four-particle cumulant was also observed at high multiplicity in collisions of small systems [10]. Measurements of p_{T}-differential flow coefficients for identified particles in high multiplicity p–Pb collisions showed a hint of a mass ordering at low p_{T}, and a baryon-meson grouping at intermediate p_{T}, which is understood as a consequence of hydrodynamic flow and partonic collectivity in heavy-ion collisions [11]. Whether these similarities with heavy-ion collisions, also observed in small collision systems, originate from the same mechanisms is yet to be understood. Phenomenological models with a hydrodynamic description of the system evolution are able to describe measurements obtained with two-particle correlations [12, 13, 14, 15, 16]. However, other alternative scenarios, including correlations of gluon fields in the initial stages of a collision, are also able to qualitatively reproduce some features of the results of flow coefficients [17, 18, 19]. Therefore, new measurements that are able to disentangle between various theoretical approaches, are of great necessity.

The study presented here shows the ALICE measurements of flow coefficients and their correlations as a function of multiplicity in a variety of collision systems [20], and new p_{T}-differential measurements of numerous identified particle species in p–Pb collisions [21]. All the results are extracted using the latest techniques to suppress non-flow contamination as much as possible, which makes the comparison to future theoretical calculations straightforward.

2. Analysis method

The analysed data samples were recorded by ALICE [22] during the LHC Run 2 period. In particular, data from Pb–Pb collisions at centre-of-mass energies $\sqrt{s_{\mathrm{NN}}} = 5.02$ TeV, Xe–Xe collisions at $\sqrt{s_{\mathrm{NN}}} = 5.44$ TeV, p–Pb collisions at $\sqrt{s_{\mathrm{NN}}} = 5.02$ TeV and pp collisions at $\sqrt{s} = 13$ TeV were selected. Minimum-bias triggered events, based on a coincidence of signals between the two arrays of the V0 detector (V0A and V0C), were used in all collision systems except for pp collisions. In this collision system, a dedicated trigger, selecting approximately 0.1% of the largest multiplicity events based on the amplitude in both arrays of the V0 detector, was applied. Events that passed the trigger threshold had 4 times larger multiplicity in the V0 acceptance than the minimum-bias average. The selected data samples were further reduced from pileup and background events. Overall, the results from the analyses with charged particles were obtained from 3.1×10^8 high-multiplicity pp, 2.3×10^8 p–Pb, 1.3×10^6 Xe–Xe and 5.5×10^7 Pb–Pb collisions, while 6.0×10^8 minimum bias p–Pb and 1.7×10^8 minimum bias pp collisions were used for the analyses with identified particles. These data samples were further divided into several event classes based on the multiplicity distribution measured with the V0A detector.

Charged tracks with a transverse momentum of $0.2 < p_{\mathrm{T}} < 3.0$ GeV/c and pseudorapidity $|\eta| < 0.8$ were used for the analyses. Particle identification of $\pi^{\pm}$, $\mathrm{K}^{\pm}$ and $\mathrm{p}(\overline{\mathrm{p}})$ was performed using a Bayesian approach [23] based on the signals from the Time Projection Chamber (TPC) and Time-of-Flight (TOF) detectors. Particles with a short life-time, $\mathrm{K}^0_{\mathrm{S}}$, $\Lambda(\overline{\Lambda})$ and ϕ, cannot be detected directly, but instead are reconstructed via their decay products on a statistical basis using the following decays: $\mathrm{K}^0_{\mathrm{S}} \rightarrow \pi^+ + \pi^-$, $\phi \rightarrow \mathrm{K}^+ + \mathrm{K}^-$ and $\Lambda \rightarrow \pi^- + \mathrm{p}$ ($\overline{\Lambda} \rightarrow \pi^+ + \overline{\mathrm{p}}$).

Measurements of multi-particle cumulants [24, 25, 26, 27] and symmetric cumulants [28] were calculated using the generic framework [28] with weighted Q_n-vectors, to correct for non-uniform acceptance and tracking inefficiencies.

Small collision systems are largely dominated by short-range few-particle correlations not related to the common symmetry plane, such as jets or resonance decays, generally denoted as “non-flow”. Correlations between particles originating from non-flow effects contaminate our measurements, and may obscure the possible signal of global collectivity involving many particles. The sub-event method [29, 30] was employed in the measurements in order to suppress this bias. In this method, particle m-tuplets are taken from different sub-events separated by a pseudorapidity gap, which ensures a long-range separation betwen particles that are correlated.

In addition, a subtraction method at the cumulant level, using data from minimum bias pp collisions, was used in p_{T}-differential measurements to remove the remaining non-flow contamination, following the equation:

$$v_2^{\mathrm{pPb,sub}}(p_{\mathrm{T}}) = \frac{d_2^{\mathrm{pPb}}(p_{\mathrm{T}}) - k \cdot d_2^{\mathrm{pp}}(p_{\mathrm{T}})}{\sqrt{c_2^{\mathrm{pPb}} - k \cdot c_2^{\mathrm{pp}}}}. \qquad (2)$$

Here, c_2 and d_2 represent the reference and the differential two-particle cumulants, respectively, and $k = \langle M\rangle^{\mathrm{pp}}/\langle M\rangle^{\mathrm{pPb}}$ is a scaling ratio of mean event multiplicities used for the estimation of non-flow effects to account for the different system sizes of pp and p–Pb collisions.

3. Results

Measurements of flow coefficients v_n of inclusive charged particles from pp, p–Pb, Xe–Xe and Pb–Pb collisions using the two particle cumulant method with a large pseudorapidity gap ($|\Delta\eta|$) to suppress non-flow effects are shown in Figure 1. Large collision systems (Pb–Pb and Xe–Xe) exhibit a strong multiplicity (N_{ch}) dependence of v_2, which is understood as a hydrodynamic response to the initial geometry of the overlapping region of the colliding nuclei at large and intermediate multiplicities. The strength of the interactions that are responsible for the transfer of the initial spatial anisotropies to the final state momentum anisotropies is reflected at low multiplicities. In addition, an ordering of flow coefficients $v_2 > v_3 > v_4$ is observed in the whole multiplicity range in both Xe–Xe and Pb–Pb collisions, except for the fluctuation-dominated region of very central collisions (with the highest multiplicity), where $v_2 \sim v_3$. In peripheral heavy-ion collisions (low multiplicities), the values of v_n become compatible with those measured in pp and p–Pb collisions. A weak N_{ch}-dependence of v_n is observed in both small and large collision systems in this region, except for v_2, which rises with multiplicity for $N_{\mathrm{ch}} \geq 50$ in Pb–Pb collisions, while the weak N_{ch}-dependence remains in pp and p–Pb collisions at similar multiplicity. An ordering of v_n is reported in small collision systems, similar to that observed in in Xe–Xe and Pb–Pb collisions.

Differential measurements of v_2 of inclusive charged hadrons $\mathrm{h}^{\pm}$ and several different species of identified particles in p–Pb collisions are reported in Figure. 2. The results are obtained with unprecedented precision as a function of p_{T} with a $|\Delta\eta| > 0.4$ separation and an additional non-flow subtraction. Similar to heavy-ion collisions, a clear mass ordering is observed for $p_{\mathrm{T}} < 2.5$ GeV/c. It suggests a strong radial expansion of particles with similar velocity, causing a push of the heavier ones to higher momenta. In the region of intermediate momenta $2.5 < p_{\mathrm{T}} \leq 6$ GeV/c, the v_2 of different particle species are grouped into two distinctive trends. Such a behaviour was also observed in heavy-ion collisions where it is understood as a consequence of parton coalescence [31] or recombination mechanism at the point of particle production [32]. Both observations and the trend, qualitatively consistent with hydrodynamic calculations [15, 16], suggest the presence of hydrodynamic collectivity in high-multiplicity p–Pb collisions.

Figure 3 presents measurements of $v_2\{m\}$ using higher order cumulants ($m > 2$) compared between small (pp and p–Pb) and large (Xe–Xe and Pb–Pb) collision systems. The existence of long-range multi-particle correlations in large collision systems is inferred from the consistency of the results from standard and subevent method ($v_2\{m\} \approx v_2\{m\}_{\mathrm{sub}}$) and of the results

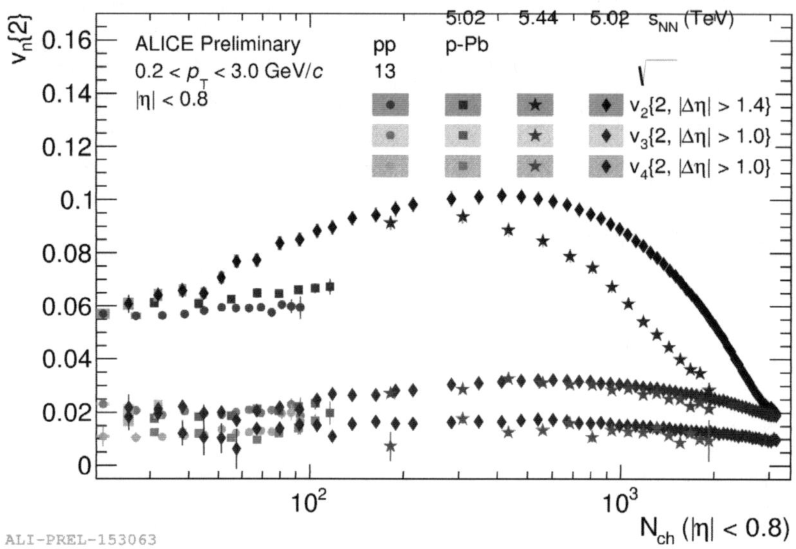

Figure 1. Multiplicity dependence of v_2, v_3 and v_4 measured using the two-particle cumulant method with a $|\Delta\eta|$ gap in small (pp and p–Pb) and large (Xe–Xe and Pb–Pb) collision systems.

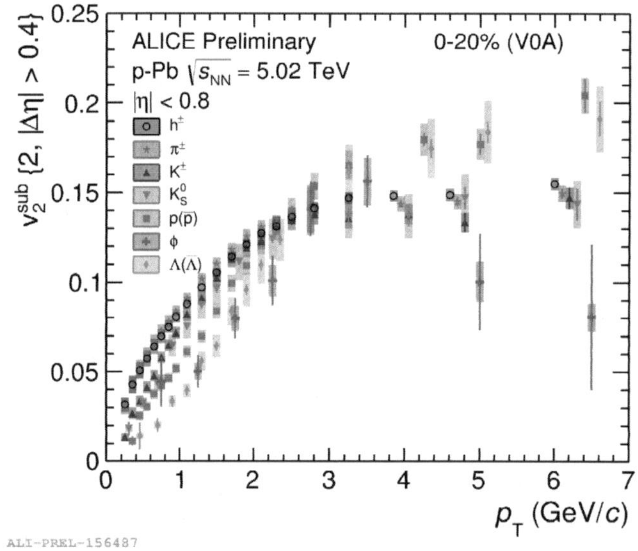

Figure 2. Transverse momentum dependence of $v_2^{\rm sub}\{2, |\Delta\eta| > 0.4\}(p_{\rm T})$ of $\rm h^{\pm}$, $\pi^{\pm}$, $\rm K^{\pm}$, $\rm K^0_S$, $\rm p(\overline{p})$, $\Lambda(\overline{\Lambda})$ and ϕ in p–Pb collisions in the $0-20\%$ V0A multiplicity class.

from cumulants of different orders ($v_2\{4\} \approx v_2\{6\} \approx v_2\{8\}$). In p–Pb collisions, the non-flow contribution to multi-particle cumulants can be further suppressed with the 3-subevent method, resulting in a decrease of the four-particle cumulant $c_2\{4\} > c_2\{4\}_{3-\rm sub}$. In turn, this leads to an increase of the flow signal $v_2\{4\}_{3-\rm sub} > v_2\{4\}$, based on the relation $v_2\{4\} = \sqrt[4]{-c_2\{4\}}$ [27]. A real-valued $v_2\{4\}$ could not be obtained in the largely non-flow dominated pp collisions. However, the contamination was successfully reduced with the 3-subevent method, leading to the first observed signal of $v_2\{4\}_{3-\rm sub}$ in pp collisions with ALICE and suggesting the presence of multi-particle correlations even in pp collisions. The collective behaviour of small systems is further supported by compatible values of different orders of multi-particle cumulants $v_2\{4\}_{3-\mathrm{sub}} \approx v_2\{6\}$, observed in both pp and p–Pb collisions.

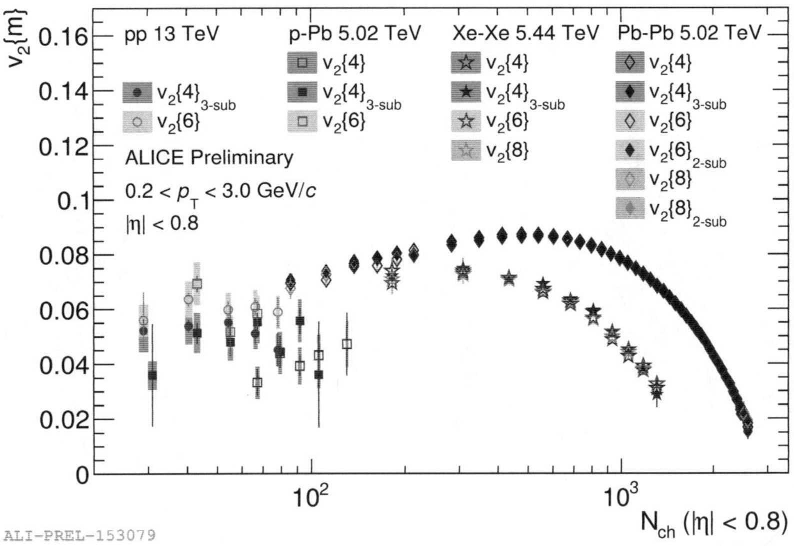

Figure 3. Multiplicity dependence of $v_2\{m\}$ for $m > 2$ in pp, p–Pb, Xe–Xe and Pb–Pb collisions.

The implications of the existence of long-range multi-particle correlations in small collision systems should be investigated with model comparisons. As it was mentioned in the Introduction, it is not yet clear which model scenario is the most suitable to describe the origin of these correlations in pp or p–Pb collisions. For that purpose, observables with a power to discriminate different model approaches are necessary. Symmetric cumulants (SC), in particular SC(3, 2) and SC(4, 2), provide access to the initial conditions and the dynamical evolution of the system [8]. Measurements of SC(m, n) with the 3-subevent method in pp, p–Pb, Xe–Xe and Pb–Pb collisions are reported in Figure 4. A positive $\mathrm{SC}(4,2)_{3-\mathrm{sub}}$ can be seen in the top panel, which is observed for all collision systems and in the entire multiplicity range. On the other hand, the $\mathrm{SC}(3,2)_{3-\mathrm{sub}}$ is negative at large and intermediate multiplicities for large collision systems, while there is a hint of a change to positive signs at $N_{\mathrm{ch}} \leq 90$. This trend seems to be followed by the results obtained in small collision systems.

4. Summary

We have presented new measurements of $v_2(p_\mathrm{T})$ using two-particle cumulants of inclusive and identified particles in high-multiplicity p–Pb collisions at $\sqrt{s_{\mathrm{NN}}} = 5.02$ TeV. In addition, the multiplicity dependence of v_n coefficients and their correlations in various collision system has been reported. The examined features of the measured results of flow coefficients in small collision systems and their striking similarity with measurements from large collision systems support the existence of collective phenomena in high-multiplicity pp and p–Pb collisions. The nature of these collective phenomena was further addressed by measurements of symmetric cumulants. The latter suggest that common mechanisms are responsible for the observations in small and large systems at similar multiplicities. All the results shown here provide an invaluable tool to further constrain the modelling of small collision systems and might allow us to understand the origin of collective effects in pp and p–Pb collisions.

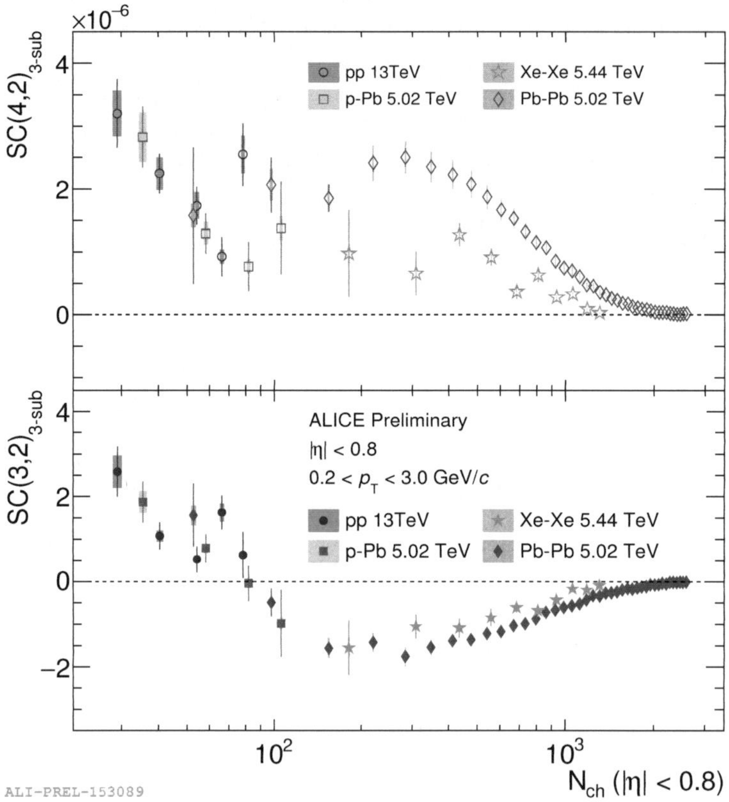

Figure 4. Multiplicity dependence of $SC(m,n)_{3-\mathrm{sub}}$ in pp, p–Pb, Xe–Xe and Pb–Pb collisions.

Acknowledgments
This proceeding are supported by the project Centre of Advanced Applied Sciences with the number: CZ.02.1.01/0.0/0.0/16-019/0000778. Project Centre of Advanced Applied Sciences is co-financed by European Union.

5. References

[1] Voloshin S and Zhang Y 1996 *Z. Phys.* **C70** 665–672 (*Preprint* hep-ph/9407282)
[2] Adam J *et al.* (ALICE) 2016 *Phys. Rev. Lett.* **116** 132302 (*Preprint* 1602.01119)
[3] Sirunyan A M *et al.* (CMS) 2019 *Phys. Lett.* **B789** 643–665 (*Preprint* 1711.05594)
[4] Acharya S *et al.* (ALICE) 2018 *JHEP* **07** 103 (*Preprint* 1804.02944)
[5] Acharya S *et al.* (ALICE) 2018 *Phys. Lett.* **B784** 82–95 (*Preprint* 1805.01832)
[6] Aaboud M *et al.* (ATLAS) 2018 *Eur. Phys. J.* **C78** 997 (*Preprint* 1808.03951)
[7] Acharya S *et al.* (ALICE) 2018 *JHEP* **09** 006 (*Preprint* 1805.04390)
[8] Adam J *et al.* (ALICE) 2016 *Phys. Rev. Lett.* **117** 182301 (*Preprint* 1604.07663)
[9] Khachatryan V *et al.* (CMS) 2010 *JHEP* **09** 091 (*Preprint* 1009.4122)
[10] Khachatryan V *et al.* (CMS) 2015 *Phys. Rev. Lett.* **115** 012301 (*Preprint* 1502.05382)
[11] Abelev B B *et al.* (ALICE) 2013 *Phys. Lett.* **B726** 164–177 (*Preprint* 1307.3237)
[12] Weller R D and Romatschke P 2017 *Phys. Lett.* **B774** 351–356 (*Preprint* 1701.07145)
[13] Mäntysaari H, Schenke B, Shen C and Tribedy P 2017 *Phys. Lett.* **B772** 681–686 (*Preprint* 1705.03177)
[14] Zhao W, Zhou Y, Xu H, Deng W and Song H 2018 *Phys. Lett.* **B780** 495–500 (*Preprint* 1801.00271)
[15] Werner K, Bleicher M, Guiot B, Karpenko I and Pierog T 2014 *Phys. Rev. Lett.* **112** 232301 (*Preprint* 1307.4379)
[16] Bozek P, Broniowski W and Torrieri G 2013 *Phys. Rev. Lett.* **111** 172303 (*Preprint* 1307.5060)
[17] Welsh K, Singer J and Heinz U W 2016 *Phys. Rev.* **C94** 024919 (*Preprint* 1605.09418)

IOP Conf. Series: Journal of Physics: Conf. Series **1271** (2019) 012002 doi:10.1088/1742-6596/1271/1/012002

[18] Dusling K, Mace M and Venugopalan R 2018 *Phys. Rev. Lett.* **120** 042002 (*Preprint* 1705.00745)
[19] Blok B and Wiedemann U A 2018 (*Preprint* 1812.04113)
[20] Gajdošová K (ALICE) 2019 *Nucl. Phys.* **A982** 487–490 (*Preprint* 1807.02998)
[21] Pacík V (ALICE) 2019 *Nucl. Phys.* **A982** 451–454 (*Preprint* 1807.04538)
[22] Aamodt K *et al.* (ALICE) 2008 *JINST* **3** S08002
[23] Adam J *et al.* (ALICE) 2016 *Eur. Phys. J. Plus* **131** 168 (*Preprint* 1602.01392)
[24] Borghini N, Dinh P M and Ollitrault J Y 2001 *Phys. Rev.* **C63** 054906 (*Preprint* nucl-th/0007063)
[25] Borghini N, Dinh P M and Ollitrault J Y 2001 *Phys. Rev.* **C64** 054901 (*Preprint* nucl-th/0105040)
[26] Borghini N, Dinh P M and Ollitrault J Y 2001 *International Workshop on the Physics of the Quark Gluon Plasma Palaiseau, France, September 4-7, 2001* (*Preprint* nucl-ex/0110016)
[27] Bilandzic A, Snellings R and Voloshin S 2011 *Phys. Rev.* **C83** 044913 (*Preprint* 1010.0233)
[28] Bilandzic A, Christensen C H, Gulbrandsen K, Hansen A and Zhou Y 2014 *Phys. Rev.* **C89** 064904 (*Preprint* 1312.3572)
[29] Jia J, Zhou M and Trzupek A 2017 *Phys. Rev.* **C96** 034906 (*Preprint* 1701.03830)
[30] Huo P, Gajdošová K, Jia J and Zhou Y 2018 *Phys. Lett.* **B777** 201–206 (*Preprint* 1710.07567)
[31] Molnar D and Voloshin S A 2003 *Phys. Rev. Lett.* **91** 092301 (*Preprint* nucl-th/0302014)
[32] Hwa R C and Yang C B 2003 *Phys. Rev.* **C67** 034902 (*Preprint* nucl-th/0211010)

IOP Conf. Series: Journal of Physics: Conf. Series **1271** (2019) 012003 doi:10.1088/1742-6596/1271/1/012003

The ALICE Inner Tracker System Upgrade

Federico Ronchetti, on behalf of the ALICE Collaboration
ALICE Team Leader at INFN, Laboratori Nazionali di Frascati
Via E. Fermi 40, 00044 Frascati, Rome, I

E-mail: federico.ronchetti@lnf.infn.it

Abstract. The phase II upgrade of the Large Hadron Collider (LHC) at CERN is planned under the assumption that the LHC will progressively increase the luminosity of heavy ion beams, eventually reaching an interaction rate of about 50 kHz in Pb-Pb collisions, i.e. instantaneous luminosities of $L = 6 \times 10^{27}$ cm^{-2}s^{-1}. In this perspective, the two main ALICE tracking devices, the TPC and the Inner Tracking system (ITS) will be both upgraded to cope with the foreseen interaction rates. In particular, the various ALICE silicon detectors will be fully replaced with a new, high-resolution, multi-layered, low-material, 12500 megapixel system allowing for very high data rates, and enhancing the vertexing and tracking capabilities at low transverse momenta (p_T).

The challenge and the uniqueness of the new ALICE ITS is the very small material budget of the silicon tracker, which will be a factor 4 to 5 smaller than existing similar systems at CERN. The present article describes the design and implementation of complex and reliable assembly procedures (focusing on the detector Outer Layers) which was crucial to push the new ITS technology into production.

1. Introduction

In order to fulfill the ALICE RUN3 (2021-23) physics program, a large area silicon pixel detector with high resolution and fast readout pixel sensors was developed under the following operating conditions:

- Material Budget: the improvement of the tracking and vertexing performance and the impact parameter resolution at low p_T can be achieved by a significant reduction of the material budget in order to minimize multiple Coulomb scattering. Specifically, the use of extremely thin Monolithic Active Pixel Sensors (MAPS) chips [1] of 50 μm thickness at maximum for the Inner tracking Layers, closest to the interaction point, will allow a $\times 7$ reduction of the thickness in comparison to the present ALICE ITS. In addition, low power consumption and a highly optimised signal and power distribution allow for the material budget of the electrical power and signal cables to be reduced by a factor of 5. Combining these new solutions it is possible to build a detector for which the material budget per detection layer is decreased down to about 0.3% radiation lengths (X_0).
- Intrinsic Spatial Resolution: the tracking performance, in particular the capability to separate secondary vertices of heavy flavour decays, is determined by the impact parameter resolution. This is a convolution of the primary vertex resolution and the track pointing resolution. Specifically, the challenging intrinsic spatial resolution of 5 μm is required to reach this performance thus requiring a pixel pitch of about 30 μm $\times$ 30 μm for the new ITS while the present pixel size is 50 μm $\times$ 425 μm.

IOP Conf. Series: Journal of Physics: Conf. Series **1271** (2019) 012003 doi:10.1088/1742-6596/1271/1/012003

- Chip Dimensions: after several years of R&D, the Tower Jazz 0.18 μm CMOS technology has been selected for the implementation of the pixel chip of the new ALICE ITS Muon Forward Tracker detectors. This allows for a maximum chip length of 30 mm in z-direction. The limitation of the chip width to 15 mm was motivated by geometrical considerations, thus a chip size of 15 mm $\times$30 mm has consequently been chosen as baseline dimension for the performance studies.
- Power Density: the maximum tolerable material budget puts severe limitations on the amount of material that can be used for detector cooling system. The power density on the sensor has thus to be brought to a minimum and does not exceed 300 mW/cm^2 in order to be compatible with the cooling capability of the support structure. This in turn requires a careful design of the power distribution as well as of the read-out architecture on the chip.
- Integration Time: in order to cope with interaction rates of up to 50 kHz for Pb-Pb collisions and 1 MHz for p-p collisions, the maximum acceptable sensor integration time is about 30 μs in order to limit pile-up effects and the consequent loss of tracking efficiency.
- Dead Time at 50 kHz Interaction Rate: a maximum dead time (equivalent to data being discarded) of 10% at 50 kHz Pb-Pb interaction rate can be tolerated. Chip buffer memories and bandwidths must be dimensioned such that they can cope with the expected occupancy level.
- Radiation Hardness: to ensure full functionality especially for the ITS Inner Layers (IL), the pixel sensors have to be tolerant against radiation levels (expected for the innermost layer) of 2700 krad. This value includes a safety factor of 4 for a collected data set corresponding to 10 nb^{-1} Pb-Pb and 6 nb^{-1} p-p collisions.
- Assembly Procedure: the highest criticality in the assembly procedure is the manipulation of very fragile low thickness silicon detectors: 50 μm for the IL and 100 μm for the Outer and Middle Layers (OL and ML, respectively). In addition, the tracking target resolution puts stringent limits on the accuracy of the positioning of each single detecting element. The goal is to reach a geometrical placement accuracy of the individual chip in the horizontal plane of the order of 50 μm or better.

This proceeding will describe many of the technical solutions that have been developed for the new ALICE ITS [2] detector. Such a practical know-how as well as the related physics analysis, will open new perspectives in view of the new generation of many different high performance silicon detectors as well as in frontier precision for high energy experiments. In addition, this project will open a range of new opportunities for applications that benefit from a high resolution tracking system, able to operate in vacuum and within a magnetic field and in many environments, including the use of hadrons (mainly protons and carbon ions) for cancer radiation treatment.

2. Outer Layers ITS Production

The efforts to establish the ITS OL and ML detector element production increased dramatically starting from mid 2016. The infrastructures at the production sites were reorganized to cope with the requirements for the assembly and the manipulation of very fragile and low-thickness (100 μm) silicon sensors with open wire bonds.

Each so-called ITS "stave" is composed by two Half-Staves (HS) where an HS is in turn composed by 7 base modules (Hybrid Integrated Circuit, HIC) containing 2 rows of 7 ALPIDE sensors.

The ALPIDE sensors are aligned and glued to the HIC printed circuit using a custom assembly machine in the HIC production centers (Bari, Strasbourg, Liverpool, Wuhan, and Pusan) and are then bonded with three Al wire per pad to ensure redundancy against wire breaking. The wire bonds are not protected: this avoids the introduction in the detector of materials that can

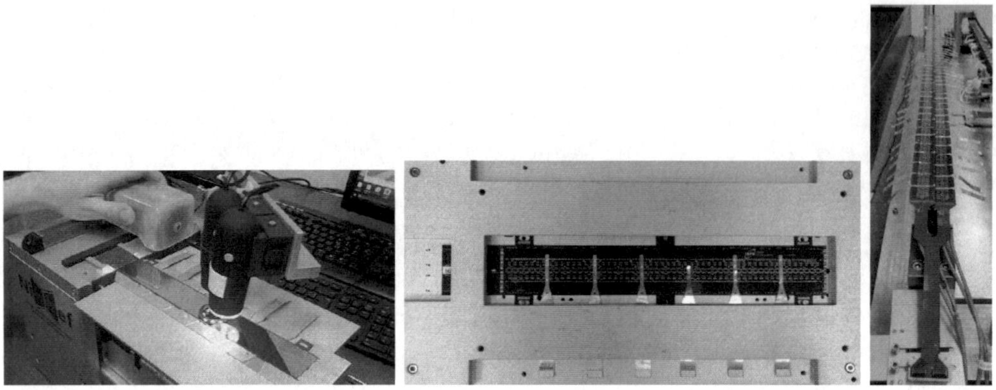

Figure 1. *An OB HIC at the reception test. Center: the HIC is contained in a carrier plate which allows transport, storage, powering, testing, and vacuum gripper placement for handling. The left side of the HIC shows the so-called "tab" used to connect the sensors to the readout system. Once the HIC is qualified for alignment and gluing, the tab has to be cut as shown in the leftmost photo. Right: the two HS assembled in a stave: the sensors are glued directly on the carbon fiber strips. The blue pipes are the embedded water cooling lines (input and outupt). The HS are read-out by soldering an extension tab which is the interface with the outside world of the 98 ALPIDE chips. It can be clearly seen that the two HS come in different flavors (upper and lower) which have a partial longitudinal overlap requiring a definite sequence of installation.*

degrade with time but on the other hand introduces a "zero tolerance" also in case of minor mishandlings. In total an HS features 98 ALPIDE sensors (7 HICs and 7×2 chips per HIC) for a total of 51 megapixels.

The two rows of sensors are readout independently using a Master-Slave scheme: the first sensor of the row acts as a master collecting the hits from the remaining 6 slaves in addition to its own. Such a setup is suitable for the ITS OL and ML as the expected occupancies are much less then those of the inner barrel which is as close as 22 mm to the interaction point. In this case the ALPIDE sensors which make up an IL stave are readout individually to cope with much higher rates.

Before usage, the HICs have to the electronically tested for integrity using the "tab" interface shown in the left and central photos of figure 1. The high density "firefly" connectors are soldered directly on the HIC printed circuit and once the HIC is tested must be cut at the level of the pad contacts in order to make the HIC alignable.

The cut is performed using a custom high-precision cutting system which rail can run the blades only 50 μm away from the silicon sensor edges as shown in the left part of figure 1. The cut HICs possess contact pads for the CLK, CTRL, and DATA lines at both ends. At this point 7 of them can be aligned and glued (chip side facing down) on water-cooled carbon fiber strips to make the HS. The HICs have reference markers which can be aligned (using the Coordinate Measuring Machine, CMM) along the longitudinal and transverse directions with an accuracy of the order of 10 μm/1.5 m. After the alignment the HICs making up an HS must have the module-to-module interconnections soldered and the termination resistors removed (except the HIC in position 7) to enable the readout lines along the HS. Each HIC position along the HS is bit encoded by 3 resistors which are removed using soldering tweezers to program the module address.

IOP Conf. Series: Journal of Physics: Conf. Series **1271** (2019) 012003 doi:10.1088/1742-6596/1271/1/012003

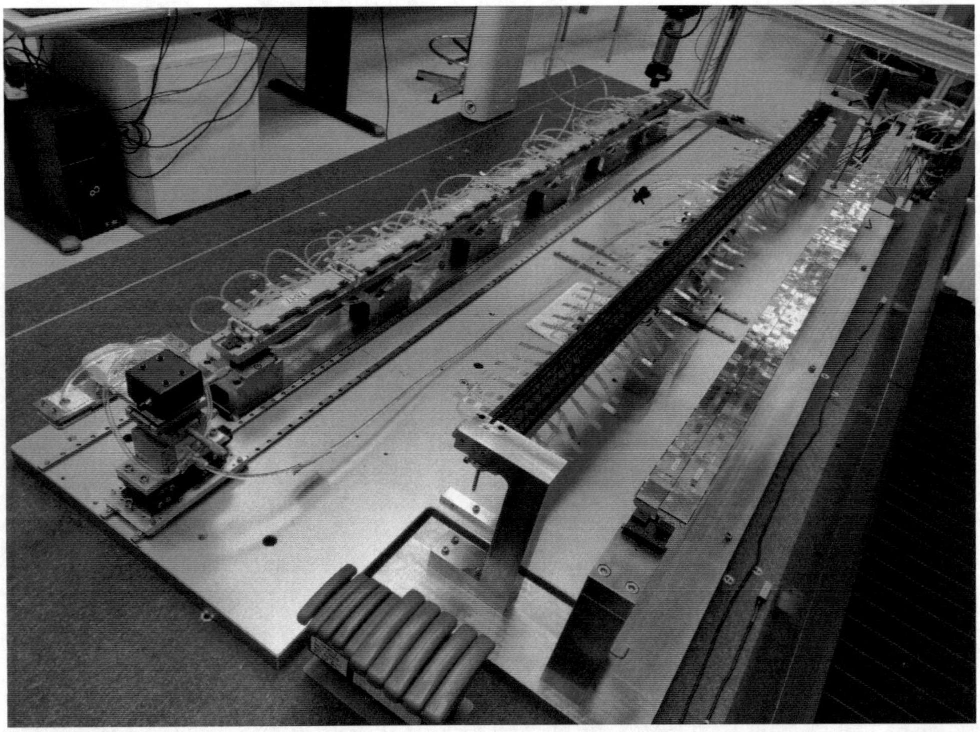

Figure 2. *Three tasks going on in parallel on the CMM jig. Station 1 on the left contains 7 HICs aligned and glued to make an HS. The vacuum grippers used to bring the HICs from the carrier plates to the gluing position can be seen as they are removed only when the Araldite 2011 glue has cured. Station 2 (in the middle) contains an already finished stave ready to be moved in the "gray area" for the installation of the PB and BB, and their folding. Station 3 on the right contains a finished stave, with the PB and BB already folded, for the final metrology.*

As mentioned above, two HS are needed to make a stave. The HS come also in two flavors: upper and lower and they must be glued to the carbon fiber support frame (Space Frame, SF) in the sequence (upper, lower) as there is a small overlap between them as shown in the right part of figure 1. The produced stave is then moved out from the CMM jig to the soldering station of the power distribution lines (Power Bus - PB - and Bias Bus - BB) and for additional electronic testing.

The CMM jig has three stations: station 1 is used to perform the HIC alignment and gluing for an HS of given flavor, while the second station is used to align and glue an already soldered HS to the support structure (SF). The third station is used to validate the finished stave metrologically before and after the installation and folding of the power and bias distribution lines as shown in figure 2.

Once the stave is assembled it is moved into the station for the installation of the power distribution system and the final tests. The two HS have separate powering called PB(BB)-U and PB(BB)-L and their relative filter boards, where L stands for lower and U stands for upper. The assembly of PB, BB, and filter board is soldered on a custom jig were the required ceramic capacitors (Murata GRM 220 μF) are first soldered to the PB. The position of soldering defines if the PB is of U or L type. The amount of tin that has to be used to solder the capacitors must be carefully controlled in order to avoid the introduction of extra thickness which can cause

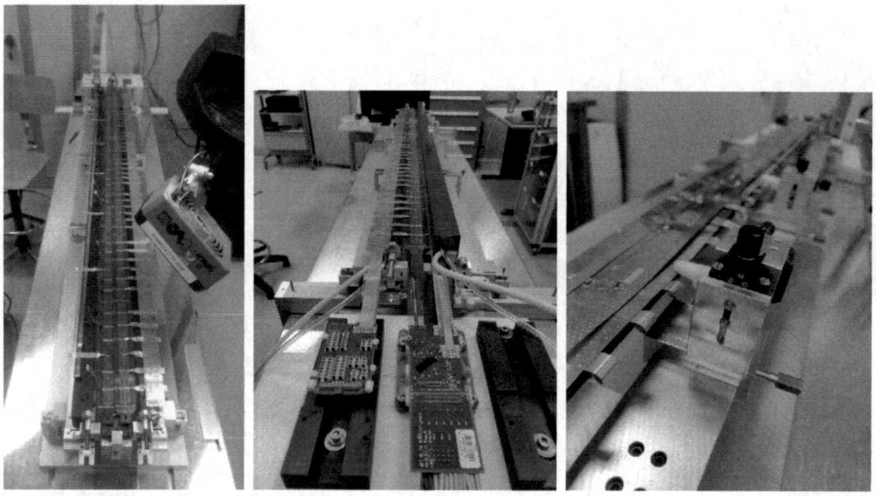

Figure 3. *The left photo shows a stave on the folding jig with the power and bias bus distributions ships ready for soldering on the handling bars. The upper PB is rolled out to allow the soldering of the BB to the corresponding HIC cross cables. Once the BB is soldered, the PB is lowered to be soldered to the remaining cross cables. The center figure shows a PB+BB assembly rotated and inserted into the U-arms of the upper half stave (keep in mind that the HS is upside-down in this position) with the filter board connected to the break-out board from the power board. In this position the HS is tested to make sure no bonds were damaged by the folding procedure. The carbon fiber handing bar is removed if the test is successful. The right photo shows the two PB+BB systems folded and the insertion of the remaining (smaller) U-arms to lock the busses into position.*

mechanical interferences once the stave is inserted into the half barrel support. The folding jig has two handling bars which can hold the PB and BB stripes in place using suction cups. These bars are fixed to the jig by means of special pivots screws which can first rotate, then step down and slide. The upper PB+BB assembly has to be inserted between the HS-U and HS-L inside special holders glued on the HS called U-arms. More U-arms are put after the folding to ensure the PB+BB system cannot move and is kept at an appropriate distance from the bonds underneath.

The stave is finally inserted into an aluminum box which is in turn closed with an envelope to keep the relative humidity under control. The Al box is inserted in a wooden box which has a suspension system to minimize mechanical shocks during the transport. The temperature must be also controlled and it shall never drop below 15° during the transport.

The production started early in 2018 and reached a sustained rate of 0.7 stave/week. The present production yield is 91% (fraction of detector grade HS).

In conclusion, the ITS Outer Layers and Middle Layers production has successfully started and ramped up in all OL (Daresbury, Frascati, Nikhef, and Torino) and ML (Berkeley) sites and is proceeding at a good rate and 95% yield. The present production rate will ensure that all the needed staves will be at CERN on time for the ITS surface commissioning (the ITS Inner Barrel construction is already finished) and for installation in the ALICE cavern during the summer of 2020.

IOP Conf. Series: Journal of Physics: Conf. Series **1271** (2019) 012003 doi:10.1088/1742-6596/1271/1/012003

3. References

[1] M. Šuljić, ALPIDE: the Monolithic Active Pixel Sensor for the ALICE ITS upgrade, 2018 *Nuovo Cimento* **C 41** 91
[2] Technical Design Report for the Upgrade of the ALICE Inner Tracking System, 2014 *J. Phys. G: Nucl. Part. Phys.* **41** 087002

IOP Conf. Series: Journal of Physics: Conf. Series **1271** (2019) 012004 doi:10.1088/1742-6596/1271/1/012004

HL-LHC prospects from ATLAS and CMS

Catrin Bernius, on behalf of the ATLAS and CMS Collaborations[1]
SLAC National Accelerator Laboratory, 2575 Sand Hill Rd, Menlo Park, CA 94025, USA
E-mail: catrin.bernius@cern.ch

Abstract. The Large Hadron Collider (LHC) has been successfully delivering proton-proton collision data at the unprecedented center of mass energy of 13 TeV. An upgrade is planned to increase the instantaneous luminosity delivered by the LHC in what is called HL-LHC, aiming to deliver a total of about 3000 fb^{-1} of data per experiment. To cope with the expected data-taking conditions ATLAS and CMS are planning major upgrades of the detectors. Additionally, ATLAS and CMS are preparing inputs to a CERN Yellow Report that aims to summarize the physics reach for HL-LHC and will be submitted as input to the European Strategy before the end of 2018. This contribution focusses on the physics reach expected for a wide range of measurements and searches at the HL-LHC for the ATLAS and CMS experiments, including Higgs coupling, di-Higgs boson production sensitivity, Vector Boson Scattering prospects as well as discovery potential for electroweak SUSY and other exotic benchmark scenarios.

1. Introduction to the HL-LHC

The ATLAS [1] and CMS [2] collaborations at the Large Hadron Collider (LHC) announced the discovery of a 125 GeV Higgs-like boson on 4 July 2012 [3, 4]. Since then, further measurements and studies by both collaborations of the mass [5], the spin and parity [6, 7] as well as the production and decay rates have found no significant deviations from the Standard Model (SM) expectations [8]. In the SM, all properties of the Higgs boson are defined once its mass is known. However, this model leaves many open questions such as the hierarchy problem, regarding the naturalness of the Higgs boson mass, and the nature of the dark matter. Therefore, precise measurements of the Higgs sector are of high priority in the future program of high-energy physics. To be able to extend the successful physics program at the LHC to high precision studies and searches for new physics Beyond the SM (BSM), much more data will need to be collected. To meet this challenge, an extensive upgrade program for the LHC [9] as well as the experiments has been implemented in several phases as shown in Figure 1. The High-Luminosity LHC (HL-LHC) is expected to collect 3000 to 4000 fb^{-1} of data, 10 times more than the integrated luminosity of the LHC Runs 1-3 combined. The HL-LHC baseline parameters with a peak luminosity of 5 to 7×10^{-34} $cm^{-2}s^{-1}$ and an average number of pp interactions (pile-up) of 140 to 200 present a very challenging environment to the existing experiments at the LHC.

IOP Conf. Series: Journal of Physics: Conf. Series **1271** (2019) 012004 doi:10.1088/1742-6596/1271/1/012004

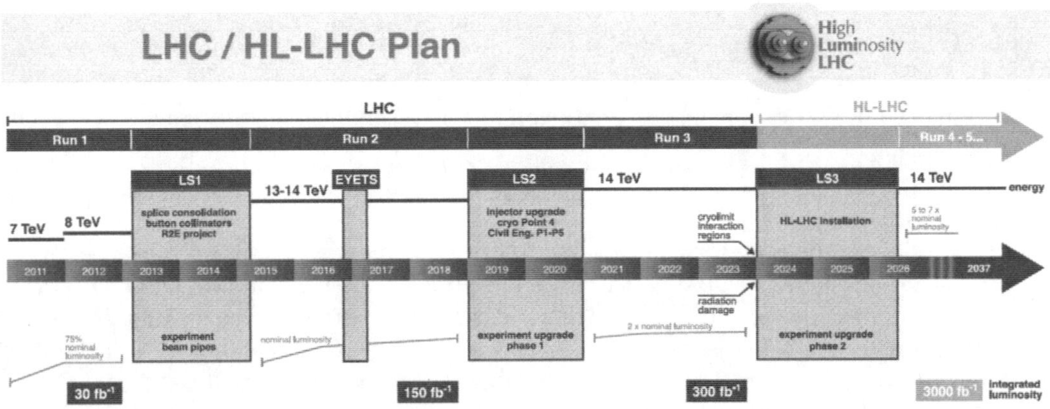

Figure 1. LHC baseline plan for the next decade and beyond in terms of collision energy (upper line) and luminosity (lower lines). The first long shutdown (LS1) 2013-14 is to allow design parameters of beam energy and luminosity. The second one, LS2 in 2018-19, is for securing luminosity and reliability as well as to upgrade the LHC Injectors. After LS3, in 2025 the machine is expected to have the High-Luminosity configuration [9].

2. Detector upgrades for Run 4

To be able to reach the physics goals set for the HL-LHC, both, the ATLAS and CMS detectors have to be upgraded in order to maintain or improve their performance and physics acceptance. A few of these upgrades will be briefly highlighted in the following. More information can be found in the respective references.

2.1. The ATLAS detector

The ATLAS detector will be upgraded to deal with the increased occupancies and data rates. The inner tracking detector will be completely replaced with a new all-silicon Inner Tracker (ITk) [10, 11], increasing the coverage up to a pseudorapidity[2] of $|\eta| = 4$. New Front-End (FE) electronics and a new read-out system in the calorimeters [12, 13] will allow for the read-out of higher resolution objects at the lowest trigger level at an increased rate. In the muon detector system [14], new inner barrel chambers will be installed for increased coverage at the trigger level. A new detector, the high-granularity timing detector [16] will also be installed in the forward region to e.g. separate jets from different *pp* interactions. The Trigger and Data Acquisition (TDAQ) [15] system will be re-designed making e.g. full granularity detector data available for hardware tracking at a 1 MHz selection rate with the software selection output being 10 kHz.

2.2. The CMS detector

The CMS detector will also undergo major upgrades to be able to cope with the HL-LHC environment. These include a replacement of the silicon strip and pixels in the tracking detector [17] to increase the granularity while reducing material as well as increasing the coverage to $|\eta| = 3.8$. The calorimeter end-caps and FE electronics in the electromagnetic barrel will be replaced for a trigger readout at 40 MHz [18, 19]. The muon system will be, amongst other

[2] ATLAS uses a right-handed coordinate system with its origin at the nominal interaction point (IP) in the centre of the detector and the z-axis along the beam pipe. The x-axis points from the IP to the centre of the LHC ring, and the y-axis points upward. Cylindrical coordinates (r, ϕ) are used in the transverse plane, ϕ being the azimuthal angle around the z-axis. The pseudorapidity is defined in terms of the polar angle θ as $\eta = -\ln\tan(\theta/2)$.

upgrades, extended with new chambers in the forward region, bringing the coverage up to $\eta = 3$ [20]. A minimum ionising particle timing detector [21] will be installed in the central region of the CMS detector to improve the particle-flow performance. Finally also the TDAQ system will undergo upgrades for hardware tracking at 40 MHz with a software selection output of 7.5 kHz [22, 23].

3. Physics prospects at the HL-LHC

Higgs boson physics will be a major component of the extensive physics program at the HL-LHC with the main measurements focussing on Higgs couplings and self-coupling, Higgs differential distributions, rare Higgs decays and heavy Higgs searches. Additionally there is strong support for a rich program in SM and flavour physics as well as for BSM searches. In the following only brief descriptions of selected projections that have been presented will be given. Far more detail can be found in the respective references as well as on the public pages of the ATLAS [24] and CMS [25] experiments. These studies of the assessment of the HL-LHC performance have been carried out as input to the HL-LHC Yellow Report (currently in preparation) as well as the European Strategy Process [26]. In the presented projections, the effects of the upgraded detectors have been taken into account by applying energy smearing, object efficiencies and fake rates to truth level quantities, and following parameterisations based on detector performance studies. An integrated luminosity of 3000 fb^{-1} and an average pile-up of 200 are assumed.

3.1. Higgs self-coupling and HH sensitivity

While many properties of the Higgs boson can be studied with the current LHC dataset, other properties, such as the Higgs self-coupling cannot be extracted from the existing dataset. Even with the large dataset expected at the HL-LHC, measurements of the Higgs self-coupling are expected to be extremely challenging. A direct measurement of the Higgs trilinear self-coupling λ_{HHH} requires the study of the Higgs boson pair productions. For a centre-of-mass energy of 14 TeV, the production cross section of pairs of 125 GeV Higgs bosons is estimated to be 39.6 fb [27].

One of the most promising channels to measure the Higgs self-coupling is the $H(\rightarrow b\bar{b})H(\rightarrow \gamma\gamma)$ final state, for which the results of the ATLAS projection are presented here, with all details of the analysis to be found in Ref. [28], while the CMS projection for the Higgs pair production can be found in Ref. [29]. This channel profits from a clean HH signal extraction thanks to the narrow mass peak of the $H(\rightarrow \gamma\gamma)$ decay. As the low branching ratio (0.13 %) for this decay leads to low expected signal rates, only the gluon-gluon production mode is studied. Following the selection procedure described in Ref. [28], the expected numbers of signal and background events, normalised to an integrated luminosity of 3000 fb^{-1}, are estimated using simulated events.

Figure 2 (right) shows the invariant mass distributions of the diphoton pairs. The dominant backgrounds are non-resonant $b\bar{b}\gamma\gamma$ and $b\bar{b}j\gamma$ final states. Other processes that also contribute significantly to the total expected background include $ZH(\rightarrow \gamma\gamma)$, $t\bar{t}H(\rightarrow \gamma\gamma)$, $c\bar{c}\gamma\gamma$, $b\bar{b}jj$, and $t\bar{t}\gamma$.

Using a cut-based analysis the estimated number of signal events is 9.544 ± 0.029, to be compared to a background level of 90.9 ± 2.0 events. The combination of these numbers gives an expected significance of 1.05σ for 3000 fb^{-1}. At 95% CL, the Higgs boson self-coupling is expected to be constrained to $-0.8 < \lambda/\lambda_{\text{SM}} < 7.7$ (see Figure 2 (right)). This is not enough on its own to claim evidence for the observation of Higgs pair production, or to determine whether the Higgs self-coupling strength is close to its SM expectation. This channel is expected to be combined with similar measurements for HH boson production in other decay channels such as $HH \rightarrow b\bar{b}b\bar{b}$ and $HH \rightarrow b\bar{b}\tau\tau$ and to be part of a combination of ATLAS and CMS results.

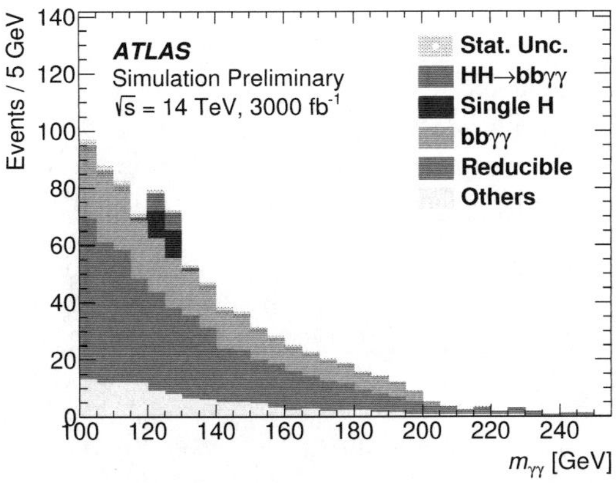

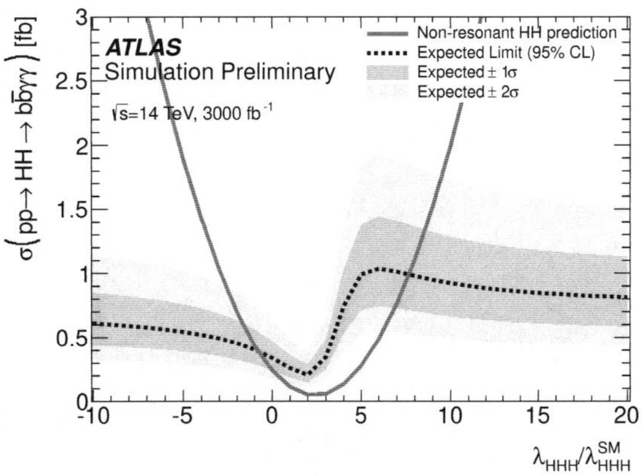

Figure 2. Left: The diphoton invariant mass distribution after all the selection cuts except for the cuts on the diphoton invariant mass for an average number of pile-up events of 200. The shaded area corresponds to the MC statistical uncertainty. Right: Expected 95% CL upper limit on the cross-section $\sigma(HH \to b\bar{b}\gamma\gamma)$ with 3000 fb^{-1} of data and neglecting systematic uncertainties, as a function of the Higgs self-coupling constant λ in units of $\lambda_{\rm SM}$. The $\pm 1\sigma$ and $\pm 2\sigma$ uncertainty bands are shown in green and yellow. The cross-section exclusion limit grows less stringent in the range $3 < \lambda/\lambda_{\rm SM} < 5$ due to the shift of the number of expected HH events towards lower values where the cross section is decreased. Figures taken from Ref. [28].

The projected sensitivity of various channels for SM HH production is described in detail in Ref. [30]. Figure 3 (left) shows the uncertainty in signal strength $\mu = \sigma/\sigma_{\rm SM}$, providing different assumptions on the systematic uncertainties which are described in detail in Ref. [30]. It can be seen that the impact of systematics differ across all channels, with the most sensitive being the $H(\to b\bar{b})H(\to \gamma\gamma)$ final state. The impact on the significance by assuming different systematic uncertainties can be seen in Figure 3 (right) at hand of the $H(\to b\bar{b})H(\to \tau\tau)$ final state. The three scenarios for the assumed uncertainties in this example are: All systematic uncertainties are kept constant with the integrated luminosity and the CMS detector performance is unchanged with respect to the reference analysis (ECFA16 S1); The theoretical uncertainties are scaled down by a factor of 2, while experimental systematic uncertainties are scaled down by the square root of the integrated luminosity until they reach a defined lower limit based on estimates of the achievable accuracy with the upgraded detector (ECFA16 S2); The statistical error only.

3.2. Standard Model physics

One of the physics goals of the HL-LHC is to perform precise SM measurements that require high accuracy and a significant amount of data. This section presents extrapolations for measurements of the top quark mass, based on results obtained mainly at $\sqrt{s} = 8$ TeV, with particular focus on the projected evolution of systematic uncertainties [31]. The top quark is the heaviest known elementary particle. Its mass, m_t, is a fundamental parameter of the SM, affecting predictions in QCD as well as the electroweak sector. It contributes significantly to the quartic Higgs coupling and affects constraints on the stability of the electroweak vacuum, as well as on models with cosmological implications.

Figure 4 summarises the expected precision on m_t for the various measurement techniques that are extensively discussed in Ref. [31]. With data collected during Run 1, most analyses were already limited by systematic uncertainties except for the J/Ψ method which is still affected

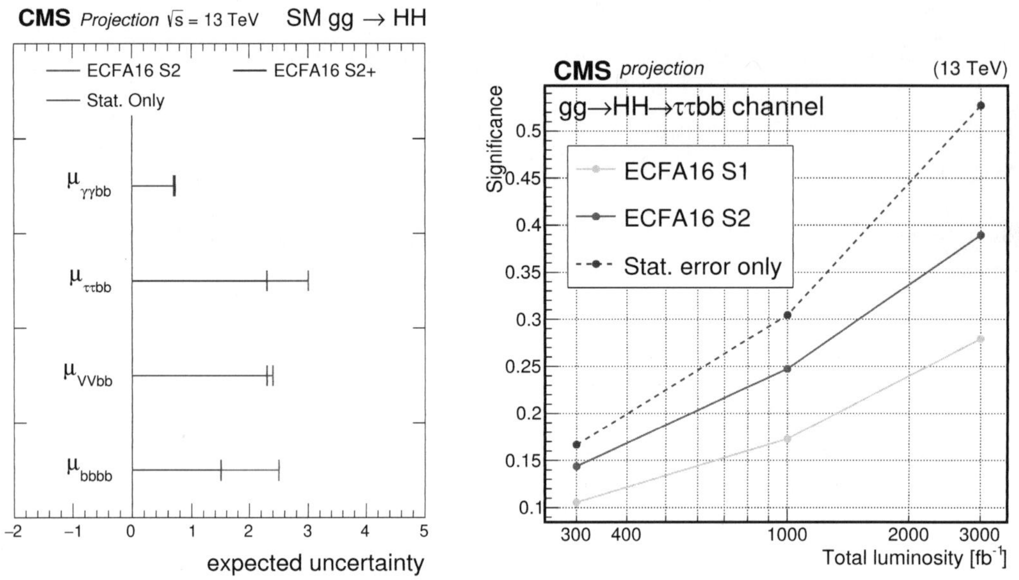

Figure 3. Left: Projection of the sensitivity to the SM $gg \rightarrow HH$ production at 3000 fb^{-1}, based on 13 TeV preliminary analyses performed with data collected in 2015. The uncertainty in the signal modifier $\mu = \sigma/\sigma_{\rm SM}$ is provided assuming different scenarios on the systematic uncertainties. Right: Projection of the sensitivity to the SM $HH \rightarrow \tau\tau b\bar{b}$ production as function of the collected luminosity, based on the 13 TeV preliminary analysis performed with data collected in 2015, under different assumptions on the systematic uncertainties. Figures taken from Ref. [30].

by a sizeable statistical uncertainty. The J/Ψ method makes use of a partial reconstruction of top quarks in leptonic final states containing $J/\Psi \rightarrow \mu^+\mu^-$ from the b quark fragmentation and determines the top quark mass through its correlation with the mass of the J/Ψ + lepton system (see Ref. [31] for a more detailed description). With 3 ab^{-1} of data, all of the considered analyses will be limited by systematic uncertainties and especially by theoretical modelling uncertainties. The most precise method, the reference method, is expected to yield an ultimate relative precision below 0.1% at HL-LHC. The other techniques are expected to be precise enough to carry significant weight in a combination with the reference method. This would make it possible to further reduce the systematic uncertainties, which are related mostly to the jet energy scale and hard process modelling.

3.3. Supersymmetry searches

In the electroweak sector, the supersymmetric partners of the Higgs, photon, Z, and W are of spin 1/2 that further mix in neutralino ($\tilde{\chi}^0_{1,2,3,4}$) and chargino ($\tilde{\chi}^\pm_{1,2}$) states, also called electroweakinos. In anomaly-mediated supersymmetry breaking (AMSB) scenarios , the supersymmetric partners of the SM W-bosons, the winos, are the lightest gaugino states. In this case, the lightest chargino ($\tilde{\chi}^\pm_1$) and the lightest neutralino ($\tilde{\chi}^0_1$), are both largely composed of the wino eigenstates and are nearly mass-degenerate. Due to this small mass difference, the $\tilde{\chi}^\pm_1$ can have a long enough lifetime such that it decays inside the detector. AMSB scenarios naturally predict a pure wino lightest supersymmetric particle (LSP), which is a dark-matter candidate. Other scenarios following from naturalness arguments suggest the absolute value of the Higgsino mass parameter μ to be near the weak scale such that the Higgsinos should be

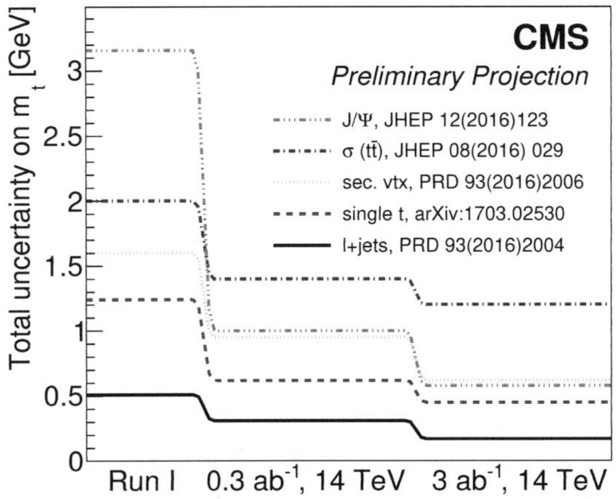

Figure 4. Total uncertainty on the top quark mass (m_t) obtained with different measurement methods and their projections to the HL-LHC. The projections for $\sqrt{s} = 14$ TeV, with 0.3 ab^{-1} or 3 ab^{-1} of data, are based on m_t measurements performed at the LHC Run 1, assuming that an upgraded detector will maintain the same physics performance despite a severe pile-up. Figure taken from Ref. [31].

light, with masses below one TeV, while the magnitude of the bino and wino mass parameters, M_1 and M_2 can be significantly larger, i.e. $|\mu| \ll M_1, M_2$. This results in the three lightest electroweakino states, $\tilde{\chi}_1^0$, $\tilde{\chi}_1^{\pm}$, $\tilde{\chi}_2^0$ being dominated by the Higgsino component. In this scenario, the three lightest electroweakino masses are separated by hundreds of MeV to tens of GeV depending on the composition of these mass eigenstates, which is determined by the specific values of M_1 and M_2. Investigating either of these scenarios, with very small mass splitting between the lightest electroweakinos, is particularly challenging at hadron colliders, both due to the small cross-sections and due to the small transverse momenta (p_T) of the final state particles. The two searches presented in here investigate electroweakino production in these two, nearly mass degenerate, scenarios assuming an integrated luminosity of 3000 fb^{-1}. In the following, only the results are presented, more details can be found in Ref. [32].

The disappearing track search investigates scenarios where the $\tilde{\chi}_1^{\pm}$ and $\tilde{\chi}_1^0$ are almost mass degenerate, leading to a long lifetime for the $\tilde{\chi}_1^{\pm}$ which decays after the first few layers of the inner detector, leaving a track in the innermost layers of the detector. Limits at 95% CL on the chargino lifetime are shown in Figure 5 as a function of the $\tilde{\chi}_1^{\pm}$ mass. The simplified models of chargino production considered include chargino pair production and chargino-neutralino production (both $\tilde{\chi}_1^{\pm}\tilde{\chi}_1^{\mp}$ and $\tilde{\chi}_1^{\pm}\tilde{\chi}_1^0$). The potential for the full HL-LHC dataset is expected to exclude at the 95% CL chargino lifetimes, assuming a wino-like (higgsino-like) LSP, of between 7 ps (10 ps) and 4 μs (1.5 μs) for light charginos with a mass of 100 GeV. Heavier wino-like (higgsino-like) charginos are expected to be excluded up to $m(\tilde{\chi}_1^{\pm}) = 1100$ GeV (750 GeV) for lifetimes of 1 ns. The discovery potential of the analysis would allow for the discovery of wino-like (higgsino-like) charginos of mass 100 GeV with lifetimes between 20 ps and 700 ns (30 ps and 250 ns), or for a lifetime of 1 ns would allow the discovery of wino-like (higgsino-like) charginos of mass up to 800 GeV (600 GeV).

The dilepton search investigates final states containing two soft muons and a large transverse momentum imbalance, which arise in scenarios where $\tilde{\chi}_2^0$ and $\tilde{\chi}_1^{\pm}$ are produced and decay via

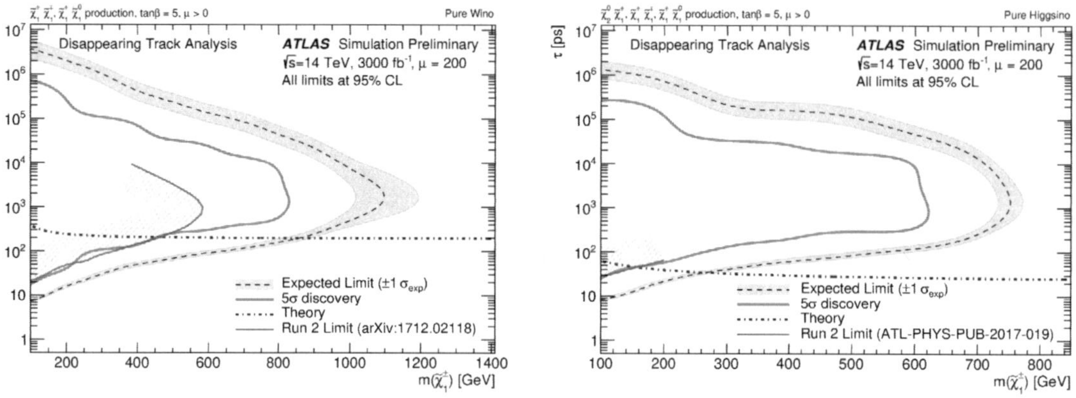

Figure 5. Expected exclusion limits at 95% CL from the disappearing track search using 3000 fb^{-1} of 14 TeV pp collision data as a function of the $\tilde{\chi}_1^{\pm}$ mass and lifetime. Simplified models including both $\tilde{\chi}_1^{\pm}\tilde{\chi}_1^{\mp}$ and $\tilde{\chi}_1^{\pm}\tilde{\chi}_1^{0}$ are considered assuming pure-wino scenarios (left) and pure-higgsino scenarios (right). The yellow band shows the 1σ region of the distribution of the expected limits. The median of the expected limits is shown by a dashed line. The red line presents the current limits from the Run 2 analysis and the hashed region is used to show the direction of the exclusion. The expected limits with the upgraded ATLAS detector would extend these limits significantly. Figures taken from Ref. [32].

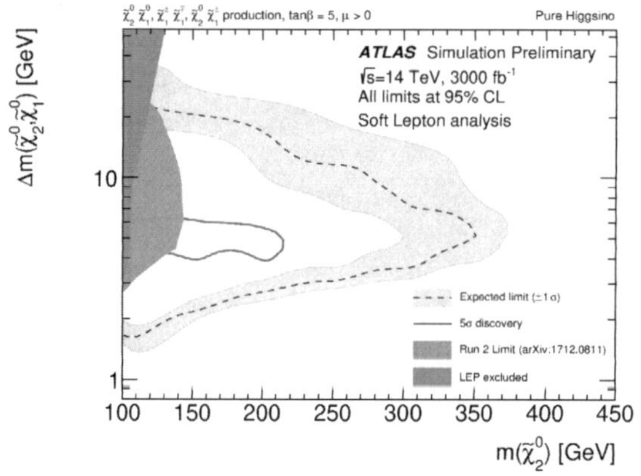

Figure 6. Expected exclusion at the 95% CL from the disappearing track and dilepton searches in the $\Delta m(\tilde{\chi}_2^0, \tilde{\chi}_1^0)$, $m(\tilde{\chi}_2^0)$ mass plane. The blue curve presents the exclusion limits from the dilepton search. The yellow contour presents the exclusion limit from the disappearing track search. The figure also presents the limits on chargino production from LEP. Figure taken from Ref. [32].

an off-shell Z or W boson respectively, $\tilde{\chi}_2^0 \to Z^*\tilde{\chi}_1^0$ and $\tilde{\chi}_1^{\pm} \to W^*\tilde{\chi}_1^0$. Due to the very small mass splitting of the electroweakinos in this scenario, a jet arising from initial-state radiation is required, to boost the sparticle system.

Figure 6 shows the 95% CL expected exclusion limits in the $\Delta m(\tilde{\chi}_2^0, \tilde{\chi}_1^0)$, $m(\tilde{\chi}_2^0)$ plane. With 3000 fb^{-1}, $\tilde{\chi}_2^0$ masses up to 350 GeV could be excluded, as well as $\Delta m(\tilde{\chi}_2^0, \tilde{\chi}_1^0)$ between 2 and 20 GeV for $m(\tilde{\chi}_2^0) = 150$ GeV. In the figure the blue curve presents the 5σ discovery potential of the search.

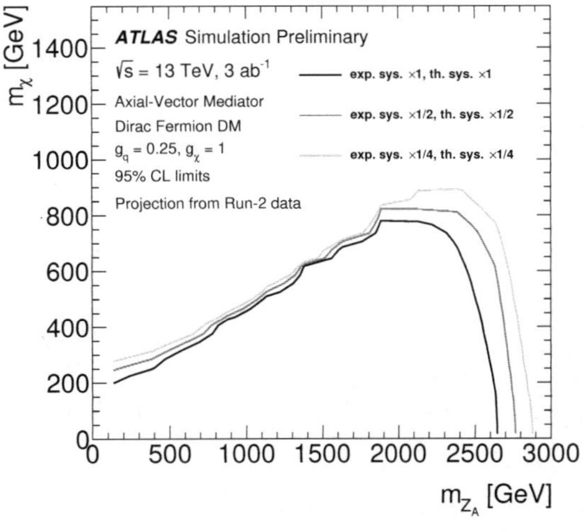

Figure 7. Expected 95% CL excluded regions on the (m_χ, m_{Z_A}) mass plane for the axial-vector simplified model with couplings $g_\chi = 1$ and $g_q = 0.25$, for a luminosity of 3000 fb^{-1}. Three contours are shown in each plot, corresponding to the three different systematic uncertainty scenarios: standard (black), reduced by a factor 2 (red) and 4 (green). Figure taken from Ref. [38].

3.4. Other BSM searches

While there are many BSM searches that will benefit from the HL-LHC dataset and the upgraded detectors, only one additional example from the BSM search sector is giving here. The final state with at least one jet and large missing transverse momentum (E_T^{miss} + jet) is a key channel for the search for dark matter at the LHC. Various theories [33, 34, 35] predict the production of pairs of Weakly Interacting Massive Particles (WIMPs) [36, 37] in association with hadronic jets from initial state radiation which can be used to discriminate signal from background events. This process is described with simplified benchmark models, which assume the existence of a massive mediator of the interaction between the initial state partons and the WIMPs. Depending on the spin-parity state of the mediator and on its couplings, different sensitivity is achieved at the LHC compared to direct and indirect detection experiments, with differing complementarity between LHC search channels, namely those with missing transverse momentum and those with resonant particles. The goal of this study [38] is to evaluate the impact of different assumed systematic uncertainty scenarios on the expected sensitivity to dark matter in the E_T^{miss} + jet channel. A simplified benchmark model is chosen to illustrate the WIMP discovery potential, where Dirac fermion WIMPs, χ, are pair-produced from the s-channel exchange of a spin-1 mediator, Z_A, with axial-vector couplings. The model is defined by four free parameters: the mediator mass (m_{Z_A}), the WIMP mass (m_χ), the coupling strength of the flavour-universal mediator-quark interaction (g_χ), and the coupling strength of the mediator-WIMP interaction (g_q). The exclusion limits, obtained with the profile likelihood method, for each of the three considered systematic uncertainty scenarios is shown in the (m_χ, m_{Z_A}) mass plane in Figure 7. The 95% CL exclusion contour for $m_\chi = 1$ GeV can be extended up to $m_{Z_A} \sim 2.65$ TeV. The excluded region that can be obtained by reducing by a factor two (four) all systematic uncertainties reaches, for low m_χ, mediator masses of about 2.77 (2.88) TeV. Small differences between systematic uncertainty scenarios are observed when approaching the region where the decay of the mediator in two WIMPs is off the mass shell ($m_{Z_A} < 2m_\chi$), due to the decrease of the signal cross-section. The results obtained for the discovery power can be found in Ref. [38].

IOP Conf. Series: Journal of Physics: Conf. Series **1271** (2019) 012004 doi:10.1088/1742-6596/1271/1/012004

4. Summary
The HL-LHC provides an unprecedented opportunity to study the Higgs boson in great detail and offers a unique ability to probe new physics beyond the Standard Model. The expected physics reach of the ATLAS and CMS detectors at the HL-LHC era and its discovery potential in the Higgs sector has been studied in detail as input to the Yellow Report that is due to be published in the beginning of 2019.

References

[1] ATLAS Collaboration 2008 *JINST* **3** S08004
[2] CMS Collaboration 2008 *JINST* **3**, S08004
[3] ATLAS Collaboration 2012 *Phys. Lett. B* **716** 1-29
[4] CMS Collaboration 2012 *Phys. Lett. B* **716** 30-61
[5] ATLAS Collaboration, CMS Collaboration 2015 *Phys. Rev. Lett.* **114** 191803
[6] CMS Collaboration 2015 *Phys. Rev. D* **92** 012004
[7] ATLAS Collaboration 2015 *Eur. Phys. J. C* **75** 476
[8] ATLAS Collaboration, CMS Collaboration 2016 *JHEP* **08** 045
[9] G. Apollinari, I. Bjar Alonso, O. Brning, P. Fessia, M. Lamont, L. Rossi, L. Tavian (editors), CERN-2017-007-M, https://cds.cern.ch/record/2284929
[10] ATLAS Collaboration, CERN-LHCC-2017-021, https://cds.cern.ch/record/2285585
[11] ATLAS Collaboration, CERN-LHCC-2017-005, https://cds.cern.ch/record/2257755
[12] ATLAS Collaboration, CERN-LHCC-2017-018, https://cds.cern.ch/record/2285582
[13] ATLAS Collaboration , CERN-LHCC-2017-019, https://cds.cern.ch/record/2285583
[14] ATLAS Collaboration, CERN-LHCC-2017-017, https://cds.cern.ch/record/2285580
[15] ATLAS Collaboration, CERN-LHCC-2017-020, https://cds.cern.ch/record/2285584
[16] ATLAS Collaboration, CERN-LHCC-2018-023, https://cds.cern.ch/record/2623663
[17] CMS Collaboration, CERN-LHCC-2017-009, https://cds.cern.ch/record/2272264
[18] CMS Collaboration, CERN-LHCC-2017-011, https://cds.cern.ch/record/2283187
[19] CMS Collaboration, CERN-LHCC-2017-023, https://cds.cern.ch/record/2293646
[20] CMS Collaboration, CERN-LHCC-2017-012, https://cds.cern.ch/record/2283189
[21] CMS Collaboration, CERN-LHCC-2017-027, https://cds.cern.ch/record/2296612
[22] CMS Collaboration, CERN-LHCC-2017-013, https://cds.cern.ch/record/2283192
[23] CMS Collaboration, CERN-LHCC-2017-014, https://cds.cern.ch/record/2283193
[24] ATLAS Collaboration, https://twiki.cern.ch/twiki/bin/view/AtlasPublic/UpgradePhysicsStudies
[25] CMS Collaboration, http://cms-results.web.cern.ch/cms-results/public-results/preliminary-results/FTR
[26] European Strategy for Particle Physics, https://europeanstrategy.cern/european-strategy-for-particle-physics
[27] D. de Florian et al., 2016 Handbook of LHC Higgs Cross Sections: 4. Deciphering the Nature of the Higgs Sector, arXiv:1610.07922 [hep-ph]
[28] ATLAS Collaboration, ATL-PHYS-PUB-2017-001, https://cds.cern.ch/record/2243387
[29] CMS Collaboration, CMS-PAS-FTR-15-002, https://cds.cern.ch/record/2063038
[30] CMS Collaboration, CMS-PAS-FTR-16-002, https://cds.cern.ch/record/2266165
[31] CMS Collaboration, CMS-PAS-FTR-16-006, https://cds.cern.ch/record/2262606
[32] ATLAS Collaboration, ATL-PHYS-PUB-2018-031, https://cds.cern.ch/record/2647294
[33] J. Abdallah et al., 2015 *Phys. Dark Univ.* **9-10** 8-23
[34] D. Abercrombie et al., 2015 Dark Matter Benchmark Models for Early LHC Run-2 Searches: Report of the ATLAS/CMS Dark Matter Forum, arXiv:1507.00966 [hep-ex]
[35] O. Buchmueller, M. J. Dolan, S. A. Malik and C. McCabe, 2015 *JHEP* **01** 037
[36] G. Steigman and M. S. Turner, 1985 *Nucl. Phys. B* **253** 375
[37] E. W. Kolb and M. S. Turner, 1990 *Front. Phys.* **69** 1
[38] ATLAS Collaboration, ATL-PHYS-PUB-2018-043, https://cds.cern.ch/record/2650050

IOP Conf. Series: Journal of Physics: Conf. Series **1271** (2019) 012005 doi:10.1088/1742-6596/1271/1/012005

Search for a Higgs boson decaying through Z_d to 4 leptons

D Boye[1,2], **X Mapekula**[1]

[1]University of Johannesburg, Department of Mechanical Engineering, South Africa, Johannesburg

[2]University of South Africa, Physics Department, South Africa Johannesburg

E-mail: diallo.boye@cern.ch

Abstract. The Standard Model (SM) is known to be incomplete (it cannot explain dark matter, dark energy, gravitational waves, matter-antimatter asymmetry, etc). The introduction of a Dark Sector via an additional $U(1)_d$ gauge symmetry added to the SM Lagrangian could be the long-awaited solution. In this model there is a dark vector boson Z_d which can mix with the SM hypercharge gauge boson. This opens the Hypercharge Portal which can mediate the fluctuation of a Z to a Z_d, or the decay of the Z_d to SM leptons. If a dark Higgs singlet also exists, then this breaks the $U(1)_d$ symmetry, allowing for Higgs mixing (parametrised by κ) between the SM and dark sector Higgs bosons, and thereby opening the Higgs portal as a discovery channel. Including dark fermionic fields in the Lagrangian allows for long-lived cold Dark Matter candidates. The various connections between the Dark and SM sectors allow descriptions of many key astro-physical phenomena. This contribution discusses an ATLAS search for the dark force boson Z_d using its production via the Higgs Portal and its decay back to SM leptons: $H \to Z_d Z_d \to 4\ell$. The biggest deviation from the Standard Model expectation is from a single event at $\langle m_{\ell\ell} \rangle \approx 20$ GeV, with a local significance of $3.2\,\sigma$ in the $2e2\mu$ channel, and the corresponding global significance is approximately $1.9\,\sigma$.

1. Introduction

The search for the Z_d dark force particle is inspired by a class of models where the Standard Model (SM) is extended by introducing hidden or dark sector states [1, 2, 3, 4, 5, 6, 7, 8, 9, 10], in a way that provides candidates for dark matter [11] and dark forces which accommodate both the indirect and the (potential) direct evidence based on astronomical observations or space platform experiments [12, 13, 14]. The hidden or dark sector is introduced in this case with an additional $U(1)_d$ dark gauge symmetry [5, 6, 7, 8, 9, 10].

If the primary route for the production of dark particles in the search, proceeds from the prior production of the Higgs particle, then we have also specialised to the Higgs Portal class of models. In this case we assume the dark sector coupling to the SM through kinetic mixing with the hypercharge boson via the kinetic mixing parameter ϵ to be very small. This is because ϵ allows for processes like $Z \longrightarrow Z_d$ and $Z_d \to \ell\ell$, which must already be constrained by the current excellent agreement between the SM and electroweak precision observables, $\epsilon \lesssim 10^{-4}$ [5, 6, 7, 8, 9, 10, 15, 16]. We introduce a Higgs level coupling between the dark sector and the SM. The $U(1)_d$ symmetry of the dark sector is broken by the introduction of a dark Higgs boson, which can mix with the SM Higgs boson [5, 6, 7, 8, 9, 10] with a coupling strength κ. The observed Higgs boson would then be the lighter partner of the new Higgs doublet, which can also decay via the dark sector. We then conceptually allow the decay $H \to h_d \to Z_d Z_d$. We also assume the dark fermions are sufficiently heavy $m_{f_d} < m_{Z_d}/2$, so that the branching ratio for the decay $Z_d \to \ell\ell$ may be taken as 100%, even though the kinetic mixing parameter ϵ has

IOP Conf. Series: Journal of Physics: Conf. Series **1271** (2019) 012005 doi:10.1088/1742-6596/1271/1/012005

been set small. The Higgs Portal has been opened by the observation of the discovered Higgs at 125 GeV [17, 18, 19] during Run 1 of the Large Hadron Collider (LHC) [20, 21]. This has ushered in a new and rich experimental program for physics beyond the SM.

This paper describes the recently published Run 2 search for the Higgs boson decaying to four leptons via two Z_d bosons using *pp* collision data at $\sqrt{s} = 13$ TeV corresponding to an integrated luminosity of 36.1±1.2 fb^{-1} collected at the CERN LHC with the ATLAS experiment [22]. The Run 1 search at $\sqrt{s} = 8$ TeV used a dataset corresponding to an integrated luminosity of 20.3±0.56 fb^{-1} [23] for $H \to Z_d Z_d \to 4\ell$ and was published in [24]. The Run 2 data both extended the search and also incorporates several extensions and improvements.

In the search, same-flavor decays of the Z_d bosons to electron and muon pairs are considered, giving the $4e$, $2e2\mu$, and 4μ final states. Final states including τ leptons are not considered. In the absence of a significant signal, a 95% CL upper limit was set on the model-independent fiducial cross section for the process $H \to Z_d Z_d \to 4\ell$. Also, considering the BSM benchmark model where the SM was extended by an additional $U(1)_d$ gauge symmetry, upper limits could be set on the branching ratio $\mathrm{BR}(H \to Z_d Z_d)$ [5, 6]. The search is restricted to the mass range where the Z_d from the decay of the Higgs boson is on-shell, i.e. $15\,\mathrm{GeV} < m_{Z_d} < m_H/2$, where $m_H = 125$ GeV. The Run 2 paper [22] also considered the search $H \to aa \to 4\mu$ where the intermediate state contains two light pseudoscalar bosons, in the range, $1\,\mathrm{GeV} < m_a < 15\mathrm{GeV}$, however this aspect is not part of this proceedings paper.

2. Experimental Setup, Monte Carlo Simulation : Signal and backgrounds

The ATLAS detector covers almost the whole solid angle around the collision point with layers of tracking detectors, calorimeters and muon chambers. Further details can be found in [25]

Signal : The trigger, event pre-selection and lepton reconstruction proceed as desribed in reference [22]. These leptons are then combined into quadruplets according to quality criteria as given below.

Table 1: Summary of the event selection taken from table 1 in [22].

Object	$\boldsymbol{H \to Z_d Z_d \to 4\ell}$
QUADRUPLET SELECTION	- Require at least one quadruplet of leptons consisting of two pairs of same-flavour opposite-charge leptons - Three leading-p_T leptons satisfy $p_T > 20$ GeV, 15 GeV, 10 GeV. - At least three muons are required to be reconstructed by combining ID and MS tracks in the 4μ channel. - Leptons in the quadruplet responsible for firing at least one trigger - $\Delta R(\ell, \ell') > 0.10$ (0.20) for all same (different) flavour leptons in the quadruplet
QUADRUPLET RANKING	– Select quadruplet with smallest $\Delta m_{\ell\ell} = \lvert m_{12} - m_{34} \rvert$
EVENT SELECTION	- Reject event if: $(m_{J/\Psi} - 0.25\ \mathrm{GeV}) < m_{12,34,14,23} < (m_{\Psi(2S)} + 0.30\ \mathrm{GeV})$ $(m_{\Upsilon(1S)} - 0.70\ \mathrm{GeV}) < m_{12,34,14,23} < (m_{\Upsilon(3S)} + 0.75\ \mathrm{GeV})$ - $m_{34}/m_{12} > 0.85$ - $115\ \mathrm{GeV} < m_{4\ell} < 130\ \mathrm{GeV}$ - $10\ \mathrm{GeV} < m_{12,34} < 64\ \mathrm{GeV}$ - $5\ \mathrm{GeV} < m_{14,32} < 75\ \mathrm{GeV}$ for $4e$ and 4μ channels

Simulated event samples are used to model the signal process, and to estimate most of the SM backgrounds. The samples are then passed through a simulation of the ATLAS detector [26] based on GEANT4 [27].

2.1. Signal

$H \to Z_d Z_d \to 4\ell$ The signal process is generated using the Hidden Abelian Higgs Model (HAHM) [9, 10] model inMadGraph5 [28] interfaced withPythia8 [29] (v8.170) for the modelling of the parton shower, hadronisation and underlying event (UE), using the CTEQ6L1 parton distribution functions (PDFs) [30] at next-to-leading order (NLO). The mass of the Z_d boson is varied for different signal hypotheses in the range of $15\ \mathrm{GeV} < m_{Z_d} < 60\ \mathrm{GeV}$ for $H \to Z_d Z_d \to 4\ell$, in 5 GeV steps. The considered Higgs boson production mode is gluon-gluon-Fusion (ggF), with a mass $m_H = 125$ GeV. The samples are normalised to the next-to-next-to-next-to-leading-order (N3LO) cross section $\sigma_{SM}(ggF) = 48.58$ pb as recommended by the Higgs cross section working group [31].

2.2. Backgrounds

$H \rightarrow ZZ^* \rightarrow 4\ell$: The POWHEG-BOX v2 MC event generator [32, 33, 34] is used to simulate Higgs production through ggF [35], vector-boson fusion (VBF) [36] and in association with a vector-boson (VH) [37] processes, using the PDF4LHC NLO PDF set [38]. For Higgs boson production in association with a heavy quark pair, events are simulated with MADGRAPH5_AMC@NLO [39], using the CT10nlo PDF set [40] for $t\bar{t}H$ and the NNPDF23 PDF set [41] for $b\bar{b}H$. For the ggF, VBF, VH, and $b\bar{b}H$ production mechanisms,PYTHIA8 [42] is used for the $H \rightarrow ZZ^* \rightarrow 4\ell$ decay as well as for parton showering, hadronisation, and multiple parton-parton interactions using the AZNLO parameter set [43]. For the showering of $t\bar{t}H$ process, HERWIG++ [44] is used with the UEEE5 parameter set [45].

$ZZ^* \rightarrow 4\ell$: The dominant $q\bar{q}$ production mechanism is modelled by POWHEG interfaced toPYTHIA8. The POWHEG-BOX v2 MC event generator [32, 33, 34] is used to simulate Higgs production through ggF [35], vector-boson fusion (VBF) [36] and in association with a vector-boson (VH) [37] processes, using the PDF4LHC NLO PDF set [38]. For Higgs boson production in association with a heavy quark pair, events are simulated with MADGRAPH5_AMC@NLO [39], using the CT10nlo PDF set [40] for $t\bar{t}H$ and the NNPDF23 PDF set [41] for $b\bar{b}H$. For the ggF, VBF, VH, and $b\bar{b}H$ production mechanisms, PYTHIA8 [42] is used for the $H \rightarrow ZZ^* \rightarrow 4\ell$ decay as well as for parton showering, hadronisation, and multiple parton-parton interactions using the AZNLO parameter set [43]. For the showering of $t\bar{t}H$ process, HERWIG++ [44] is used with the UEEE5 parameter set [45].

EWK6 $\rightarrow 4\ell+2X$: Higher-order electroweak processes (with cross sections proportional to α^6 at leading order) include triboson production and vector-boson scattering, which lead to four leptons in the final state, with two additional particles (quarks, neutrinos, or electrons and muons). These processes are modelled by SHERPA 2.1 with the CT10 PDFs.

$Z + (t\bar{t}/J/\psi/\Upsilon) \rightarrow 4l$: Production of a Z boson, in association with either a quarkonium state ($b\bar{b}$ or $c\bar{c}$) that decays to leptons, or $t\bar{t}$ production with leptonic decays of the prompt W bosons. The processes involving quarkonia are modelled using PYTHIA8 with the CTEQ6L1 PDFs. The $t\bar{t}Z$ process is modelled at leading order with MADGRAPH5 [46] interfaced to PYTHIA8, using the NNPDF 2.3 PDFs [41] and the A14 tune [47].

Reducible background: Processes like Z + jets, $t\bar{t}$ and WZ, produce less than four prompt leptons but can contribute to the selection through objects (e.g. jets) misidentified as leptons. Z + jets events are modeled using SHERPA 2.2. Semi- and fully-leptonic production of $t\bar{t}$ is generated with POWHEG interfaced to PYTHIA6 [48] for parton shower and hadronisation. The WZ production is modelled by POWHEG interfaced to PYTHIA8 and the CT10 PDFs.

3. Analysis procedure

3.1. Backgrounds to $H \rightarrow XX \rightarrow 4\ell$

The main background contributions come from the $H \rightarrow ZZ^* \rightarrow 4\ell$ and $ZZ^* \rightarrow 4\ell$ processes. These backgrounds are suppressed by the requirements on the lepton invariant masses. They account for 63% and 19% of the total prediction, respectively. Other sources of background with smaller contributions come from higher-order electroweak processes (with cross-sections proportional to α^6 at leading order). These include triboson production and vector-boson scattering, which lead to four leptons in the final state, with two additional particles (quarks, neutrinos, or electrons and muons). They are approximately 17% and 19% of the total prediction for the high- and low- mass selections, respectively. The production of $Z + (J/\psi/\Upsilon) \rightarrow 4\ell$ and $Z + t\bar{t} \rightarrow 4\ell$ constitute another minor source of background, accounting for approximately 1% of the total. All the background processes described above are estimated from simulation and normalised to the theoretical calculations of their cross sections.

3.2. Sytematics and uncertainties

The identification and quantification of systematics and uncertainties are important in the statistical procedure to identify the significance in the case of discovery and the limit in the case of exclusion.

Luminosity and pile-up: The uncertainty on the integrated luminosity is 3.2%, affecting the overall normalisation.

Lepton-related uncertainties: Uncertainties associated with leptons arise from the reconstruction and identification efficiencies [49, 50], as well as lepton momentum scales and resolutions [51, 50]. The combined effect of all these uncertainties results in an overall normalisation uncertainty in the signal and background of less than 10% (5% in muon final states).

MC background modelling: Uncertainties on the factorisation and renormalisation scales, the parton shower, the choice of PDF and the hadronisation and underlying event model affect those backgrounds normalised with their theory cross sections. Uncertainties on $H \rightarrow ZZ^* \rightarrow 4\ell$ are found to be between 3% and 9% depending on the Higgs boson production mode, while for Standard Model $q\bar{q}/gg \rightarrow ZZ^*$ processes these sources of uncertainties add up to 5%.

Signal modelling: Several sources of systematic uncertainty affect the theoretical modelling of the signal acceptance. Uncertainties originating from the choice of PDFs, the factorisation and renormalisation scales, the modelling of parton shower, hadronisation, and underlying event account for a total effect of 9% [52, 53].

Less than 4 background events are predicted in the signal region for the $H \rightarrow Z_d Z_d \rightarrow 4\ell$ search, so the dominant uncertainty in the background prediction is the statistical uncertainty.

4. Results

The m_{12} and m_{34} distributions of the selected events are shown in figure 1. The lower triangle comes from the labeling of the di-lepton masses which defines m_{12} as the di-lepton mass closer to the mass of Z-boson and m_{34} the other di-lepton mass, this is equivalent to $m_{12} > m_{34}$ in the 15 GeV $< m_{12,34} <$ 60GeV mass range, and the green shaded area with non-crossed points represents the signal region with cuts applied as in table 1.

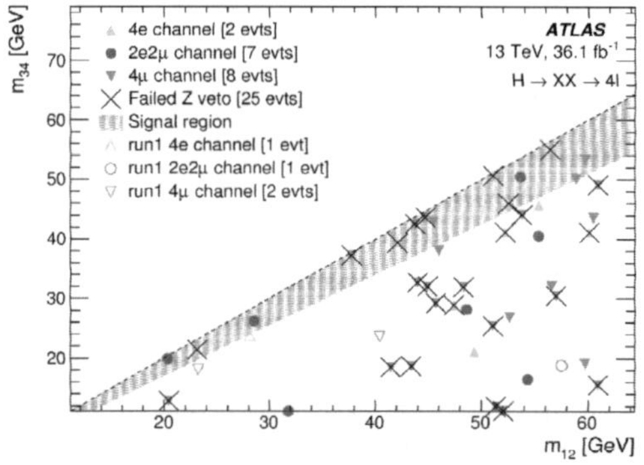

Figure 1: Leading dilepton mass (m_{12}) vs subleading dilepton mass (m_{34}) for events passing the selections (the crossed-through points fail the Z veto) [22]. The events passing the selection used in a similar analysis performed in run 1 are also shown [54].

IOP Conf. Series: Journal of Physics: Conf. Series **1271** (2019) 012005 doi:10.1088/1742-6596/1271/1/012005

The crossed points are events that fail the Z veto. This cut is applied onto the alternative-pairing masses m_{32} and m_{14} (relevant only to the $4e$ and 4μ channels), that they are less than 75 GeV. It is applied to mitigate against any small contributions from SM processes, even in the case of misidentified pairing, where there is Z-boson production with large cross sections.

It can be seen that 6 events appear in the signal region where 3.9 ± 3 are expected.

The biggest deviation from the Standard Model expectation is from a single event at $\langle m_{\ell\ell} \rangle \approx 20$ GeV, with a local significance of $3.2\,\sigma$ in the $2e2\mu$ channel. The corresponding global significance is approximately $1.9\,\sigma$, estimated using the likelihood test statistic tail probability approximation described in Ref. [55].

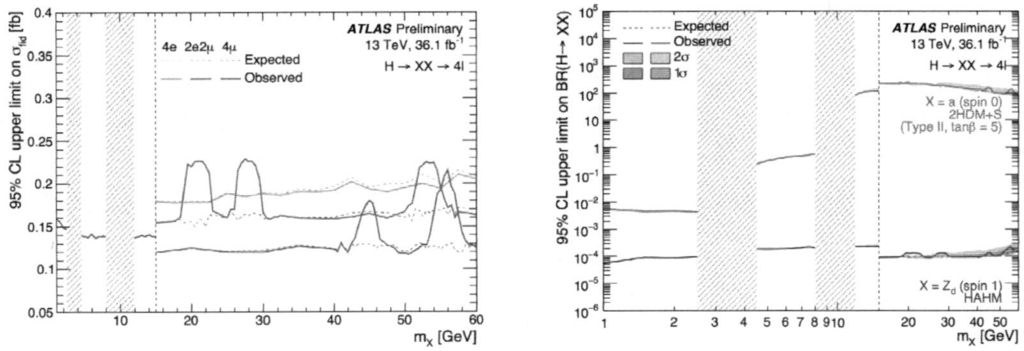

Figure 2: 95% CL upper limits of the process $H \to XX \to 4\ell$ on (left) model-independent fiducial cross sections,. and (right) branching fractions $\mathrm{BR}(H \to Z_d Z_d)$ for the benchmark model. The section from 1 to 15 GeV is not discussed in this proceedings article, see [22].

As there is no statistically significant evidence for the signal processes of $H \to Z_d Z_d \to 4\ell$, the results are interpreted in terms of the $Z_d Z_d$ benchmark model [5, 6]. The model provided the signal simulation for computing the reconstruction efficiencies in the fiducial phase spaces defined to mimic the analysis selections described above [22]. Upper limits on the fiducial cross sections should be applicable to any models of 125 GeV Higgs-boson decays to four leptons via two intermediate, on-shell, narrow, promptly-decaying bosons. The fiducial cuts are applied to the four leptons in this decay. These efficiencies are used to compute 95% CL upper limits on the cross sections in the fiducial phase space using the CL_s frequentist formalism [56] with the profile-likelihood test statistic. The results are shown at the left in figure 2.

Model-dependent acceptances for the fiducial phase spaces are computed per channel for the $H \to Z_d Z_d \to 4\ell$ search. The acceptances are used in a combined statistical model to compute upper limits on $\sigma_H \cdot \mathrm{BR}(H \to Z_d Z_d \to 4\ell)$. These cross section limits are converted into limits on the branching ratios of $H \to Z_d Z_d$ by using the theoretical branching ratios $Z_d \to \ell\ell$ from the benchmark model [5, 6], and assuming for σ_H the SM cross section[1] for Higgs boson production at $\sqrt{s} = 13$ TeV [53]. The limits on these branching ratios are shown at the right in figure 2 for the $H \to Z_d Z_d \to 4\ell$ search. The branching ratio limit $H \to Z_d Z_d$ is improved over the Run 1 result of Ref. [57] by about a factor of four which corresponds to the increase in both luminosity and Higgs cross section between Run 1 and Run 2.

Figure 3 shows the local p-value for the of the under the background-only hypothesis for the process $H \to XX \to 4\ell$ in the high-mass range, using the indicated profile likelihood ratio as test statistic. No globally significant excess is observed.

5. Conclusion

This proceedings article presents a dedicated search for exotic decays of the Standard Model Higgs boson with a mass of 125 GeV to two new spin-one particles, $H \to Z_d Z_d$, which in all

[1] Assumes the presence of BSM decays of the Higgs boson does not alter the SM Higgs-boson production cross section.

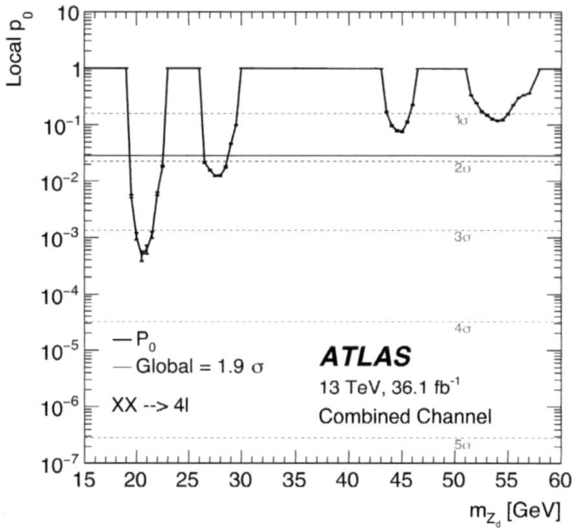

Figure 3: Observed local p-values under the background-only hypothesis for the process $H \to XX \to 4\ell$ for $15\,\text{GeV} < m_X < 60\,\text{GeV}$. The profile-likelihood ratio $(-2\log(L(\mu = 0, \hat{\theta})/L(\hat{\mu}, \hat{\theta})))$ is used as the test statistic. The most significant excess corresponds to a local significance of 3.2σ and a global significance of 1.9σ in $2e2\mu$ channel.

cases decay promptly to two lepton pairs. Six events are observed for a background prediction of 3.9±3. The results are expressed in terms of upper limits on the branching ratios $\mathrm{BR}(H \to Z_d Z_d)$ as a function of Z_d mass for the HAHM (Hidden Abelian Higgs Model) [5, 6] benchmark model. Limits are also provided on model-independent fiducial cross sections.

6. Acknowledgments

We thank CERN and all associated staff for the very successful operation of the LHC. The support of the National Research Foundation (NRF) and Department of Science and Technology, both of South Africa, is acknowledged. Similar acknowledgements apply for all participating institutions in the ATLAS Collaboration. Support from WLCG is acknowledged.

References

[1] Fayet P 2004 *Phys.Rev.* **D 70** 023514 (*Preprint* hep-ph/0403226)
[2] Finkbeiner D P and Weiner N 2007 *Phys Rev* **D 76** 083519 (*Preprint* astro-ph/0702587)
[3] Arkani-Hamed, N et al 2009 *Phys.Rev.* **D 79** 015014 (*Preprint* 0810.0713)
[4] Dudas, E et al 2012 *J. High Energy Phys.* **1210** 123 (*Preprint* 1205.1520)
[5] Curtin D, Essig R, Gori S and Shelton J 2015 *J. High Energy Phys.* **1502** 157 (*Preprint* 1412.0018)
[6] Curtin D *et al.* 2014 *Phys.Rev.* **D 90** 075004 (*Preprint* 1312.4992)
[7] Davoudiasl H, Lee H S, Lewis I and Marciano W J 2013 *Phys.Rev.* **D 88** 015022 (*Preprint* 1304.4935)
[8] Davoudiasl H, Lee H S and Marciano W J 2012 *Phys.Rev.* **D 85** 115019 (*Preprint* 1203.2947)
[9] Wells J D 2008 (*Preprint* 0803.1243)
[10] Gopalakrishna S, Jung S and Wells J D 2008 *Phys.Rev.* **D 78** 055002 (*Preprint* 0801.3456)
[11] Clowe D *et al.* 2006 *Astrophys.J.* **648** L109–L113 (*Preprint* astro-ph/0608407)
[12] Adriani O , et al (PAMELA Collaboration) 2009 *Nature* **458** 607–609 (*Preprint* 0810.4995)
[13] Chang J et al (ATIC Collaboration) 2008 *Nature News* **456** 362
[14] Aguilar M et al (AMS Collaboration) 2013 *Phys. Rev. Lett.* **110**(14) 141102
[15] Hook A, Izaguirre E and Wacker J G 2011 *Adv.High Energy Phys.* **2011** 859762 (*Preprint* 1006.0973)
[16] Hoenig I, Samach G and Tucker-Smith D 2014 *Phys.Rev.* **D 90** 075016 (*Preprint* 1408.1075)
[17] Englert F and Brout R 1964 *Phys. Rev. Lett* **13** 321–323
[18] Higgs P W 1964 *Phys. Rev. Lett* **13** 508–509
[19] Guralnik G S, Hagen C R and Kibble T W B 1964 *Phys. Rev. Lett* **13** 585–587
[20] ATLAS Collaboration 2012 *Phys.Lett.* **B716** 1–29 (*Preprint* 1207.7214)
[21] CMS Collaboration 2012 *Phys.Lett.* **B716** 30–61 (*Preprint* 1207.7235)
[22] ATLAS Collaboration 2018 *J. High Energy Phys.* **1806** 166 (*Preprint* 1802.03388)

[23] ATLAS Collaboration 2013 *Eur.Phys.J.* **C73** 2518 (*Preprint* 1302.4393)
[24] Wells J D 2015 (*Preprint* 1505.07645)
[25] ATLAS Collaboration 2009 (*Preprint* 0901.0512)
[26] ATLAS Collaboration 2010 *Eur. Phys. J. C* **70** 823 (*Preprint* 1005.4568)
[27] Agostinelli S et al (GEANT4 Collaboration) 2003 *Nucl. Instrum. Meth.* **A506** 250–303
[28] Alwall, J et al 2011 *J. High Energy Phys.* **1106** 128 (*Preprint* 1106.0522)
[29] Sjostrand T, Mrenna S and Skands P Z 2008 *Comput.Phys.Commun.* **178** 852–827 (*Preprint* 0710.3820)
[30] Pumplin, J et al 2002 *JHEP* **07** 012 (*Preprint* hep-ph/0201195)
[31] LHC Higgs Cross Section Working Group and Heinemeyer, S et al CERN, Geneva, 2013 *CERN-2013-004*
[32] Nason P 2004 *JHEP* **11** 040 (*Preprint* hep-ph/0409146)
[33] Frixione S, Nason P and Oleari C 2007 *JHEP* **11** 070 (*Preprint* 0709.2092)
[34] Alioli S, Nason P, Oleari C and Re E 2010 *JHEP* **06** 043 (*Preprint* 1002.2581)
[35] Hamilton K, Nason P, Re E and Zanderighi G 2013 *JHEP* **10** 222 (*Preprint* 1309.0017)
[36] Nason P and Oleari C 2010 *JHEP* **02** 037 (*Preprint* 0911.5299)
[37] Luisoni G, Nason P, Oleari C and Tramontano F 2013 *JHEP* **10** 083 (*Preprint* 1306.2542)
[38] Butterworth J *et al.* 2016 *J. Phys.* **G43** 023001 (*Preprint* 1510.03865)
[39] Alwall, J et al 2014 *JHEP* **07** 079 (*Preprint* 1405.0301)
[40] Lai, H-L et al 2010 *Phys. Rev.* **D82** 074024 (*Preprint* 1007.2241)
[41] Ball RD et al, (NNPDF Collaboration) 2013 *Nucl. Phys. B* **867** 244–289
[42] Sjöstrand T, Mrenna S and Skands P Z 2008 *Comput. Phys. Commun.* **178** 852–867 (*Preprint* 0710.3820)
[43] ATLAS Collaboration 2014 *JHEP* **09** 145 (*Preprint* 1406.3660)
[44] Bahr M *et al.* 2008 *Eur. Phys. J.* **C58** 639–707 (*Preprint* 0803.0883)
[45] Seymour M H and Siodmok A 2013 *JHEP* **10** 113 (*Preprint* 1307.5015)
[46] Alwall, J et al 2011 *JHEP* **06** 128 (*Preprint* 1106.0522)
[47] ATLAS Collaboration 2014 ATL-PHYS-PUB-2014-021 URL http://cdsweb.cern.ch/record/1966419
[48] Sjöstrand T, Mrenna S and Skands P Z 2006 *JHEP* **05** 026 (*Preprint* hep-ph/0603175)
[49] ATLAS Collaboration 2017 *Eur. Phys. J. C* **77** 195 (*Preprint* 1612.01456)
[50] ATLAS Collaboration 2016 *Eur. Phys. J. C* **76** 292 (*Preprint* 1603.05598)
[51] ATLAS Collaboration 2016 ATL-PHYS-PUB-2016-015 URL https://cds.cern.ch/record/2203514
[52] LHC Higgs cross section working group, Dittmaier, S et al 2011 *CERN-2011-002* (*Preprint* 1101.0593)
[53] LHC Higgs cross section working group, Dittmaier, S et al 2012 *CERN-2012-002* (*Preprint* 1201.3084)
[54] ATLAS Collaboration 2015 *Phys. Rev.* **D92** 092001 (*Preprint* 1505.07645)
[55] Gross E and Vitells O 2010 *Eur. Phys. J.* **C70** 525–530 (*Preprint* 1005.1891)
[56] Read A L 2002 *Journal of Physics G* **28** 2693 URL http://stacks.iop.org/0954-3899/28/i=10/a=313
[57] ATLAS Collaboration 2015 *Phys. Rev. D* **92** 092001 (*Preprint* 1505.07645)

IOP Conf. Series: Journal of Physics: Conf. Series **1271** (2019) 012006 doi:10.1088/1742-6596/1271/1/012006

Constraints on Standard Model physics from colliders

Kristof Schmieden, on behalf of the ATLAS and CMS collaborations
CERN[1]

E-mail: kristof.schmieden@cern.ch

Abstract. Effective field theories provide an elegant approach to study possible new physics scenarios, by comparing precision measurements with Standard Model predictions. The latest results on multi-boson processes from the ATLAS and CMS collaborations at the LHC, which are sensitive to anomalous gauge couplings and thus can be used to constrain EFT operators, are presented. The consistency of the Standard Model can also be tested by a global electroweak fit to precision observables in the electroweak sector. We present here the precision measurement of the W boson mass as well as the new results on the determination of the electroweak mixing angle and the top quark mass.

1. Introduction

The Standard Model of particle physics (SM) is tremendously successful in understanding the fundamental structure of matter. The precision of theoretical predictions are improving constantly and so are experimental results, confirming the theory predictions to an ever increasing accuracy. Yet despite the huge success of the Standard Model, it is since long known that it is incomplete. Big open questions not answered to date are the nature of neutrinos and dark matter, the origin of the baryon asymmetry in the universe as well as the quantum level description of gravity, to name a few. Without conclusive experimental evidence of new physics in high energy particle processes the question arises at which energies and in which processes physics beyond the Standard Model will appear. These proceedings give an overview of two approaches used by the ATLAS [1] and CMS [2] collaborations at the LHC to find effects from physics beyond the Standard Model in high energy particle collisions.

Section 2 will discuss the usage of an effective field theory (EFT) approach to model non Standard Model contributions to gauge boson couplings. The latest results from the ATLAS and CMS collaborations on the electroweak production of vector boson as well as multi-boson production will be presented, along with the interpretation of the results in terms of limits on contributions of EFT operators. The second approach of finding effects beyond the Standard Model relies on probing the consistency of the Standard Model to the best achievable precision. This is done by performing a global fit of all electroweak parameters of the Standard Model using the most precise experimental measurements, and is detailed in section 3.

2. Constraining Effective Field theory operators using measurements of vector boson production

Effective field theories are expansions in inverse powers of the energy scale of new interactions assuming perturbative coupling coefficients. For a detailed introduction to EFTs see e.g. [3, 4]. Essentially an effective field theory is a systematic Taylor expansion of all occurring variables.

$$f(x) = f(0) + f'(0)x + f''(0)x^2 + ... \tag{1}$$

At low energies (e.g. $\sqrt{s} \approx m_Z$) only simple observables are accessible, like the properties of the Z boson (mass, width, angular distributions) which can be converted into the respective couplings. Contributions from new physics could potentially lead to a rescaling of those couplings. At higher energies $\sqrt{s} \gg m_Z$ however, more and more observables become accessible. Modifications to those are described by higher orders of the Taylor expansion which has to be truncated at some order. The minimal useful order of the Taylor expansion is derived from the most minimal meaningful extension to the Standard Model possible. These are operators of dimension 6, which do not introduce new states and solely lead to modified couplings. In the most recent analyses, however, operators up to dimension 8 are considered. The EFT Lagrangian can then be written as

$$\mathcal{L}_{EFT} = \mathcal{L}_{SM} + \sum_{i=WWW,W,B,\Phi W,\Phi B} \frac{c_i}{\Lambda^2}\mathcal{O}_i + \sum_{j=1,2} \frac{f_{S,j}}{\Lambda^4}\mathcal{O}_{S,j} + \sum_{j=0..9} \frac{f_{T,j}}{\Lambda^4}\mathcal{O}_{T,j} + \sum_{j=0..7} \frac{f_{M,j}}{\Lambda^4}\mathcal{O}_{M,j} \tag{2}$$

where the first term corresponds to dimension 6 operators. Those are sensitive to triple gauge boson couplings (TGC), amongst which only the WWZ and $WW\gamma$ couplings are non-zero at tree level in the SM. The remaining three terms correspond to dimension 8 operators. Those operators contribute to quartic gauge boson couplings (QGC) as well as neutral TGCs. The EFT formalism facilitates a model independent, well defined parametrization of new physics above the mass reach of a given experiment, thus providing an ideal benchmark to assess the impact of SM measurements on the search for new physics phenomena. A historically used alternative to the EFT formalism is referred to as the framework of effective vertex functions [5] in which anomalous coupling vertices are added explicitly to the SM Lagrangian together with complex form factors modifying their contributions. Those form factors are usually labeled h_i^V, $\Delta g, \Delta\kappa, \lambda, \tilde{\kappa}$ and $\tilde{\lambda}$ in literature and some will be presented later on. However, as the estimation of those form factors is model dependent, recent LHC analyses are encouraged to calculate limits on anomalous couplings in term of limits on EFT operators.

Experimentally two approaches are distinguished when studying multi-gauge boson interactions.

(i) Inclusive cross-section measurements of di- and tri-boson final states.

(ii) Measurement of the electroweak production cross-section of single gauge bosons.

Representative Feynman diagrams for both processes are shown in figure 1. Both the ATLAS and CMS collaborations perform a variety of measurements using either approach. A recent, but not exhaustive, summary of measurements of di-boson final states from the CMS collaboration is shown in figure 2(left). A summary of the results on electroweak (EW) production of vector bosons in vector boson fusion (VBF) and vector boson scattering (VBS) analyses is presented in figure 2(right). In the following a selection of analyses will be presented which is by no means exhaustive. Usually corresponding analyses are performed by both the ATLAS and CMS collaborations. For a complete list of measurements performed please refer to [6] and [7], respectively. Overall no significant deviations from the SM predictions have been observed so far.

Selected results from studies of inclusive multi-boson production are presented in section 2.1, followed by selected results of measurements studying the electroweak production of single vector bosons in section 2.2.

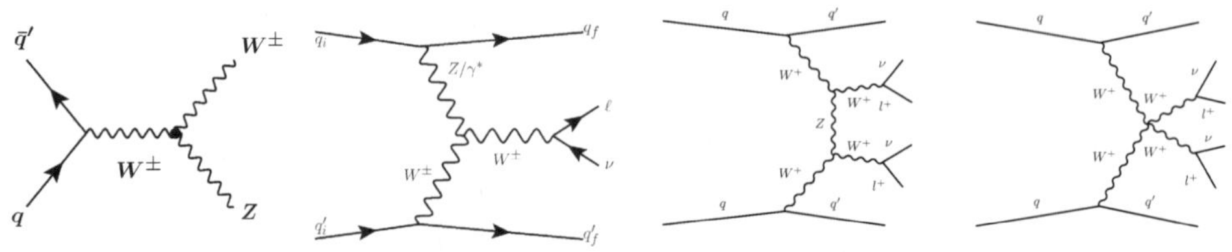

Figure 1. Feynman diagrams showing from left to right: the tree-level process for WZ production, electroweak production of a W boson, WW boson pair production involving TGCs and QGC.

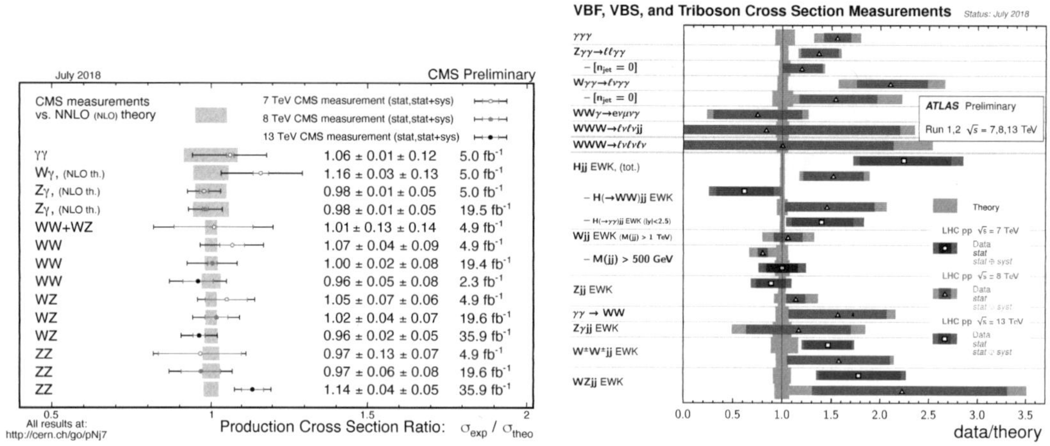

Figure 2. Overviews of di-boson production cross-section measurements from the CMS collaboration (left) [7] as well as VBF, VBS and tri-boson production cross-section measurements from the ATLAS collaboration (right) [6].

2.1. Inclusive production of multiple gauge bosons

The measurement of the differential production cross-section of a Z boson together with a photon is performed by ATLAS using $36\,\text{fb}^{-1}$ of $13\,\text{TeV}$ pp collisions data [8]. The invisible Z boson decay into neutrinos is utilized in this analysis. Studying the Z boson decay into neutrinos has several advantages over processes where the Z boson decays into hadrons or charged leptons. The hadronic decay channel is contaminated with a large multi-jet background. A larger Z boson branching ratio into neutrinos relative to that into charged leptons provides an opportunity to study this process for larger photon transverse energies (E_{T}^{γ}), where the sensitivity bosonic couplings is higher. In addition, the neutrino channel is sensitive to anomalous neutrino dipole moments, although a larger integrated luminosity than that available to this study would be required to significantly improve upon the LEP results [9]. Events featuring photons with $E_{\text{T}}^{\gamma} > 600\,\text{GeV}$ are used to set limits on anomalous triple gauge couplings (aTGC). Uncertainties on aTGC limits estimated from the $Z\gamma$ final state were reduced by a factor 4 to 7 within last two years, profiting from the increased LHC beam energy and a larger data set. This is visible in figure 3, where the parameters $h_{3,4}^{\gamma}$, calculated using the framework of effective vertex functions, are compared between the ATLAS measurement performed in 2016 using 20.3fb^{-1} of $8\,\text{TeV}$ data [10] and the one from 2018 presented here.

The only TGC between W and Z bosons allowed in the SM is the WWZ coupling. This can be studied using WZ final states to search for any additional anomalous WWZ couplings. Such

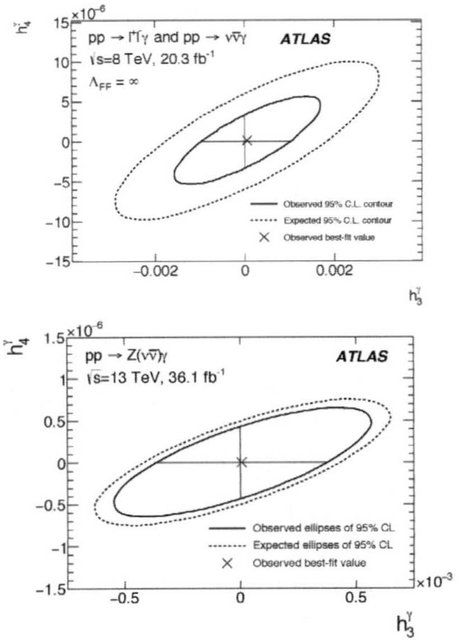

Figure 3. 95% C.L. limits on h_3 and h_4, extracted from the study of $Z(\bar{\nu}\nu)\gamma$ events published in 2016 [10] (top) and 2018 [8] (bottom).

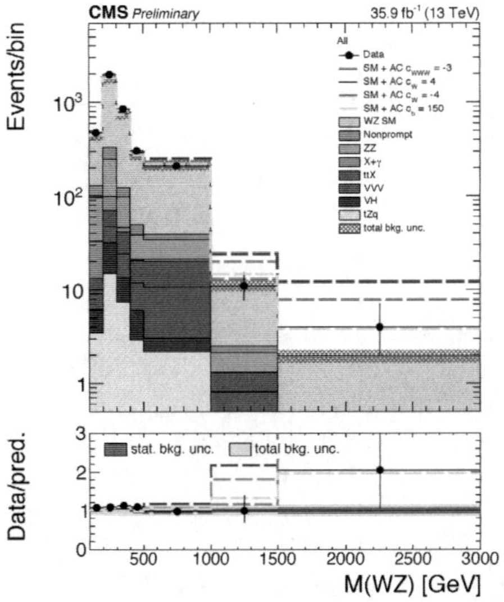

Figure 4. Measured invariant mass distribution of WZ boson pairs (black markers) compared to the SM expectation (shaded histograms) and predictions with modified coupling parameters (dashed lines)[12].

analyses have been performed by both the ATLAS [11] and CMS [12] collaborations yielding similar results. Anomalous couplings changing the WWZ vertex would result in an altered invariant mass distribution of the selected WZ events. Measurement and predictions for various modified coupling scenarios are shown in figure 4. As no deviation from the SM expectation is observed limits are set on the EFT parameters contributing to anomalous WWZ couplings and shown in table 1 together with results from various other analyses.

Finally, the three-boson final state is sensitive to QGCs and a total of 16 dimension 8 EFT operators. Events containing $WW\gamma$ or $WZ\gamma$ boson triplets have been studied by the ATLAS collaboration [17]. Two final state signatures are considered: $e\nu\mu\nu\gamma$ which is enriching $WW\gamma$ events and $e\nu jj\gamma$ enriching $WZ\gamma$ events. One of the variables that is most sensitive to the EFT operators is the photon transverse energy, which is shown for the $e\nu jj\gamma$ selection in figure 5 together with the limits set on all accessible dimension 8 operators.

2.2. Electroweak production of vector bosons

The fusion of vector bosons is a particularly important process for measuring particle properties, such as the couplings of the Higgs boson, as well as for searching for new particles beyond the SM. It is a sensitive probe of anomalous couplings of vector bosons. Several analyses from the ATLAS and CMS collaborations study EW production of single vector bosons as well as di-vector boson production. In *pp* collisions, a characteristic signature of these processes is the production of two high-momentum jets of hadrons at small angles with respect to the incoming proton beams.

Several topologies are studied using dedicated analyses: single W or Z boson production, same sign $W^{\pm}$ pair production and $W^{\pm}Z$ pair production. EW production of single vector bosons is sensitive to the WWZ TGC, as can be seen from figure 1. A variety of analyses have been performed using W or Z bosons in the final state [15, 16, 18–22]. The results are

Table 1. Overview of expected and observed 1-D confidence intervals at 95% confidence level for various analyses presented throughout section 2.

Analysis	Parameter	(expected) [TeV^{-2}]	(observed) [TeV^{-2}]
WZ [12]	c_w/Λ^2	[-3.3, 2.0]	[-4.1, 1.1]
WV [13]	c_w/Λ^2	[-6.0, 6.7]	[-5.1, 5.8]
WV [14]	c_w/Λ^2	-	[-2.0, 5.7]
Zjj [15]	c_w/Λ^2	[-12.6, 14.7]	[-8.4, 10.1]
Wjj [16]	c_w/Λ^2	[-39, 37]	[-33, 30]
WZ [12]	c_{www}/Λ^2	[-1.8, 1.9]	[-2.0, 2.1]
WV [13]	c_{www}/Λ^2	[-3.6, 3.6]	[-3.1, 3.1]
WV [14]	c_{www}/Λ^2	-	[-2.7, 2.7]
Zjj [15]	c_{www}/Λ^2	[-3.7, 3.6]	[-2.6, 2.6]
Wjj [16]	c_{www}/Λ^2	[-16, 13]	[-13, 9]
WZ [12]	c_B/Λ^2	[-130, 170]	[-100, 160]
WV [13]	c_B/Λ^2	[-22, 23]	[-19, 20]
WV [14]	c_B/Λ^2	-	[-14, 17]
Wjj [16]	c_B/Λ^2	[-200, 190]	[-170, 160]
Wjj [16]	$c_{\tilde{W}/\Lambda^2}$	[-720, 720]	[-580, 580]
Wjj [16]	$c_{\tilde{W}WW/\Lambda^2}$	[-14, 14]	[-11, 11]

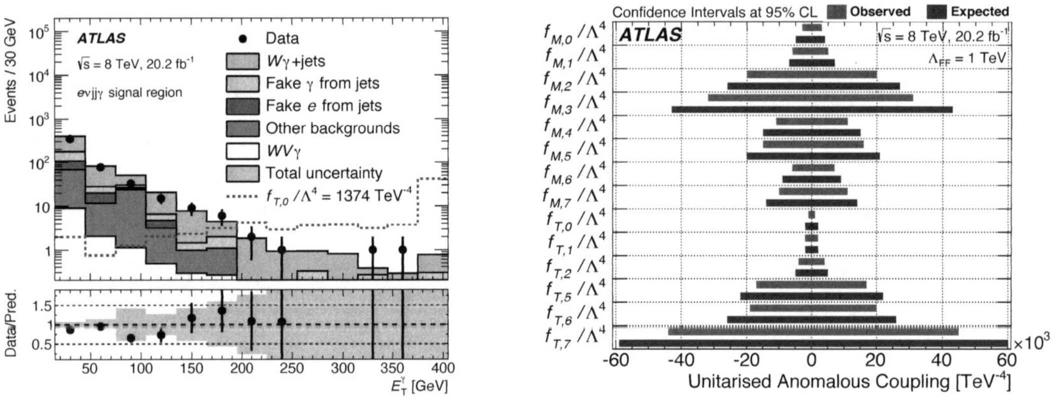

Figure 5. The left plot shows the $e\nu jj\gamma$ signal region of the tri-boson analysis [17], which is sensitive to quartic gauge couplings. Extracted limits on dimension 8 EFT operators are shown in the right plot.

summarized in figure 6. Those are turned into limits on EFT parameters or limits on anomalous coupling form factors in the framework of effective vertex functions. The most stringent limits are derived from the Zjj final state, whereas more operators have been studied in the Wjj final state. A summary of the limits on the EFT operators from several analyses is given in table 1.

The EW production of two vector bosons has been studied in several recent analyses as well [23, 24]. Of particular interest is the EW production of two same charge W bosons, i.e. the scattering process of two W bosons, to which tripple and quartic gauge couplings contribute as well as Higgs boson exchange. The latter leads to large cancelations between the Feynman diagrams involved, restoring the unitarity of the SM. Hence this process is a fundamental probe of the electroweak symmetry breaking and was first observed by the CMS collaboration in 2017

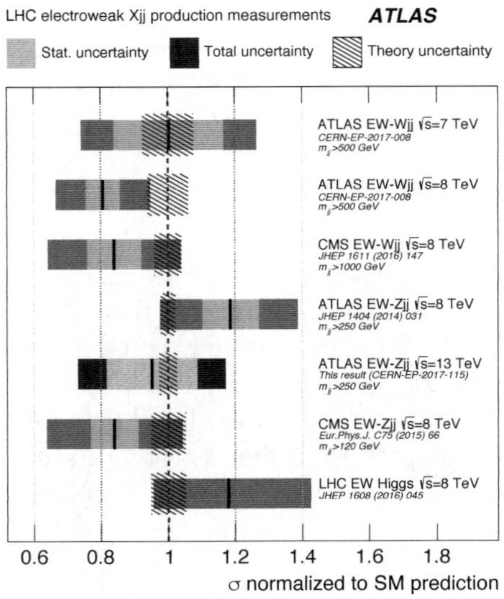

Figure 6. Overview of recent results on cross-section measurements for electroweak Vjj production from the ATLAS and CMS collaborations [6].

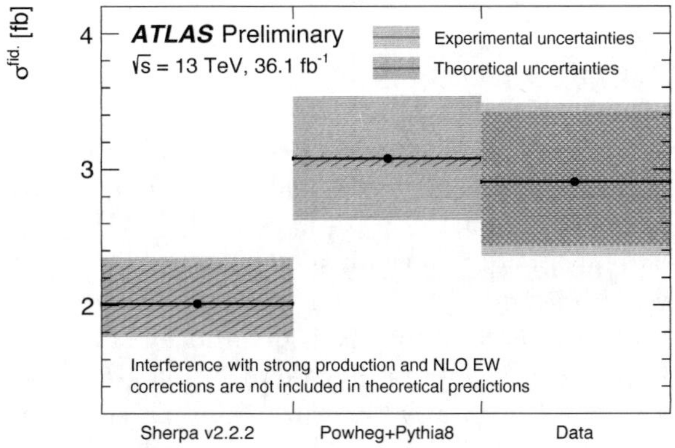

Figure 7. Predicted WW boson pair production cross-sections by Sherpa and Powheg+Pythia (blue) in comparison to the measurement (orange) [23].

[25] with a significance of $5.5\,\sigma$ ($5.7\,\sigma$ expected), and recently by the ATLAS collaboration with a significance of $6.9\,\sigma$ ($4.6\,\sigma$ expected) [23]. The measurements have been used to set limits on various dimension 8 EFT operators. The same sign WW production process is the only di-boson process to date for which EW and QCD NLO corrections have been computed [26]. Including NLO EW correction shifts the predicted cross-section by -15% in the fiducial region and significantly reduces the uncertainties from around 10% to 2%. Significant differences in the cross-section prediction by the Sherpa [27] and Powheg [28] event generators have been observed in [23] which is shown in figure 7.

The EW production of a W boson together with a Z boson is another key probe of the gauge symmetry in the electroweak sector of the SM as it is directly sensitive to the vector boson self-coupling. This process has recently been observed by the ATLAS collaboration for the first time [29] with a significance of $5.6\,\sigma$ ($3.3\,\sigma$ expected). The measured fiducial cross-section of the EW production of the WZ boson pair is measured to be

$$\sigma_{meas.}^{\text{fid.,EW}} = 0.57^{+0.14}_{-0.13}(\text{stat})^{+0.05}_{-0.04}(\text{syst})^{+0.04}_{-0.03}(\text{th})\text{fb}.$$

Currently only LO EW predictions are available for this process. Sherpa v2.2.2 predicts a cross-section of $\sigma_{Sherpa}^{\text{fid.,EW}} = 0.32 \pm 0.03\,\text{fb}$, significantly below the measurement. Triple and quartic gauge couplings contribute to the WZ pair production and hence this process can be used to set limits on several EFT parameters. This was done in a recent analysis by CMS [24] which provided limits on several dimension 8 operators. In addition the analysis exhibits sensitivity to the decay of a hypothetical charged Higgs boson into a WZ pair. As no evidence for such a process has been observed, limits were set accordingly.

All of the analyses involving multi-boson final states are limited by the size of the available datasets. Hence significant improvements can be expected with the analysis of the full LHC run-2 data set and even more so using the future data from LHC run-3 and beyond. An estimate of the evolutions of the cross-section uncertainty for the vector boson scattering processes has been presented in [30]. Analyses discussed in these proceedings use an integrated luminosity of $36\,\text{fb}^{-1}$ at most. A factor 3 improvement on the uncertainty can be expected until the end of run-3, assuming an integrated luminosity of $300\,\text{fb}^{-1}$. At the end of the HL-LHC data taking a reduction of the uncertainty by a factor of ≈ 8 with respect to the current analyses is expected.

3. Global electroweak fit

Simultaneous fitting of all SM electroweak parameters to theory calculations is a unique way of combining experimental electroweak precision observables with the most accurate theoretical calculations available in a global way. This has proven very successful in the past in predicting e.g. the masses of the top quark and the Higgs boson. It is also a comprehensive test of the consistency of the SM as illustrated in figure 8(top), which was released with the latest update of the results from the GFitter group [31]. Recent theory calculations used in the global fit include full two loop EW terms (NNLO) and even partial 3 and 4 loop terms. The latest updates of the experimental inputs include an updated measurement of the W boson mass, the top quark mass as well as the weak mixing angle and the Higgs mass.

The results of the global fit impressively confirm the consistency of the SM within the current experimental uncertainty. The calculated $\chi^2_{min}/\text{d.o.f.} = 18.6/15$ yields a probability of 23%. The largest potential for improvement lies in the W boson mass measurement, where the experimental uncertainty (13 MeV) and the uncertainty from the fit (7 MeV) have the same order of magnitude and the global fit is pulled by the measurement. Hence this observable should be assigned the highest priority for experimental improvements. The only, long lasting, tension visible in figure 8(left) is found between the asymmetry measurements $A^{0,b}_{FB}$ and $A_l(SLD)$ from LEP and SLD. A new generation of analyses at the LHC measuring the leptonic electroweak mixing angle may eventually resolve the tension.

Measurements for three input parameters to the global electroweak fit will be presented, namely the measurements of m_W, the weak mixing angle and the top quark mass.

3.1. Measurement of the W boson mass

When studying inclusive W boson decays two variables are sensitive to its mass. The transverse momentum distribution of the decay lepton p^l_{T} exhibits a Jacobean peak at $m_W/2$ at tree level. The distribution of the transverse mass, which is defined as $m_T = \sqrt{2p^l_{\text{T}}p^{miss}_{\text{T}}(1-\cos\Delta\phi)}$, has an endpoint corresponding to the mass of the W boson. Here p^{miss}_{T} is the missing transverse momentum of the event and $\Delta\phi$ is the difference in azimuth between the decay lepton and the missing transverse energy, which is associated to the neutrino from the W boson decay. Various effects modify the reconstructed p^l_{T} and m_T distributions, the most prominent ones being QED initial state radiation, the unknown transverse momentum of the W boson due to QCD effects as well as the detector response. The latest measurement of the W mass was released by the ATLAS collaboration in 2018 [32]. In this analysis both sensitive distributions were fitted independently to templates generated by simulation in 14 different event categories, separated in bins of η^l, by the W boson charge as well as electron and muon final states. Both, p^l_{T} and m_T, exhibit different sensitivity to additional effects. The m_T distribution is sensitive to the hadronic recoil resolution, but exhibits only a low sensitivity to the transverse momentum of the W boson. In contrast, the p^l_{T} distribution is directly sensitive to p^W_{T} but less dependent on the hadronic recoil measurement. The combination of all measurement channels yields a W boson mass of

$$m_W = 80370 \pm 7(\text{stat.}) \pm 11(\text{exp.syst.}) \pm 14(\text{mod.syst.})\,\text{MeV} = 80370 \pm 19\,\text{MeV}\,. \quad (3)$$

Results for individual W boson charges are shown along results from other experiments in Figure 9 (left). The largest uncertainty arises from the modeling of the W boson transverse momentum distribution, which is theoretically very difficult to describe. Hence the more precise

IOP Conf. Series: Journal of Physics: Conf. Series **1271** (2019) 012006 doi:10.1088/1742-6596/1271/1/012006

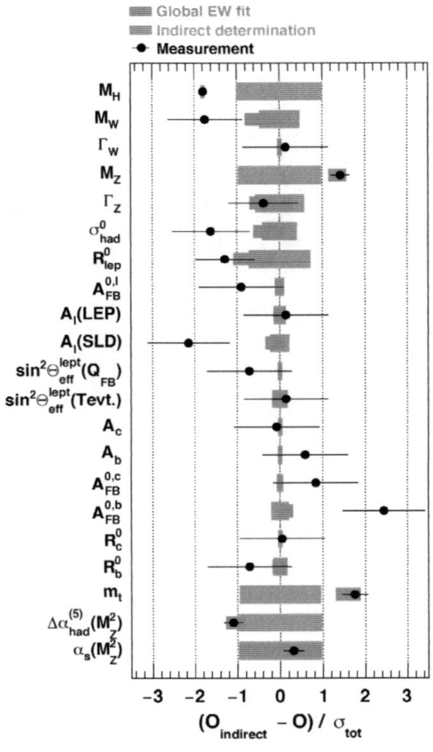

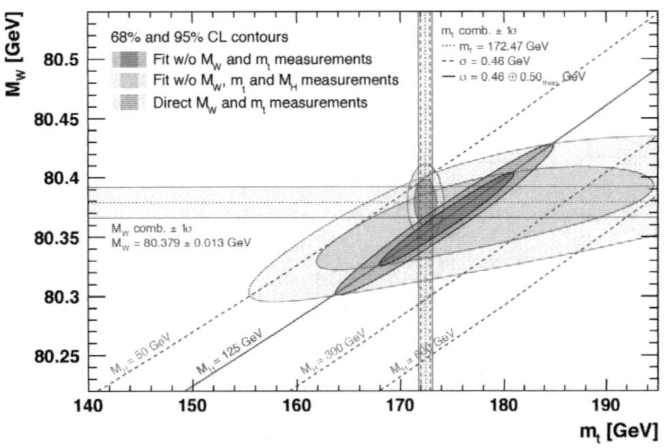

Figure 8. GFitter results [31]. (Left): comparison of input measurements to fit results and indirect determinations in units of the total uncertainties. The latter corresponds to a fit without using the constraint from the corresponding input measurement. (Top): Contours at 68% and 95% CL obtained from scans of m_W versus m_t for the global fit including (blue) and excluding (grey) the m_H measurement. Direct measurements and their $1\,\sigma$ errors are shown as green bands (ellipses show $1\,\sigma$ and $2\,\sigma$ errors).

measurement of the transverse momentum of the Z boson is transformed into the expected p_{T}^{W} distribution using predictions of the ratio $\sigma_Z/\sigma_W(p_{\mathrm{T}})$. Interestingly, different predictions of this ratio disagree significantly, in particular for low values of $p_{\mathrm{T}}^{W,Z}$, as is presented in [33]. A measurement of p_{T}^{W} in bins of 5 GeV with an accuracy of 1% is needed to significantly reduce this uncertainty. This could be achieved utilizing the data recorded under low pileup conditions by the ATLAS and CMS experiments, which amounts to $\approx 500\,\mathrm{pb}^{-1}$. Expected uncertainties for various PDFs in conjunction with the analysis of $200\,\mathrm{pb}^{-1}(1\,\mathrm{fb}^{-1})$ of low pileup data are presented in figure 9 (right), where a significant reduction of uncertainties is expected using a small amount of low pileup data. Other major contributions to the uncertainty on m_W arise from the uncertainties on the lepton reconstruction efficiency and calibration as well as uncertainties due to PDFs. Results from the most precise measurement of inclusive W, Z production performed using the 7 TeV LHC dataset were used to determine the PDF set that best describes the data used for the measurement of m_W to be the CT10 PDF. The uncertainty due to PDFs were assesed using this and comparisons to two additional PDF sets.

3.2. Measurement of the leptonic weak mixing angle

The full event information of the inclusive Z boson production and subsequent decay into lepton pairs can be described by the 5-dimensional differential cross-section which in turn can be

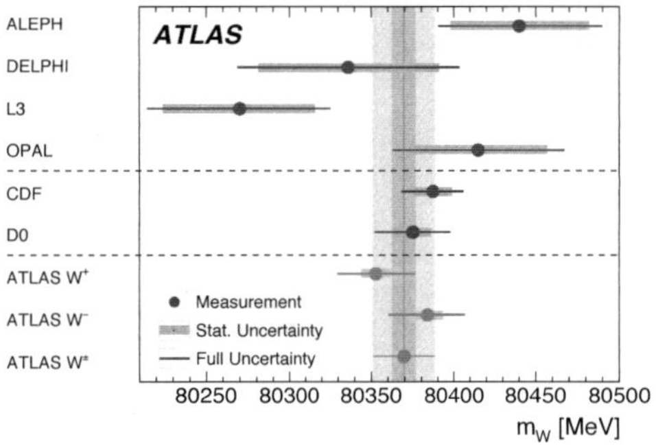

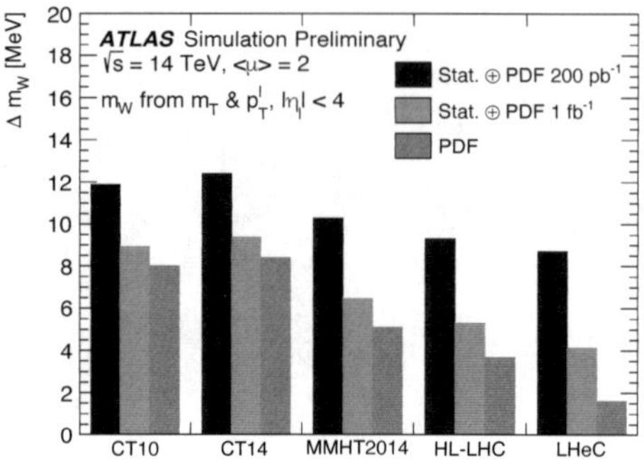

Figure 9. Results from the ATLAS m_W measurements are shown along with previous results in the left figure [32]. Estimates of the expected uncertainty on m_W are shown for various PDF sets assuming $200\,\mathrm{pb}^{-1}(1\,\mathrm{fb}^{-1})$ of low pileup data in the right figure [34].

decomposed into 1+8 othortogonal polynomials

$$\frac{d\sigma}{dp_{\mathrm{T}}^{ll}dy^{ll}dm^{ll}d\cos\theta d\phi} = \frac{3}{16\pi}\frac{d\sigma^{U+L}}{dp_{\mathrm{T}}^{ll}dy^{ll}dm^{ll}} \cdot$$

$$\Big\{\tfrac{1}{2}A_0\left(1-3\cos^2\theta\right) + A_1\sin 2\theta\cos\phi + \tfrac{1}{2}A_2\sin^2\theta\cos 2\phi + A_3\sin\theta\cos\phi + A_4\cos\theta$$

$$+A_5\sin\theta\sin 2\phi + A_6\sin 2\theta\sin\phi + A_7\sin\theta\sin\phi\Big\} \tag{4}$$

with θ, ϕ being the polar and azimuth angles of the final state lepton. The coefficients A_i are themselves functions of p_{T}^{ll}, y^{ll} and m^{ll} and have been measured by ATLAS [35]. A_4 is directly related to the forward-backward asymmetry in the full phase space via the relation $A_{FB} = A_4 \cdot 3/8$, where $A_{FB} = (\sigma_{\cos\theta>0} - \sigma_{\cos\theta<0})/\sigma$. A_4 and hence A_{FB} are sensitive to the effective weak mixing angle $\sin^2\theta_{\mathrm{W}}^{eff}$. The coefficient A_3 also exhibits some sensitivity to $\sin^2\theta_{\mathrm{W}}^{\mathrm{eff}}$ but is not used in the presented analyses.

The CMS collaboration measured the double differential forward-backward asymmetry $A_{FB}(m^{ll}, y^{ll})$ [36] independently in electron and muon final states. The double differential measurement allows to simultaneously constrain the PDF used in the analysis and extract $\sin^2\theta_{\mathrm{W}}^{\mathrm{eff}}$, which significantly reduces the uncertainty due to the PDFs. The analysis uses templates generated for various values of $\sin^2\theta_{\mathrm{W}}^{\mathrm{eff}}$ using Powheg in conjunction with the NNPDF3.0 PDF set. Those templates are fitted to measured distributions yielding the value of $\sin^2\theta_{\mathrm{W}}^{\mathrm{eff}}$ found in data. One hundred PDF replicas were used to generate weights according to their agreement with the measured A_{FB} distribution. Using the constraints on the PDF reduces the uncertainty from the PDF from $57\cdot10^{-5}$ to $31\cdot10^{-5}$. The largest systematic uncertainty arises from the limited size of the simulated event sample, contributing an uncertainty of $15(33)\cdot10^{-5}$ in the muon (electron) channels, respectively. The largest theory uncertainties arise from variations of the renormalisation and factorization scales in the event simulation. The final result for the effective leptonic weak mixing angle after combining the electron and muon channels is shown in table 2, together with results from ATLAS and the Tevatron experiments.

ATLAS uses $A_4(p_{\mathrm{T}}^{ll}, y^{ll}, m^{ll})$ to extract $\sin^2\theta_{\mathrm{W}}^{\mathrm{eff}}$. A_4 was measured in three independent channels using muons in the final state, electrons with both electrons being within $|\eta| < 2.4$ and the so called central-forward channel where one electron lies within $|\eta| < 2.4$ and one electron within $2.5 < |\eta| < 3.2$. A linear parametrization of A_4 in terms of $\sin^2\theta_{\mathrm{W}}^{\mathrm{eff}}$ was derived using the improved Born approximation (IBA) including NLO EW effects. The impact of

Table 2. Recent measurements of $\sin^2\theta^l_{\text{eff}}$.

Experiment	$\sin^2\theta^l_{\text{eff}}$	stat. $[10^{-5}]$	syst. $[10^{-5}]$	theo. $[10^{-5}]$	PDF $[10^{-5}]$
CMS [36]:	0.23101 ± 0.0053	36	18	16	31
ATLAS [37]:	0.23140 ± 0.0037	21	16	-	24
D0 + CDF [38]:	0.23148 ± 0.0033	27	05	-	18

NLO EW contributions on the analysis was studied in great detail as those could potentially break the factorization of the cross-section into the polynomials shown in equation 4. For this purpose the results from calculations using the IBA where transferred to the EW LO Powheg simulation by means of form factors. Though the factorization is shown to be preserved, the impact on the extraction of $\sin^2\theta^{\text{eff}}_{\text{W}}$ amounts to about the same size as the uncertainty from the A_4 measurement. Hence NLO EW effects are included throughout the analysis. Details are presented in Ref. [37]. The Tevatron collaborations CDF and D0 recently released an updated result of a similar measurement, utilizing the single differential measurement of $A_{FB}(m^{ll})$ as sensitive variable [38]. The result is presented in table 2. All results are in agreement with each other, and are also consistent with the combined result from the LEP and SLD experiments.

3.3. Measurements of the top mass

The measured mass of the top quark m_t is another important input to the global electroweak fit. Two distinct methods are used to measure the top mass. In the direct measurement the full event kinematic is reconstructed and variables sensitive to m_t are compared to simulated templates. Essentially this method measures a parameter in the simulation, often called $m^{\text{MC}}_{\text{top}}$. The difficulty in this approach lies in the conversion of the measured parameter of the simulation to the theoretically well defined pole mass of the top quark and this leads to ambiguities. On the other hand the indirect measurement of m_t exploits the dependence of the $t\bar{t}$ production cross-section on the top pole mass. Essentially the mass measurement is turned into a cross-section measurement and $\sigma_{t\bar{t}}$ is compared to theory calculation in order to extract m_t. This allows to unambiguously measure the top pole mass within the renormalisation scheme adopted for the cross-section measurement. However, this approach is largely sensitive to the PDF used in the cross-section predictions which in turn leads to significant uncertainties.

The latest top mass measurement from the ATLAS collaboration [39] uses the direct measurement approach yielding $m_t = 172.08 \pm 0.39(stat.) \pm 0.82(syst.)$ GeV. The latest result from the CMS collaboration [40] uses both, the indirect and direct mass measurements, yielding results of $m_t^{\text{MC}} = 172.33 \pm 0.14(stat.)^{+0.66}_{-0.72}(syst.)$ GeV and $m^{\text{pole}}_{t,\text{NNPDF3.0}} = 172.4 \pm 1.6(\text{fit} + \text{PDF} + \alpha_s)^{+1.3}_{-2.0}(scale)$ GeV. A large variety of top mass measurements have been released by the ATLAS and CMS collaborations, utilizing various final states and measurement methods. An overview is given in figure 10 (left). The latest compilation of all available top pole mass results was released in March 2018 [41] and is presented in figure 10 (right).

4. Summary

Over the past two years impressive progress has been made studying the SM gauge couplings as well as the EW parameters at the LHC experiments. New multi-boson processes became experimentally accessible, and limits on anomalous gauge couplings are narrowed down bit by bit. The first measurement of the W boson mass at the LHC was released as well as precision measurements on EW observables. No evidence of physics beyond the SM was observed and all results confirm the consistency of the SM to unprecedented accuracy. Many analyses, in

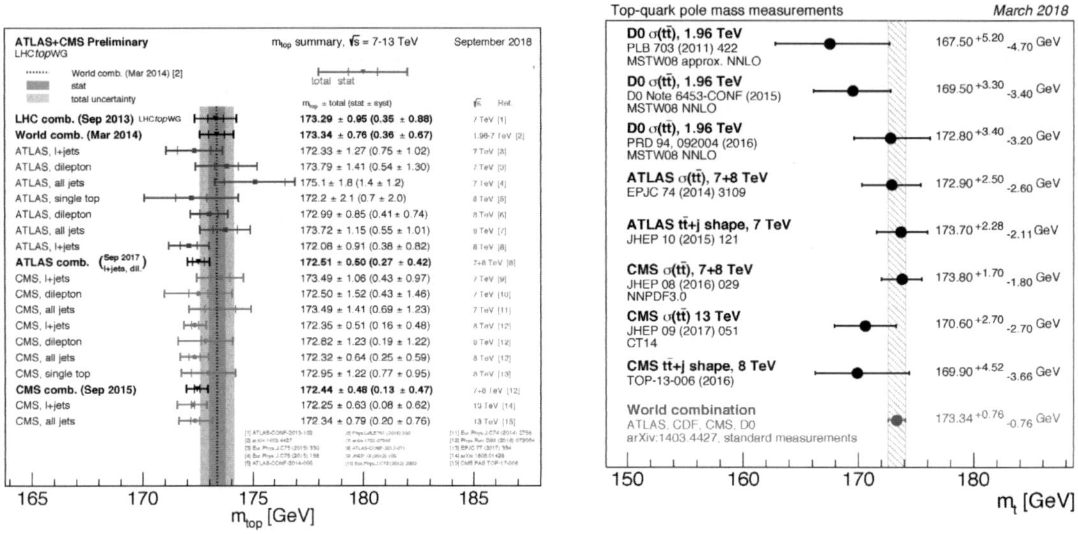

Figure 10. An overview of top quark mass measurement results performed at ATLAS and CMS is given in the left figure [42]. Various top quark pole mass measurements are compared to the world combination in the right figure.

particular those targeting multi-gauge boson final states, are limited by the available statistics. Hence significant improvements are expected from results using the full dataset recorded during the LHC run-2, and even more so once the LHC run-3 data will be available.

References

[1] ATLAS Collaboration. "The ATLAS Experiment at the CERN Large Hadron Collider". In: *JINST* 3 (2008), S08003.

[2] CMS Collaboration. "The CMS experiment at the CERN LHC". In: *JINST* 3 (2008), S08004.

[3] C. Degrande et al. "Effective Field Theory: A Modern Approach to Anomalous Couplings". In: *Annals Phys.* 335 (2013), pp. 21–32. arXiv: `1205.4231 [hep-ph]`.

[4] C. Degrande. "A basis of dimension-eight operators for anomalous neutral triple gauge boson interactions". In: *JHEP* 02 (2014), p. 101. arXiv: `1308.6323 [hep-ph]`.

[5] U. Baur and E. L. Berger. "Probing the weak-boson sector in $Z\gamma$ production at hadron colliders". In: *Phys. Rev. D* 47 (11 June 1993), pp. 4889–4904.

[6] ATLAS Collaboration. *ATLAS public standard model results.* 2019. URL: `https://twiki.cern.ch/twiki/bin/view/AtlasPublic/StandardModelPublicResults` (visited on 03/03/2019).

[7] CMS Collaboration. *CMS public standard model results.* 2019. URL: `http://cms-results.web.cern.ch/cms-results/public-results/publications/SMP/index.html` (visited on 03/03/2019).

[8] ATLAS Collaboration. "Measurement of the $Z\gamma \to \nu\bar{\nu}\gamma$ production cross section in pp collisions at $\sqrt{s} = 13$ TeV with the ATLAS detector and limits on anomalous triple gauge-boson couplings". In: *JHEP* (2018). arXiv: `1810.04995 [hep-ex]`.

[9] A. Gutierrez-Rodriguez et al. *Studies on the anomalous magnetic and electric dipole moments of the tau-neutrino in pp collisions at the LHC.* arXiv: `1712.02439 [hep-ph]`.

[10] ATLAS Collaboration. "Measurements of $Z\gamma$ and $Z\gamma\gamma$ production in pp collisions at $\sqrt{s} = 8$ TeV with the ATLAS detector". In: *Phys. Rev. D* 93 (2016), p. 112002. arXiv: 1604.05232 [hep-ex].

[11] ATLAS Collaboration. *Measurement of $W^{\pm}Z$ boson pair-production in pp collisions at $\sqrt{s} = 13$ TeV with the ATLAS Detector and confidence intervals for anomalous triple gauge boson couplings.* ATLAS-CONF-2016-043. 2016. URL: https://cds.cern.ch/record/2206093.

[12] CMS Collaboration. "Measurements of the pp→WZ inclusive and differential production cross section and constraints on charged anomalous triple gauge couplings at $\sqrt{s} = 13$ TeV". In: *submited to JHEP* (2019). arXiv: 1901.03428 [hep-ex].

[13] ATLAS Collaboration. "Measurement of $WW/WZ \rightarrow \ell\nu qq'$ production with the hadronically decaying boson reconstructed as one or two jets in pp collisions at $\sqrt{s} = 8$ TeV with ATLAS, and constraints on anomalous gauge couplings". In: *Eur. Phys. J. C* 77 (2017), p. 563. arXiv: 1706.01702 [hep-ex].

[14] CMS Collaboration. "Search for anomalous couplings in boosted $WW/WZ \rightarrow \ell\nu q\bar{q}$ production in proton–proton collisions at $\sqrt{s} = 8$ TeV". In: *Phys. Lett. B* 772 (2017), p. 21. arXiv: 1703.06095 [hep-ex].

[15] CMS Collaboration. "Electroweak production of two jets in association with a Z boson in proton–proton collisions at $\sqrt{s} = 13$ TeV". In: *Eur. Phys. J. C* 78 (2018), p. 589. arXiv: 1712.09814 [hep-ex].

[16] ATLAS Collaboration. "Measurements of electroweak Wjj production and constraints on anomalous gauge couplings with the ATLAS detector". In: *Eur. Phys. J. C* 77 (2017), p. 474. arXiv: 1703.04362 [hep-ex].

[17] ATLAS Collaboration. "Study of $WW\gamma$ and $WZ\gamma$ production in pp collisions at $\sqrt{s} = 8$ TeV and search for anomalous quartic gauge couplings with the ATLAS experiment". In: *Eur. Phys. J. C* 77 (2017), p. 646. arXiv: 1707.05597 [hep-ex].

[18] ATLAS Collaboration. "Observation of electroweak $W^{\pm}Z$ boson pair production in association with two jets in pp collisions at $\sqrt{s} = 13$ TeV with the ATLAS detector". In: *Phys. Lett.* (2018). arXiv: 1812.09740 [hep-ex].

[19] ATLAS Collaboration. "Measurement of the cross-section for electroweak production of dijets in association with a Z boson in pp collisions at $\sqrt{s} = 13$ TeV with the ATLAS detector". In: *Phys. Lett. B* 775 (2017), p. 206. arXiv: 1709.10264 [hep-ex].

[20] ATLAS Collaboration. "Studies of $Z\gamma$ production in association with a high-mass dijet system in pp collisions at $\sqrt{s} = 8$ TeV with the ATLAS detector". In: *JHEP* 07 (2017), p. 107. arXiv: 1705.01966 [hep-ex].

[21] CMS Collaboration. "Measurement of the cross section for electroweak production of $Z\gamma$ in association with two jets and constraints on anomalous quartic gauge couplings in proton–proton collisions at $\sqrt{s} = 8$ TeV". In: *Phys. Lett. B* 770 (2017), p. 380. arXiv: 1702.03025 [hep-ex].

[22] CMS Collaboration. "Measurement of electroweak production of a W boson and two forward jets in proton–proton collisions at $\sqrt{s} = 8$ TeV". In: *JHEP* 11 (2016), p. 147. arXiv: 1607.06975 [hep-ex].

[23] ATLAS Collaboration. *Observation of electroweak production of a same-sign W boson pair in association with two jets in pp collisions at $\sqrt{s} = 13$ TeV with the ATLAS detector.* ATLAS-CONF-2018-030. 2018. URL: https://cds.cern.ch/record/2629411.

[24] CMS Collaboration. *Measurement of electroweak WZ boson production and search for new physics in $WZ +$ two jets events in pp collisions at $\sqrt{s} = 13$ TeV.* 2019. arXiv: 1901.04060 [hep-ex].

[25] CMS Collaboration. "Observation of Electroweak Production of Same-Sign W Boson Pairs in the Two Jet and Two Same-Sign Lepton Final State in Proton–Proton Collisions at 13 TeV". In: *Phys. Rev. Lett.* 120 (2018), p. 081801. arXiv: 1709.05822 [hep-ex].

[26] B. Biedermann, A. Denner, and M. Pellen. "Large Electroweak Corrections to Vector-Boson Scattering at the Large Hadron Collider". In: *Physical Review Letters* 118 (Nov. 2016).

[27] S. Schumann and F. Krauss. "A Parton shower algorithm based on Catani-Seymour dipole factorisation". In: *JHEP* 03 (2008), p. 038. arXiv: 0709.1027 [hep-ph].

[28] S. Frixione, P. Nason, and C. Oleari. "Matching NLO QCD computations with Parton Shower simulations: the POWHEG method". In: *JHEP* 11 (2007), p. 070. arXiv: 0709.2092 [hep-ph].

[29] ATLAS Collaboration. *Observation of electroweak $W^{\pm}Z$ boson pair production in association with two jets in pp collisions at $\sqrt{s}$ = 13 TeV with the ATLAS detector.* ATLAS-CONF-2018-033. 2018. URL: https://cds.cern.ch/record/2630183.

[30] ATLAS Collaboration. *HL-LHC prospects for diboson resonance searches and electroweak vector boson scattering in the $WW/WZ \to \ell\nu qq$ final state.* ATL-PHYS-PUB-2018-022. 2018. URL: https://cds.cern.ch/record/2645269.

[31] J. Haller et al. "Update of the global electroweak fit and constraints on two-Higgs-doublet models". In: *Eur. Phys. J.* C78.8 (2018), p. 675. arXiv: 1803.01853 [hep-ph].

[32] ATLAS Collaboration. "Measurement of the W-boson mass in pp collisions at $\sqrt{s} = 7$ TeV with the ATLAS detector". In: *Eur. Phys. J. C* 78 (2018), p. 110. arXiv: 1701.07240 [hep-ex].

[33] ATLAS Collaboration. *Prospects for the measurement of the W-boson transverse momentum with a low pileup data sample at $\sqrt{s}$ = 13 TeV with the ATLAS detector.* ATL-PHYS-PUB-2017-021. 2017. URL: https://cds.cern.ch/record/2298152.

[34] ATLAS Collaboration. *Prospects for the measurement of the W-boson mass at the HL- and HE-LHC.* ATL-PHYS-PUB-2018-026. 2018. URL: https://cds.cern.ch/record/2645431.

[35] ATLAS Collaboration. "Measurement of the angular coefficients in Z-boson events using electron and muon pairs from data taken at $\sqrt{s} = 8$ TeV with the ATLAS detector". In: *JHEP* 08 (2016), p. 159. arXiv: 1606.00689 [hep-ex].

[36] CMS Collaboration. "Measurement of the weak mixing angle using the forward-backward asymmetry of Drell–Yan events in pp collisions at 8 TeV ". In: *Eur. Phys. J. C* 78 (2018), p. 701. arXiv: 1806.00863 [hep-ex].

[37] ATLAS Collaboration. *Measurement of the effective leptonic weak mixing angle using electron and muon pairs from Z-boson decay in the ATLAS experiment at $\sqrt{s}$ = 8 TeV.* ATLAS-CONF-2018-037. 2018. URL: https://cds.cern.ch/record/2630340.

[38] T. A. Aaltonen et al. "Tevatron Run II combination of the effective leptonic electroweak mixing angle". In: *Phys. Rev.* D97.11 (2018), p. 112007. arXiv: 1801.06283 [hep-ex].

[39] ATLAS Collaboration. "Measurement of the top quark mass in the $t\bar{t} \to$ lepton+jets channel from $\sqrt{s} = 8$ TeV ATLAS data and combination with previous results". In: *Eur. Phys. J.* (2018). arXiv: 1810.01772 [hep-ex].

[40] CMS Collaboration. "Measurement of the $\mathrm{t\bar{t}}$ production cross section, the top quark mass, and the strong coupling constant using dilepton events in pp collisions at $\sqrt{s} = 13$ TeV". In: *Submitted to: Eur. Phys. J.* (2018). arXiv: 1812.10505 [hep-ex].

[41] "First combination of Tevatron and LHC measurements of the top-quark mass". In: (2014). arXiv: 1403.4427 [hep-ex].

IOP Conf. Series: Journal of Physics: Conf. Series **1271** (2019) 012006 doi:10.1088/1742-6596/1271/1/012006

[42] LHC top working group. *LHC top physics summary plots.* 2019. URL: https://twiki.cern.ch/twiki/bin/view/LHCPhysics/LHCTopWGSummaryPlots (visited on 03/03/2019).

IOP Conf. Series: Journal of Physics: Conf. Series **1271** (2019) 012007 doi:10.1088/1742-6596/1271/1/012007

Heavy Ion Results from ATLAS and CMS

David Krofcheck

The University of Auckland
For the CMS and ATLAS Collaborations

d.krofcheck@auckland.ac.nz

Abstract. The heavy-ion programmes in the ATLAS and CMS experiments at the Large Hadron Collider aim to probe and characterise properties of the quark-gluon plasma created in relativistic nuclear collisions. This work presents selected results of collective effects, system size effects, and quarkonia production in p-p, p-Pb, Pb-Pb, and Xe-Xe collisions.

1. Introduction

The ATLAS[1] and CMS Collaborations[1] perform measurements in heavy ion (HI) collision programmes, in addition to proton-proton (p-p) discovery physics. The goal of the HI programmes is to discover and explain the properties of Quark-Gluon Plasma (QGP) created in these collisions. Smaller system sizes are also studied as cases in which QGP creation is not favoured. Recent results from Pb-Pb at centre-of-mass energies of 2.76 TeV and 5.02 Tev, Xe-Xe at 5.44 TeV, and p-Pb at 8.16 TeV are presented. Both HI programmes use p-p reactions as reference collisions to explore initial-state and final-state collision effects.

2. Collision System Size Effects

Most of the current HI programme uses the Pb-Pb system to probe hot and dense matter nuclear. More recent measurements are used to explore the Xe-Xe reaction, which provides a tool for exploring system size dependence on the production of hadrons.

2.1 Pb-Pb and Xe-Xe

Figure 1 shows the pseudorapidity density ($dN_{ch}/d\eta$) of charged hadrons produced in nuclear collisions as a function of rapidity (y) at different energies-per-nucleon [2]. The plot shows the expected growth in charged hadron production as both the LHC centre-of-mass energy increases, and the size of the collision region increases when going from Xe-Xe- to Pb-Pb reactions.

IOP Conf. Series: Journal of Physics: Conf. Series **1271** (2019) 012007 doi:10.1088/1742-6596/1271/1/012007

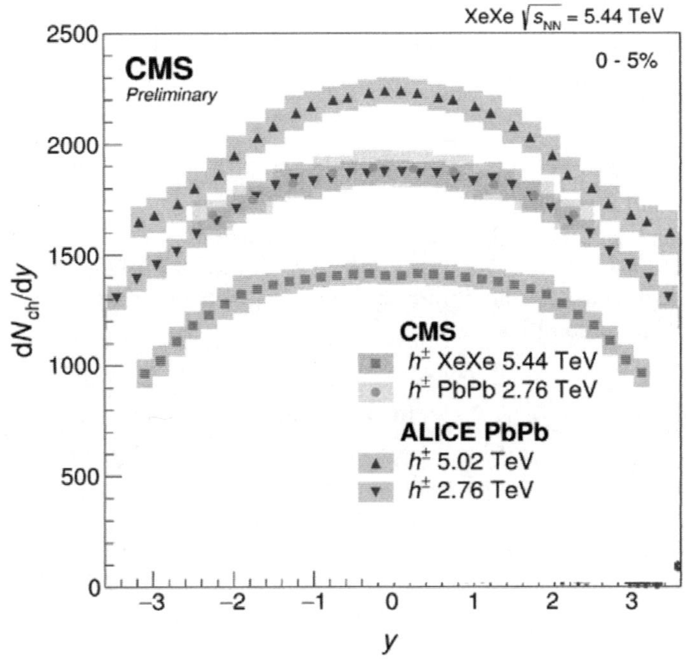

Figure 1. Average and symmetrised dNch/dy as a function of rapidity [2].

2.2 System size scaling

A study of the number of hadronic particles produced as a function of the average number of participants (<Npart>) is shown on the left side of Figure 2, for both Pb-Pb and Xe-Xe collisions. For fixed and large <Npart> no system size scaling is observed.

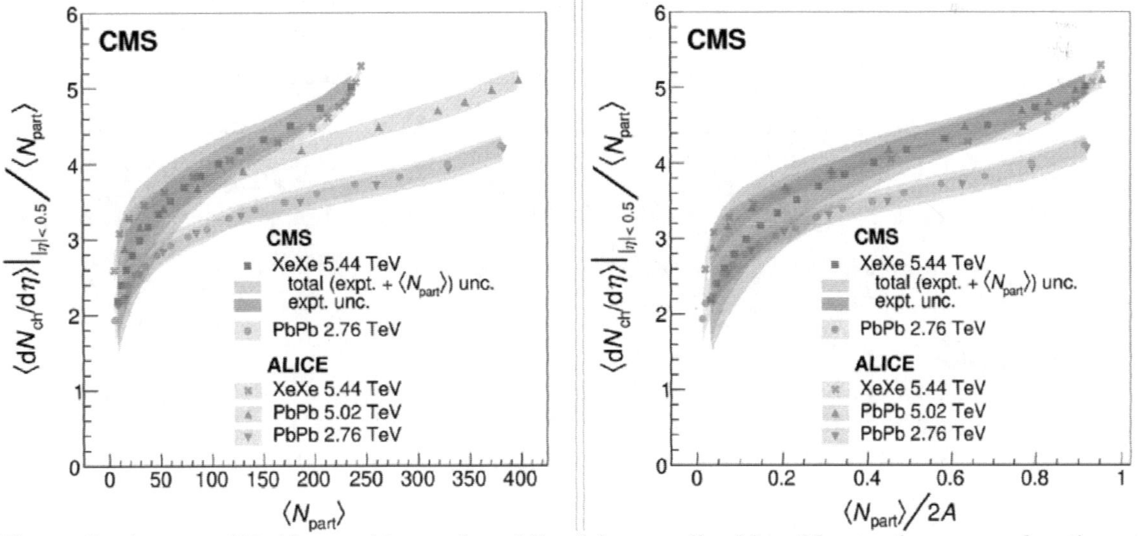

Figure 2. Average dNch/dη at mid-pseudorapidity (η) normalised by <Npart>, shown as a function of <Npart> (left), and <Npart>/2*A* (right), where *A* is the atomic number of the nuclei [2].

The per-participant multiplicity for Xe-Xe and Pb-Pb collisions for the same <Npart> and similar energies but different geometry or centrality are different. This is particularly evident for the most central (largest <Npart>) collisions. However, as in Figure 2 (right), where <Npart>/2A is used as a

IOP Conf. Series: Journal of Physics: Conf. Series **1271** (2019) 012007 doi:10.1088/1742-6596/1271/1/012007

substitute for centrality, the per-participant charged-hadron multiplicity for different colliding nuclei are equal within uncertainties when the geometry (centrality) and the energy of the compared systems are the same [2].

2.3 Charged particle suppression

Measurement of the nuclear suppression factor R_{AA} indicates how the QGP suppresses the observed charged hadron particle production. The data in Figure 3 show similar numbers of <Npart> but for different centralities in the Xe-Xe and Pb-Pb systems. The Xe-Xe system exhibits slightly stronger suppression in the most central collisions.

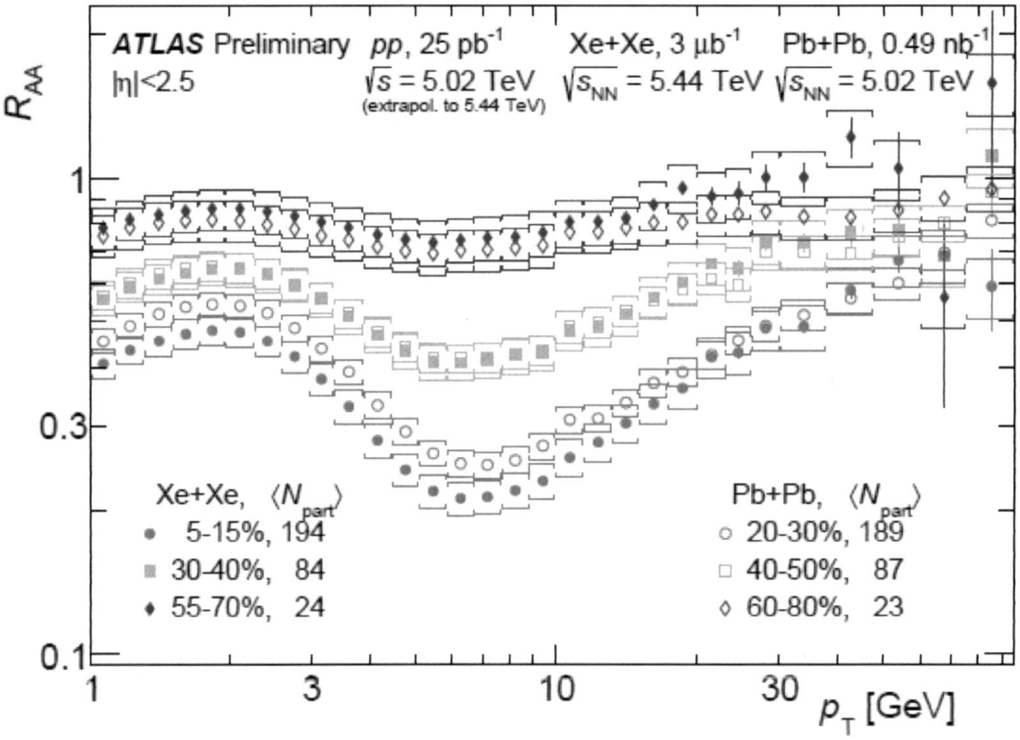

Figure 3. The nuclear modification factor R_{AA} for Xe-Xe and Pb-Pb systems [3].

3. Collective effects

Strong collective flow behaviour is exhibited in high energy nucleus-nucleus collisions. Studies of smaller systems such as p-Pb collisions and high multiplicity p-p collisions also reveal collective flow. Current research is aimed at measuring the flow of heavy quarks in small systems.

3.1 Charm and strange quark elliptic flow

In Figure 4 the background corrected V_2, called V_2^{sub}, per constituent quark for mesons and baryons is presented. For particle transverse kinetic energy per constituent quark values less than 1GeV, the V_2^{sub} of prompt J/ψ mesons is consistent with prompt D^0, K_S^0 and Λ scaling along with the D^0meson. There is a suggestion of prompt J/Ψ mesons, which consist of two charm quarks, breaking the scaling at higher transverse kinetic energies. This is a hint that heavy quarks show weaker collective behavior in compressed nuclear matter.

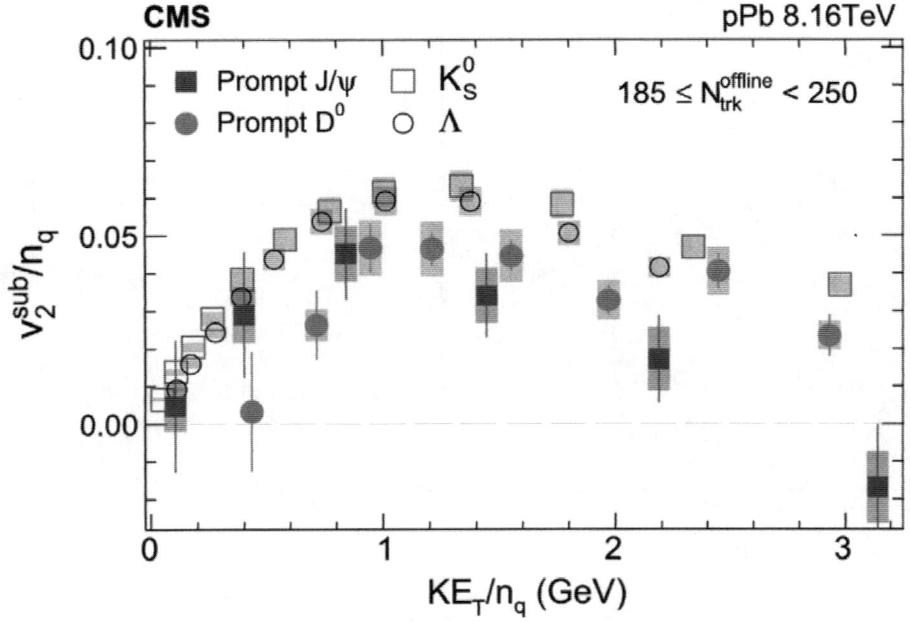

Figure 4. Background subtracted constituent quark normalized elliptic flow (V_2^{sub} / nq) as a function of normalized traverse kinetic energy [4].

3.2 Multiparticle correlations in azimuthal distributions

Multiparticle azimuthal correlations produced in heavy ion collisions extend over a considerable range in pseudorapidity. The observed azimuthal correlations are characterized by Fourier harmonics, with V_2 and V_3 referred to as elliptic and triangular flow, respectively [5]. The ratio between the four-particle and two-particle harmonics provides information on the relative importance of the global geometry and the fluctuation-driven asymmetries.

The first ever small system four-particle measurements of V_3 are shown in Figure 5. For the small system formed in p-Pb (left side of Figure 5) the four-particle and two-particle harmonics are very similar. This is consistent with the origin of these harmonics coming from the same initial state fluctuation. For the larger system Pb-Pb (right side of Figure 5) the elliptic flow harmonic ratio is larger than the triangular flow harmonic. This result if expected if the global collision geometry dominates the Pb-Pb results [5].

IOP Conf. Series: Journal of Physics: Conf. Series **1271** (2019) 012007 doi:10.1088/1742-6596/1271/1/012007

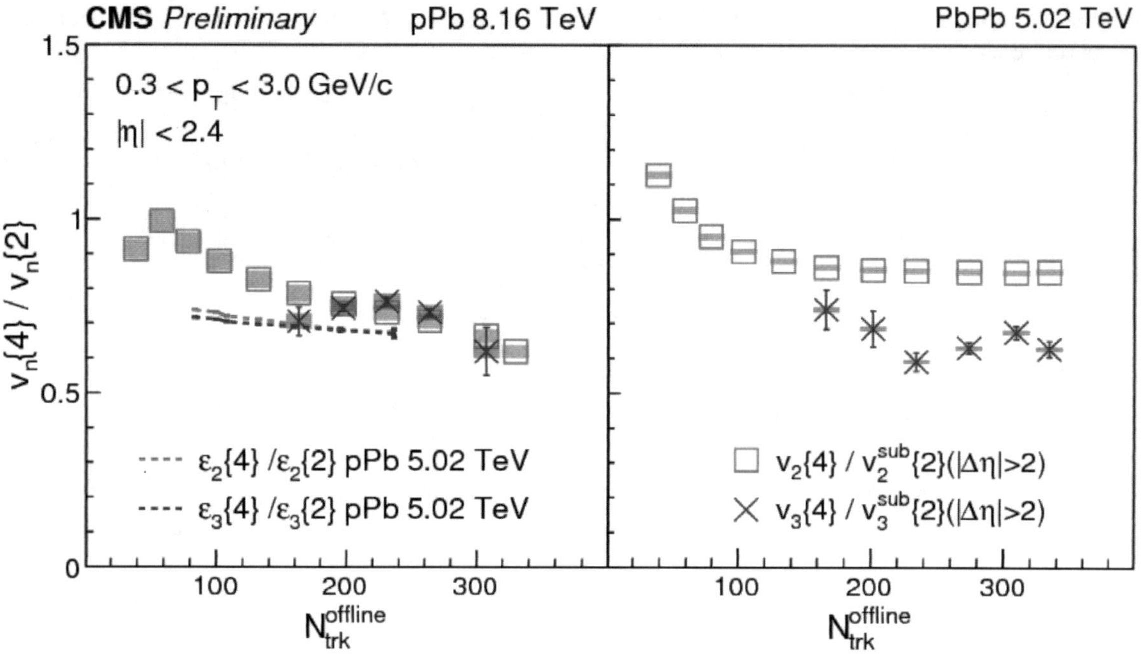

Figure 5. The ratios of four- and two-particle harmonics for pPb at 8.16 TeV (left), and for PbPb at 5.02 TeV (right) [5].

3.3 Dijet asymmetry

Dijets are sprays of particles produced in nucleus-nucleus collisions. If the dijets are approximately back-to-back then the phenomenon of jet quenching, the difference between thetransverse energies of the two jets, can be observed.

New results on the dijet asymmetry (defined as $X_J = P_T^2/P_T^1$, in which P_T^1 is the leading jet transverse momentum, and P_T^2 is the sub-leading jet transverse momentum) for Xe-Xe and Pb-Pb collisions are presented across four centrality bins in Figure 6 [3]. There appears to be very little difference in the behaviour of X_J as the size of the nuclear overlap changes. The jet quenching is uniform for central (0-10%) collisions and potentially evolves smoothly towards a non-uniform shape in the most peripheral (60-80%) collisions.

IOP Conf. Series: Journal of Physics: Conf. Series **1271** (2019) 012007 doi:10.1088/1742-6596/1271/1/012007

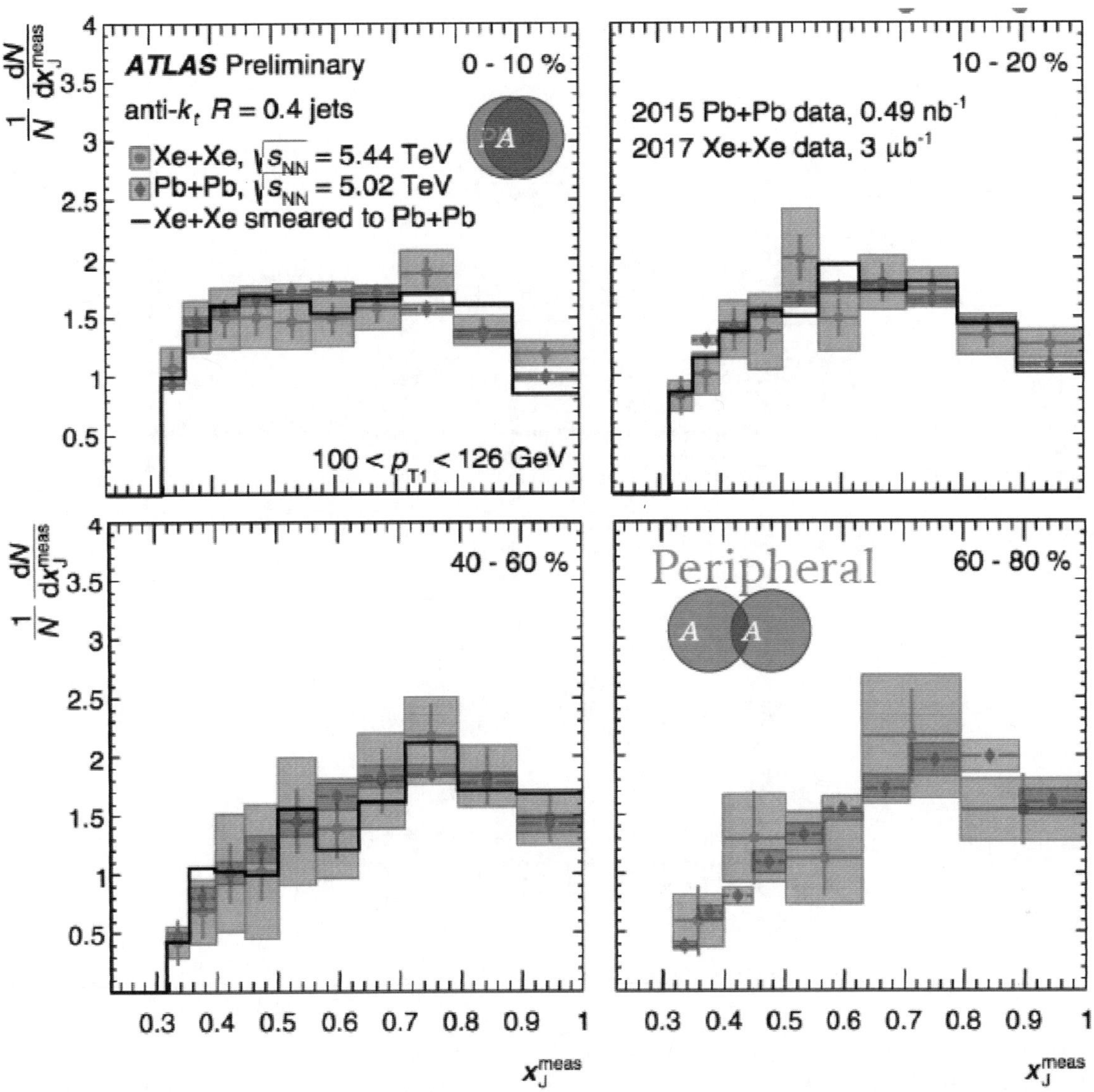

Figure 6. The normalised dijet asymmetry X_J in Xe-Xe and Pb-Pb collisions for four centrality bins [3].

Photons act as calibration probes in HI collisions as they essentially lose no energy as they traverse the QGP. One of the jets can be replaced by a high energy gamma ray. When comparing p-p collisions to Pb-Pb collisions the jet-gamma asymmetry parameter (defined as $X_{J\gamma} = P_T^{jet}/P_T^{\gamma}$) behaves in a substantially different way than X_J defined for Figure 6. The $X_{J\gamma}$ parameter for p-p reactions is compared to two centrality choices in Pb-Pb collisions at 5.02 TeV in Figure 7. Since the gamma-ray balances the jet momentum in p-p reactions we see a prominent peak at $X_{J\gamma} = 1$. In the Pb-Pb collisions we see a similar peak in peripheral collisions, but this peak smoothly evolves with increasing centrality. It nearly disappears for the most central (0-10%) Pb-Pb collisions. The jet momentum is quenched by the QGP in the central Pb-Pb collisions.

IOP Conf. Series: Journal of Physics: Conf. Series **1271** (2019) 012007 doi:10.1088/1742-6596/1271/1/012007

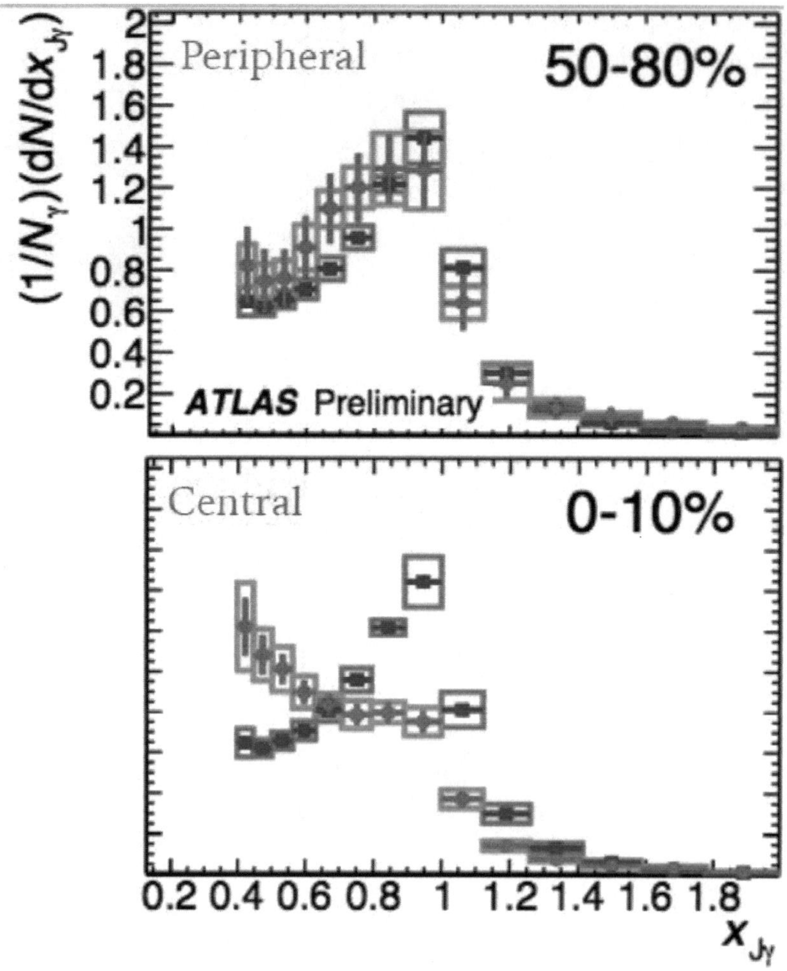

Figure 7. The jet-gamma asymmetry $X_{J\gamma}$ in p-p and Pb-Pb collisions at 5.02 TeV for two centrality bins [6].

4. Quarkonia and heavy quarks

Quarkonia and heavy quarks are also useful probes of QGP produced in HI collisions. The relevant observable quantity is the nuclear suppression factor (R_{AA}). The J/ψ meson is composed of charm and anti-charm quarks. As shown in Figure 8, suppression of J/ψ production in Pb-Pb collisions gets larger as the collision centrality increases, e.g., more QGP is produced in the most central collisions (0-10%).

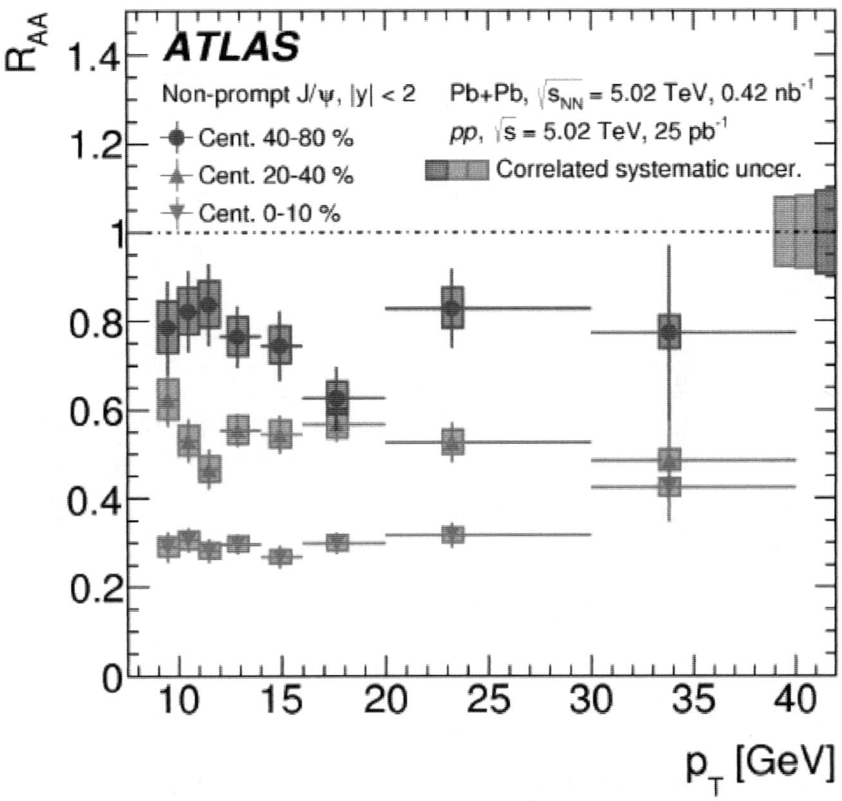

Figure 8. The nuclear modification factor R_{AA} as a function of p_T for prompt J/ψ production in Pb-Pb collisions at 5.02 TeV, shown for three centrality bins [7].

The J/ψ meson has an excited state called the ψ(2S). The less tightly bound ψ(2S) is more sensitive to the high temperature QGP. This is illustrated in Figure 9 (left) where the R_{AA} is plotted for both the J/ψ and ψ(2S) mesons, as a function of participant particles in the Pb-Pb collision [8]. Prompt ψ(2S) are suppressed uniformly across the number of participants when compared to prompt J/ψ in Pb-Pb collisions at 5.02 TeV.

On the right side of Figure 9 is presented the nuclear suppression factor R_{pPb} for the p-Pb collisions at 5.02 TeV. p-Pb collisions are used to observe final state interactions. The ψ(2S) is expected to be suppressed by the same amount as the J/ψ meson, but the data in the negative rapidity region suggests that the ψ(2S) is more suppressed [8]. This may be due to the larger size of the ψ(2S) meson.

The ϒ meson also has a third excited state accessible in HI collisions. These mesons are labelled ϒ(1S), ϒ(2S), and ϒ(3S) in Figure 10 [9]. Similar to the J/ψ meson and the ψ(2S) the three upsilon mesons, consisting of bottom and anti-bottom quarks, show a sequential suppression that increases as the collision centrality increases. The ϒ(3S) has the smallest R_{AA} observed for any hadron. Superimposed upon the data are ideal fluid hydrodynamics calculations [9].
Strangeness enhancement in HI collisions is accompanied by heavy quark creation. This motivates the search for strange neutral B mesons; e.g., $B_S^0 \rightarrow J/\Psi\ \varphi \rightarrow \mu^+ \mu^- K^+ K^-$ in A-A collisions. In Figure 11 (left) is an invariant mass yield plot of B_S^0 mesons[10]. In Figure 11 (right) is the nuclear suppression R_{AA} for B mesons, with and without a strange quark, moving through QGP. Within current uncertainties, the results are consistent with models of strangeness enhancement and a suppression as observed for the B+ mesons.

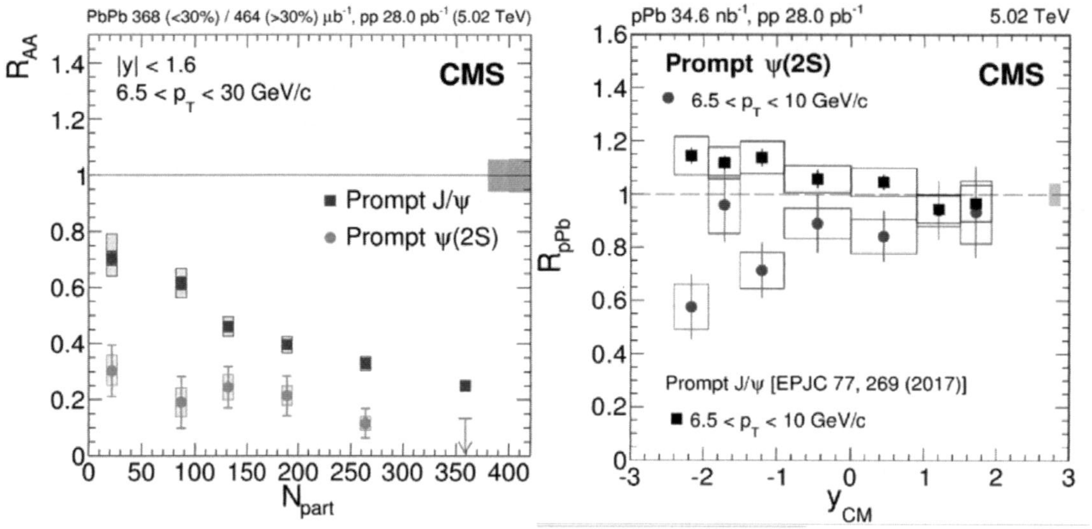

Figure 9. Nuclear suppression factors R_{AA} for J/ψ and ψ(2S) mesons in Pb-Pb(left) and in p-Pb (right) collisions plotted versus the number of participating particles (N_{part}), and the rapidity (y_{CM}), respectively [8].

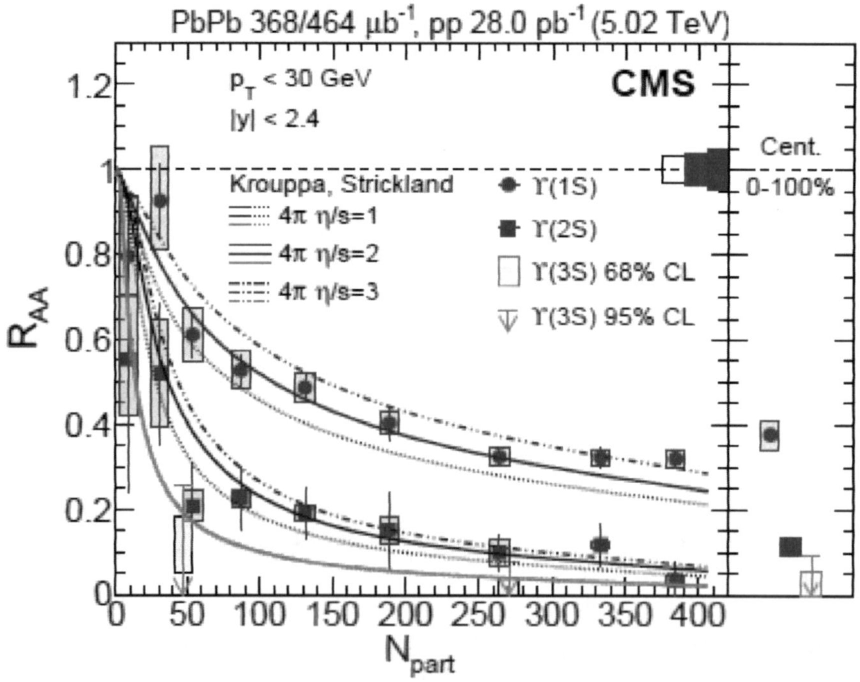

Figure 10. The nuclear modification factor R_{AA} as a function of p_T for ϒ(1S), ϒ(2S), and ϒ(3S) production in Pb-Pb collisions at 5.02 TeV, shown as a function of the number of participating nucleons [9]. Ideal fluid model calculations by Krouppa and Strickland are also presented [9].

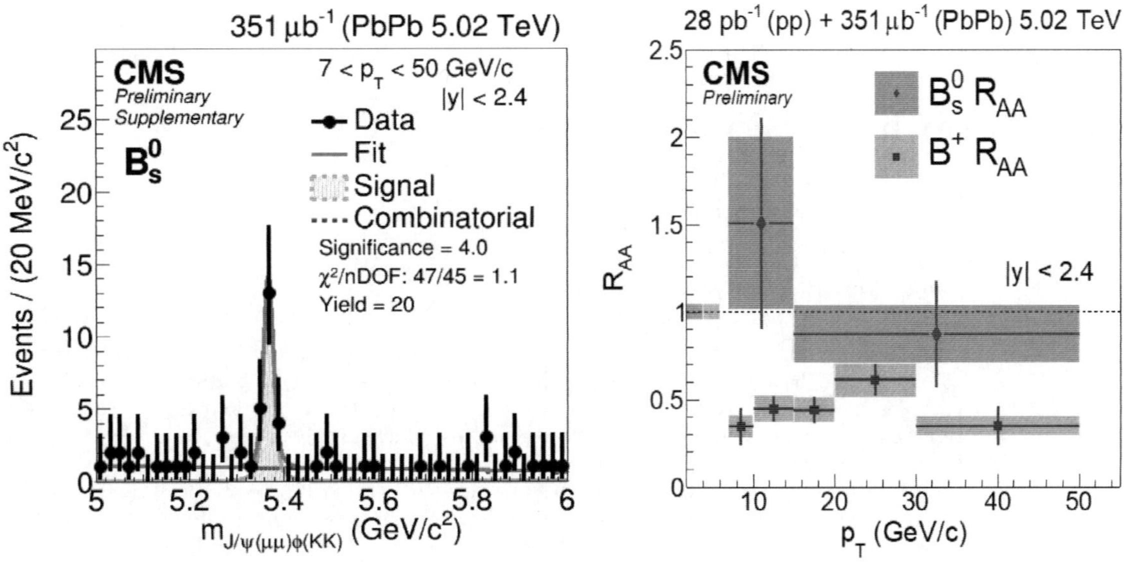

Figure 11. The invariant mass distribution for B_S^0 decay in Pb-Pb collisions (left). Nuclear suppression factor for B measons from Pb-Pb collisions (right) [10].

5. Conclusions

The heavy ion reaction programmes at both ATLAS and CMS are focussed on the study of QGP. A variety of particles and observables are used to investigate the earliest stages of nucleus-nucleus collision. System size effects, particle correlations, jet quenching, quarkonia, and heavy quark nuclear suppression provide insight to the earliest moments in relativistic nucleus-nucleus collisions.

References

[1] ATLAS Collaboration, *JINST* 3 (2008) S08003. https://doi.org/10.1088/1748-0221/3/08/S08003
CMS Collaboration, *JINST* 3 (2008) S08003 https://doi.org/10.1088/1748-0221/3/08/S08004
[2] CMS Collaboration 2019 (preprint http://arxiv.org/abs/1902.03603)
JHEP **08** (2011) 141, http://dx.doi.org/10.1007/JHEP08(2011)141
ALICE Collaboration http://dx.doi.org/10.1103/PhysRevLett.106.032301
http://dx.doi.org/10.1103/PhysRevLett.116.222302
[3] ATLAS Collaboration, (ATLAS-CONF-2018-007)
[4] CMS Collaboration, 2019 , *Phys. Lett.* **B 791**, 172 doi:10.1016/j.physletb.2019.02.018
[5] CMS Collaboration, 2018 (CMS PAS HIN-17-004)
[6] ATLAS Collaboration, (ATLAS-CONF-2018-009)
[7] ATLAS Collaboration,2018 *Eur. Phys. J.* **C78**, **762** doi:10.1140/epjc/s10052-018-6219-9
[8] CMS Collaboration, 2018 *Eur. Phys. J.* **C78**, 509 10.1140/epjc/s10052-018-5950-6
CMS Collaboration, 2019 *Phys. Lett.* **B790** 509 10.1016/j.physletb.2019.01.058
[9] B. Kroupa and M. Strickland, 2016 Predictions for Bottomonia Suppression in 5.023 TeV Pb-Pb Collisions, *Universe* **2**(3) https://doi.org/10.3390/universe2030016
CMS Collaboration, 2019 *Phys. Lett.* **B790** 270 https://doi.org/10.1016/j.physletb.2019.01.006
[10] CMS Collaboration, 2018 (preprint https://arxiv.org/abs/1810.03022)
CMS Collaboration, 2017 *Phys. Rev. Lett* **119** 152301
https://doi.org/10.1103/PhysRevLett.119.152301

IOP Conf. Series: Journal of Physics: Conf. Series **1271** (2019) 012008 doi:10.1088/1742-6596/1271/1/012008

Results on heavy ion physics at LHCb

Giacomo Graziani, on behalf of the LHCb Collaboration
INFN, Sezione di Firenze
E-mail: Giacomo.Graziani@fi.infn.it

Abstract. In the last years, the LHCb experiment established itself as an important contributor to heavy ion physics by exploiting some of its specific features. Production of particles, notably heavy flavour states, can be studied in *p*-*p*, *p*-Pb and Pb-Pb collisions at LHC energies in the forward rapidity region (pseudorapidity between 2 and 5), providing measurements which are highly complementary to the other LHC experiments. Moreover, owing to its forward geometry, the detector is also well suited to study fixed-target collisions, obtained by impinging the LHC beams on gas targets with different mass numbers. In this configuration, *p*-A collisions can be studied at the relatively unexplored scale of $\sqrt{s_{\rm NN}} \sim 100$ GeV, also providing valuable inputs to cosmic ray physics. An overview of the measurements obtained so far by the LHCb ion program is presented.

1. Introduction

The LHCb experiment [1] has been conceived with the main goal of studying heavy flavour physics in *p*-*p* collisions at the LHC, exploiting the unprecedented yield of *b*-quark pairs, which are mainly produced at small angles with respect to the direction of the colliding proton beams. The detector is therefore designed as a forward spectrometer covering the pseudorapidity region $2 < \eta < 5$ and providing excellent vertexing, tracking and particle identification capabilities for the reconstruction of heavy flavour decays. Another key feature is the online selection system, consisting of a hardware level with high output bandwidth (up to 1 MHz), followed by a software level providing high flexibility.

Though heavy ion physics was not among the original motivations for the experiment, the detector capabilities offer some unique possibilities also for this field:

- the forward acceptance, with a fully instrumented detector, highly complementary to the other LHC experiments;
- the excellent reconstruction performance, unrivaled at the LHC, for exclusive heavy flavour states down to null transverse momentum ($p_{\rm T}$), disentangling charmed particles produced promptly from those coming from *b*-hadron decays;
- the possibility to operate the detector in fixed target mode, by injecting small amount of noble gas (He, Ne and Ar) in the LHC vacuum [2] and studying beam-gas collisions, for which the forward geometry of the detector is very well suited.

On the other hand, the most central Pb-Pb collisions can't be properly reconstructed due to the high track density in the forward region. LHCb is therefore more suited for smaller collision systems like *p*-Pb, but can contribute also to study peripheral Pb-Pb collisions. It has to be noted that studies in small collision systems have become increasingly important to the

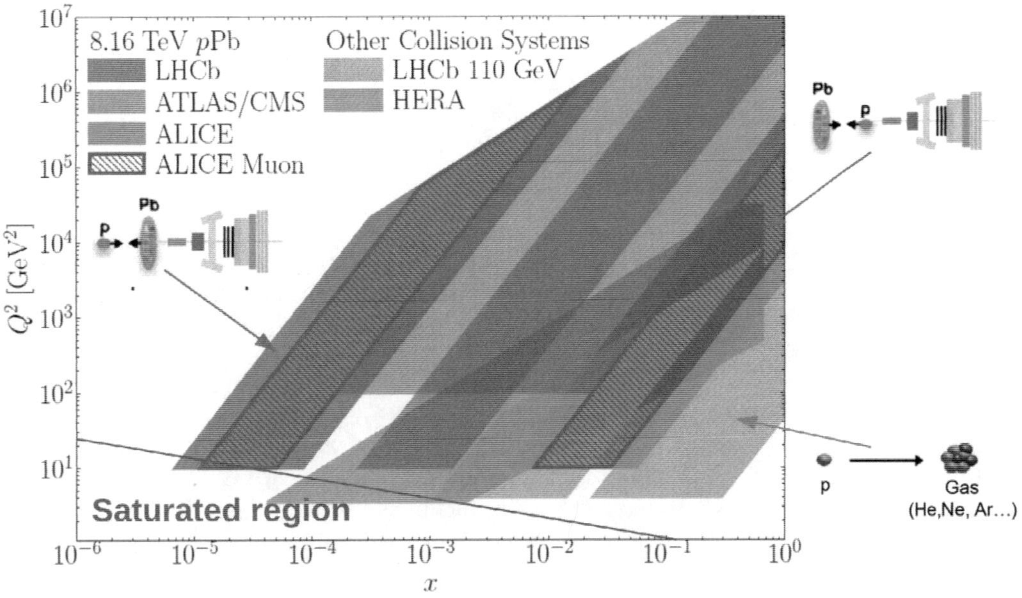

Figure 1. Kinematic reach corresponding to the acceptance of the four LHC experiments in p-Pb collisions. The kinematic regions accessible in the fixed-target configuration at LHCb and in e-p collisions at HERA are also shown.

interpretation of relativistic heavy ion collisions after the discovery, performed at the LHC, of sizeable collective behaviour even in p-p collisions [3].

A comparison among the kinematic reaches of LHCb and the other LHC experiments in p-A collisions, in terms of the Bjorken-x value of the nucleon in the nuclear target and the squared parton-parton invariant mass Q^2, is shown in Figure 1. Two regions are covered in p-Pb collisions, depending on the orientation of the proton and lead beams. The so-called *forward* configuration, when the proton beam points toward the detector (i.e., it enters the detector region from its vertex detector), corresponds to values of x down to 10^{-5}, where gluon saturation is expected to occur. In the *backward* configuration, when the Pb beam points toward the detector, measurements are sensitive to the anti-shadowing region up to $x \sim 0.1$.

In fixed target collisions, the nucleon-nucleon centre-of-mass (c.m.) energy reaches 110 GeV for proton beams of 6.5 TeV energy. Measurements at this energy scale, intermediate between SPS fixed target experiments and beam-beam collisions at LHC, provides an additional test bed to explore the energy evolution in the dynamics of nuclear matter. The detector acceptance corresponds to mid and backward rapidities in the c.m. frame: $-2.8 < y^* < 0.2$. As also depicted in Figure 1, this gives access to large values of x in the target nucleon, where nuclear PDFs are modified by the EMC effect and where the contribution of a possible intrinsic heavy quark content in the nucleon could be substantiated.

2. p-Pb collisions

The experiment collected a first dataset of p-Pb collisions at $\sqrt{s_{\mathrm{NN}}} = 5.02$ TeV in 2013, corresponding to integrated luminosities of 1.1 and 0.5 nb^{-1} in the forward and backward configuration, respectively. After this succesfull experience, larger luminosities were delivered to LHCb during the 2016 p-Pb run at $\sqrt{s_{\mathrm{NN}}} = 8.16$ TeV, with 13.6 and 30.8 nb^{-1} recorded. The results published so far by LHCb focus on production of heavy flavour states, whose modification

IOP Conf. Series: Journal of Physics: Conf. Series **1271** (2019) 012008 doi:10.1088/1742-6596/1271/1/012008

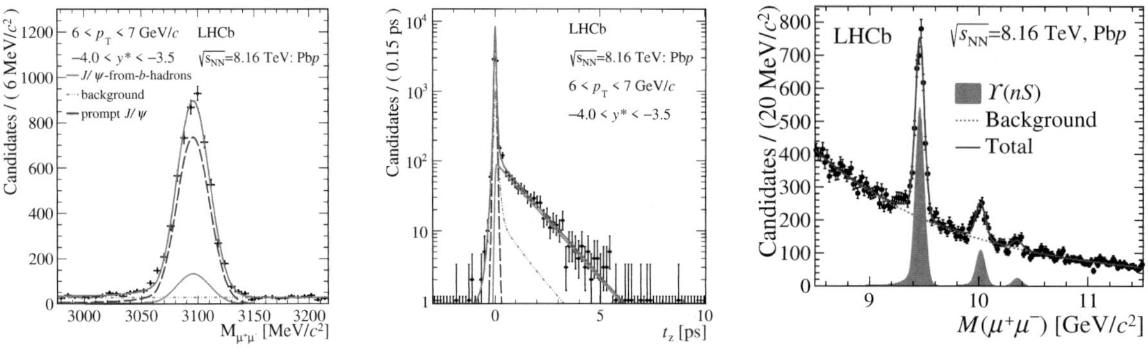

Figure 2. Distributions of (left) reconstructed mass and (middle) pseudo-proper time $t_z \equiv (z_{J/\psi} - z_{PV}) \times (M/p_z)_{J/\psi}$ for the $J/\psi \to \mu^+\mu^-$ decay candidates from the 8 TeV Pb-p sample [4] in the rapidity bin $-4.0 < y^* < -3.5$ ($z_{J/\psi}$ and z_{PV} are the reconstructed longitudinal position of the J/ψ decay vertex and the interaction primary vertex, respectively). The result of a fit to determine prompt signal, from-b signal and background fractions is overdrawn. In the right plot. mass distribution for the $\Upsilon(nS)$ candidates from the same sample in $-5.0 < y^* < -2.5$ [5].

with respect to p-p collisions constitutes a major probe for the hot and dense matter, known as quark-gluon plasma (QGP), produced in Pb-Pb collisions at LHC. In p-Pb collisions QGP is not expected to be formed and the system is considered to be a reference for the understanding of cold nuclear matter (CNM) effects.

Figure 2 illustrates the ability to distinguish promptly produced J/ψ mesons from those produced in b-hadron decays in reconstructed $J/\psi \to \mu^+\mu^-$ decays. The nuclear modification factor $R_{pPb} \equiv \sigma_{p\mathrm{Pb}}/(A_{\mathrm{Pb}}\, \sigma_{pp})$, where $\sigma_{p\mathrm{Pb}}$ and σ_{pp} are the production cross-sections in the two collision systems and $A_{\mathrm{Pb}} = 208$ is the lead mass number, can be measured separately for the two components. The result for the prompt component obtained from the 8 TeV sample [4] is shown in Figure 3 as a function of p_{T} and the c.m. rapidity y^*. A clear suppression with respect to p-p collisions is observed at forward rapidity and low p_{T}, compatible with the expected effect from nuclear PDF modifications (shadowing) as computed in the framework of NRQCD factorisation using several collinear nuclear PDF sets with the HELAC-Onia package [6, 7]. However, the rapidity dependence can also be well explained by the coherent energy loss model [8]. The result at forward rapidity is also compatible with the latest calculations based on the Color Glass Condensate model [9]. More measurements, notably on Drell-Yan production, are needed to disentangle these physical effects [10].

If final-state effects in CNM can't be excluded from the J/ψ result, they are definetely needed to explain the different modifications of quarkonia states. The first measurement of $\psi(2S)$ production in the 5 TeV sample [11] indicated a larger suppression with respect to J/ψ, in agreement with results from the other LHC experiments [12–15]. A more recent result [5] shows evidence for different suppression among the three $\Upsilon(nS)$ states, which are cleanly observed in the 8 TeV sample (see Figure 2). The result, shown in Figure 4, is nicely described in the framework of the "comovers" model [16], where dissociation of the quarkonia states is attributed to interaction with final-state particles which are close in phase-space. This relatively large effect needs to be taken into account in the interpretation of the spectacular suppression of Υ states recently observed by CMS in Pb-Pb collisions [17].

Production of open charm states, D^0 mesons [18] and Λ_c^+ baryons [19], has also been measured with large statistics already in the 5 TeV sample. Data are precise enough to constrain nuclear PDFs, assuming that initial-state effects dominate the observed nuclear modification.

IOP Conf. Series: Journal of Physics: Conf. Series **1271** (2019) 012008 doi:10.1088/1742-6596/1271/1/012008

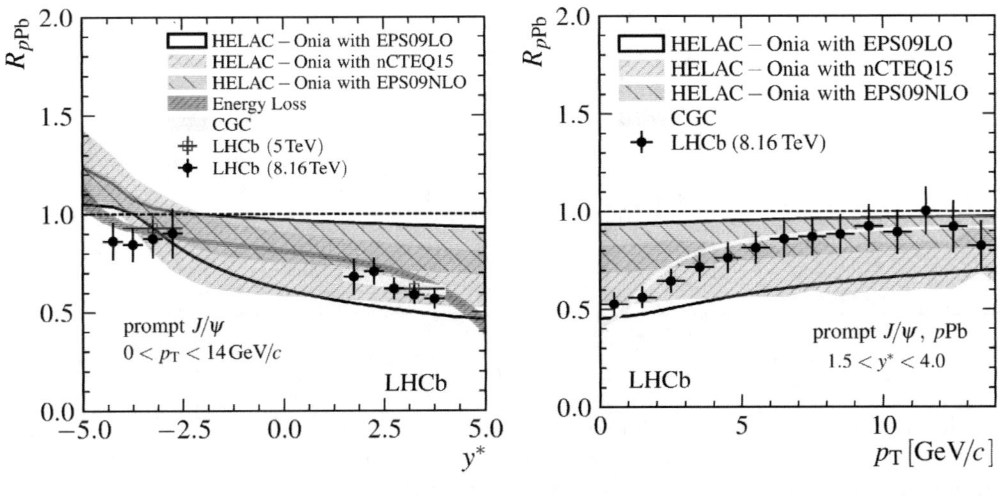

Figure 3. Nuclear modification factor R_{pPb} for prompt J/ψ production as a function of (left plot) y^* and (right plot) p_T for the forward configuration [4].

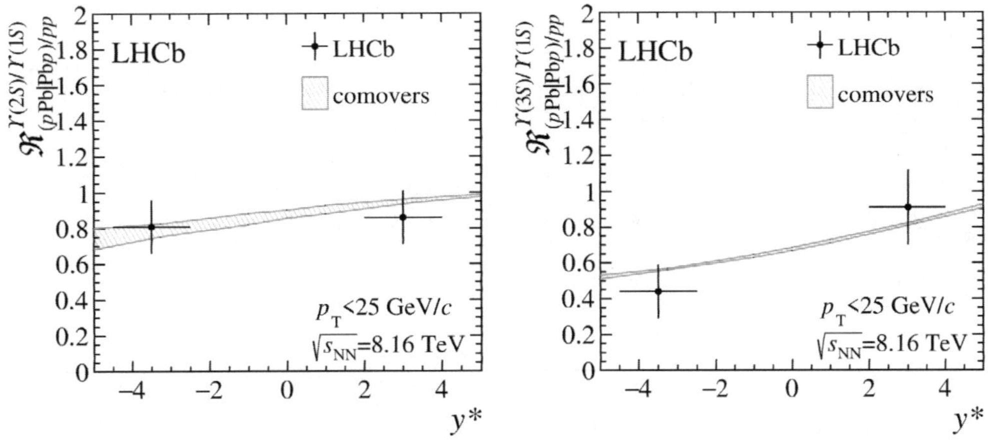

Figure 4. Ratio between Υ nuclear modification factors as a function of rapidity for (left plot) $\Upsilon(2S)/\Upsilon(1S)$ and (right plot) $\Upsilon(3S)/\Upsilon(1S)$ [5].

The Λ_c^+/D^0 ratio is an important input to the hadronisation phenomenology, since baryon enhancement in heavy ion collisions, notably at low p_T, is expected from production via coalescence and is affected by the thermal properties of the nuclear medium. LHCb measures a ratio between the Λ_c^+ and D^0 prompt production around 0.3, with no evidence of strong dependence on the rapidity or transverse momentum (see Figure 5). Results are in substantial agreement with HELAC-onia computations where the only nuclear effects are due to PDF modifications and largely cancel in the ratio.

For the first time, exclusive b-hadron decays have been reconstructed in nuclear collisions [21]. Clean samples of four decay modes ($B^+ \to \overline{D}^0\pi^+, B^+ \to J/\psi K^+, B^0 \to D^-\pi^+, \Lambda_b^0 \to \Lambda_c^+\pi^-$) have been obtained from the 8 TeV p-Pb dataset, down to p_T values below the hadron mass. The B^+ modes have been used to measure the nuclear modification factor as a function of y^* and p_T. The results, shown in Figure 6, confirm the suppression pattern observed with detached J/ψ, consistently with nuclear PDF effects. The ratio between prompt production of Λ_b^0 baryons

IOP Conf. Series: Journal of Physics: Conf. Series **1271** (2019) 012008 doi:10.1088/1742-6596/1271/1/012008

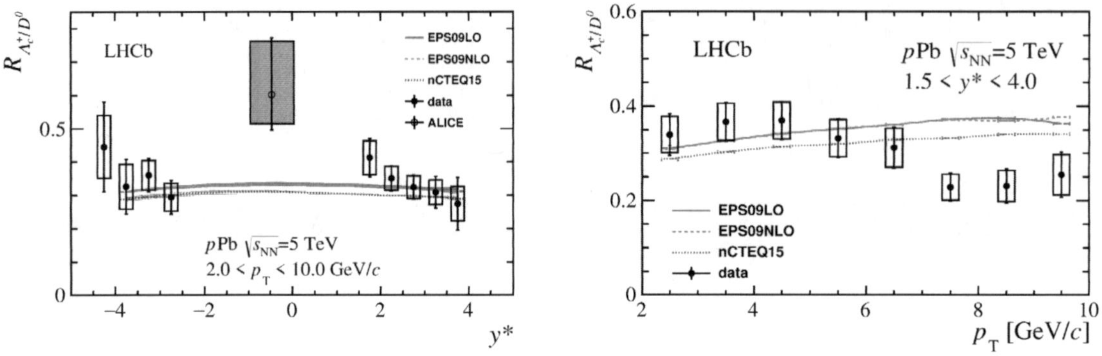

Figure 5. Ratio between Λ_c^+ and D^0 production in p-Pb collisions at 5 TeV, as a function of (left plot) y^* and (right plot) p_{T} for the forward configuration [19]. The result from ALICE [20] at mid rapidities is also shown for comparison.

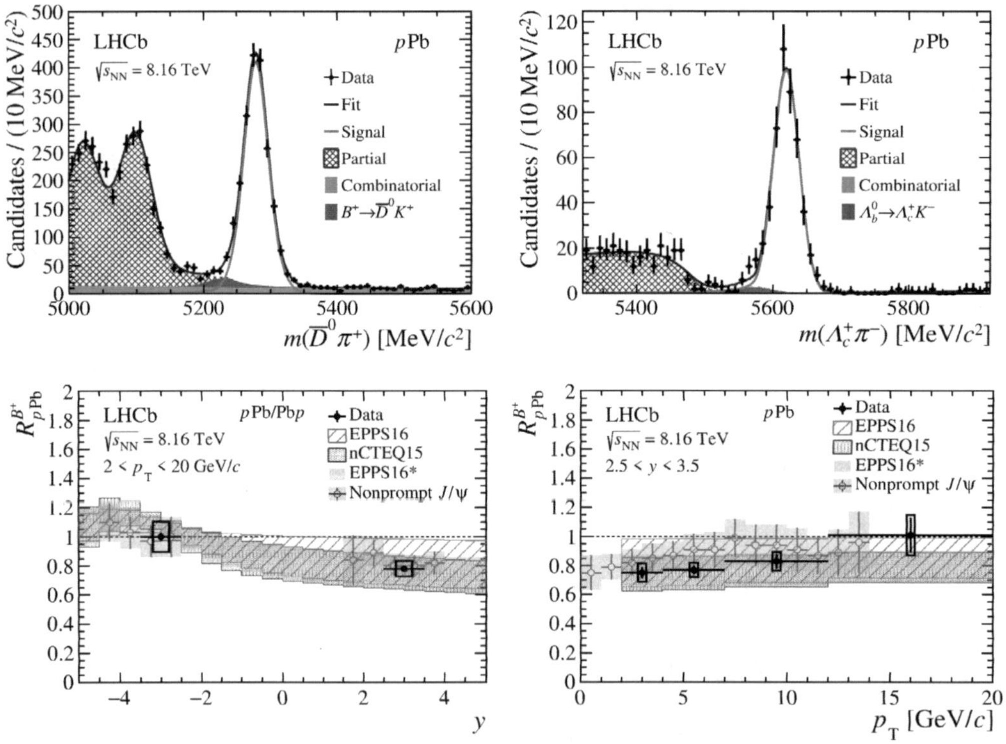

Figure 6. In the upper plots, reconstructed mass distribution of candidates for the (left) $B^+ \to \overline{D}{}^0\pi^+$ and (right) $\Lambda_b^0 \to \Lambda_c^+\pi^-$ decays in the p-Pb forward configuration data at 8 TeV. In the lower plots, the resulting nuclear modification factor R_{pPb} for B^+ meson production is plotted as a function of (left plot) y^* and (right plot) p_{T} for the forward configuration [21]. Results are compared with the measurement for J/ψ from b-hadron decays [4] and with HELAC-onia calculations using different nuclear PDF sets.

and B^0 mesons is also measured and found consistent with the p-p measurement.

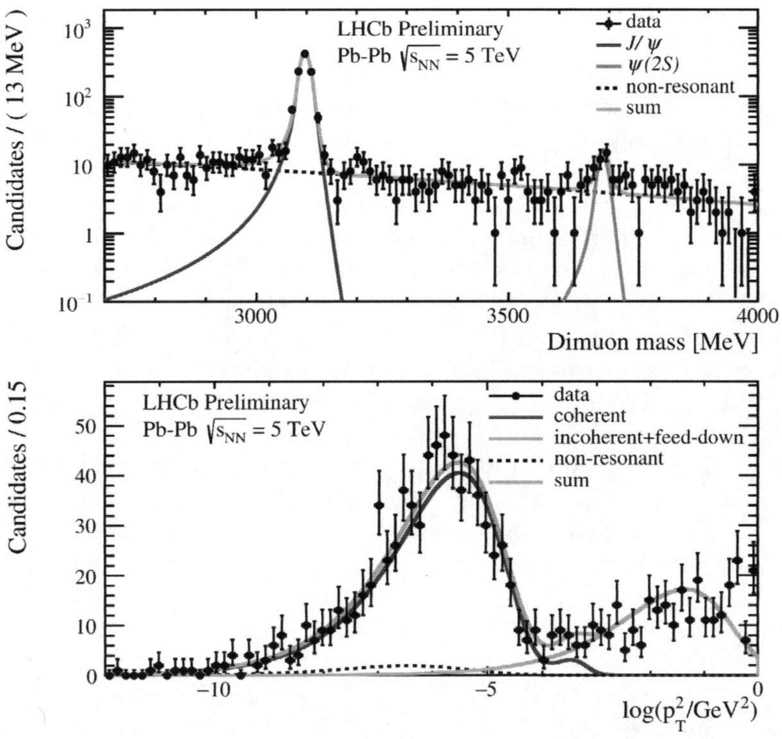

Figure 7. Candidates for J/ψ exclusive production in Pb-Pb collisions at 5 TeV [22], reconstructed from $J/\psi \to \mu^+\mu^-$ decays. In the upper plot, the reconstructed mass distribution is shown in an extended mass range where, beside the clean J/ψ peak, a small signal for $\psi(2S)$ can also be seen. In the lower plot, the $\log(p_T^2)$ distribution in the J/ψ mass range (3096.9 ± 65 MeV) is fitted with templates for coherent and incoherent production.

3. Pb-Pb collisions

A first small sample of Pb-Pb collisions was recorded by LHCb during the 2015 run, corresponding to about 10 μb^{-1}. The tracking detector performance in this challenging environment was found to be satisfactory for events of centrality above 50%.

The first preliminary physics result [22] has been obtained from ultraperipheral collisions (UPC), where hadron photoproduction is enhanced by the large photon flux from the lead nuclei. The observation of photoproduction of heavy flavour states, providing a hard scale for perturbative QCD calculations, is particularly interesting to explore the gluon density down to the saturation region at $x \sim 10^{-5}$. The exclusive production of J/ψ is cleanly observed (see Figure 7). The excellent p_T resolution allows to distinguish coherent and uncoherent production, whose p_T distributions are found to be well described by templates obtained with the STARlight generator [23].

The accuracy of the result is limited by the size of the data sample, but this analysis demonstrates the LHCb potential for physics in Pb-Pb UPC. During the run performed in november 2018, an integrated luminosity of 210 μb^{-1} was collected, providing the possibility for a precision measurement of exclusive J/ψ and $\psi(2S)$ production, and possibly to access rarer channels like Υ production and light-by-light scattering.

4. Fixed-target collisions

Samples of beam-gas collisions with proton and lead beams of different energy, and with the three possible targets (He, Ne and Ar) have been collected by LHCb during the LHC Run 2, as summarized in Figure 8. The largest sample, p-Ne collisions at $\sqrt{s_{\rm NN}} = 69$ GeV, corresponds to an integrated luminosity of about 100 $\rm nb^{-1}$. A sample of Pb-Ne collisions at the same energy has been recently collected during the 2018 Pb-Pb run.

The first physics results, demonstrating the potential of this novel program, have been obtained from some of the first samples collected in 2015 ane 2016. The first measurement of charm production [24] is based on the 7.6 ± 0.5 $\rm nb^{-1}$ of p-He collisions at 87 GeV, and a few $\rm nb^{-1}$ of p-Ar data at 110 GeV. Order of 10^3 D^0 mesons and 10^2 J/ψ mesons are reconstructed from both samples. These data provide the first determination of the $c\overline{c}$ cross-section at this relatively unexplored energy scale (see Figure 9). The rapidity dependence of the production is found to agree with predictions based on collinear factorisation not including a contribution from intrinsic charm, as shown in Figure 9. Therefore, these measurements, which are sensitive to x values up to ~ 0.5, do not favour the large intrinsic charm contributions predicted by some models in this kinematic range [25, 26]. The larger available datasets, and possibly the use of a hydrogen target in the future, will allow for more accurate constraints.

The antiproton production in the p-He sample at 110 GeV [28] has also been measured. This study is motivated by the recent precision measurements performed in space, notably by AMS-02 [29], of the antiproton content in cosmic rays, which is sensitive to possible exotic contributions like dark matter annihilation. For antiprotons above 10 GeV, the largest uncertainty on the expected flux of antiprotons from known sources, namely production in collisions between primary cosmic rays and the interstellar medium, is due to the limited knowledge of the corresponding production cross-sections. LHCb performed the first $\overline{p}$ production measurement in p-He collisions, which are responsible for about 40% of the expected cosmic $\overline{p}$ flux, in the range of $\overline{p}$ momentum between 12 and 110 GeV. The results, shown in Figure 10, are significantly more precise than the spread among the predictions from different phenomenological models,

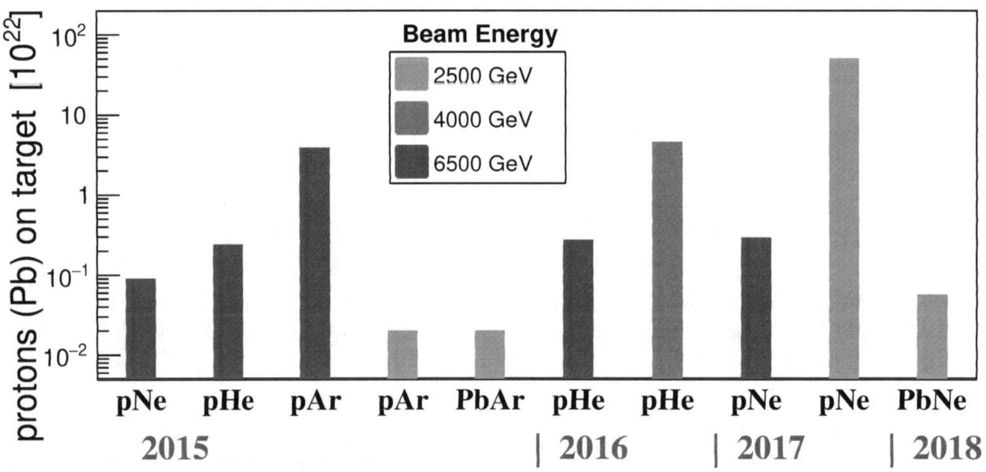

Figure 8. Summary of collected fixed-target physics samples at LHCb. The data size is given in terms of delivered protons (ions) on target (POT). For a nominal target pressure of 2×10^{-7} mbar, 10^{22} POT correspond to an integrated luminosity of about $5\,\rm nb^{-1}$ per meter of gas, though the actual target pressure and the data taking efficiency vary among samples.

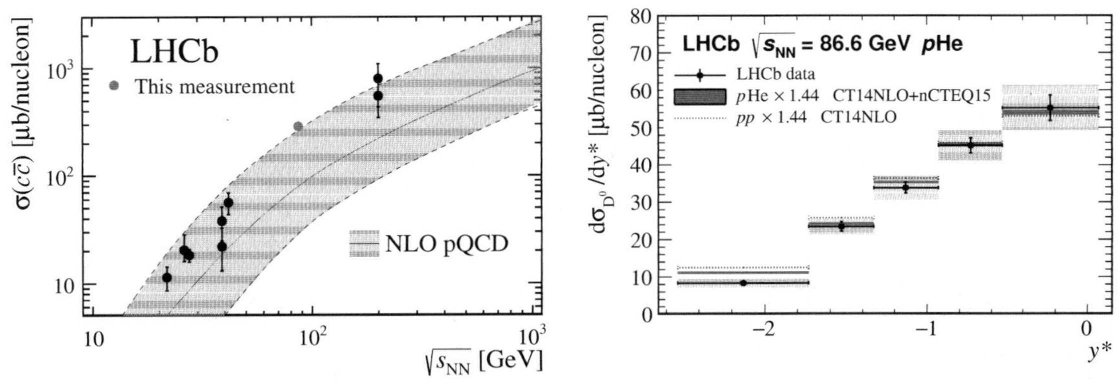

Figure 9. In the left plot, the result for the $c\bar{c}$ production cross-section per nucleon, obtained from p-He collisions at 69 GeV [24], is compared with other experimental results and with a NRQCD calculation [27]. The right plot shows the rapidity dependence of the D^0 production cross-section in the same sample. The result is compared with HELAC-onia calculations, not accounting for intrinsic charm, which are rescaled to reproduce the integrated value of the measured cross-section.

and are contributing to improve the models for cosmic secondary $\overline{p}$ and the resulting constraints on dark matter contributions [30, 31].

For both results, the integrated luminosity of the sample is estimated from the yield of elastically scattered electrons from the target atoms. The yield is determined by selecting events with a single low-pt track in the detector, identified as an $e^{\pm}$. The background to this normalization channel is due to soft diffractive collisions with a single track reconstructed in the detector and is expected to be symmetric in charge, so that it can be estimated and subtracted using the e^+ candidates, as illustrated in Figure 10. The integrated luminosity is determined using this method with an accuracy of 6% [28], dominated by the systematic uncertainty on the electron reconstruction efficiency.

5. Conclusions and Prospects

The results obtained so far by the LHCb collaboration from heavy ion collisions demonstrate the capability of the experiment to provide unique contributions to this field, exploiting a variety of collision systems and energy scales, where exclusive final states, notably in the heavy flavour sector, can be reconstructed with high efficiency and purity.

The rich datasets collected so far provide many additional possibilities that are being investigated. In p-Pb collisions, studies on direct photons, both inclusive and in gamma-jet events, are underway, with unique sensitivity to the saturation region [33]. The size of the sample collected in 2018 at 8 TeV is expected to provide access to new channels, as χ_c and η_c quarkonia states and Drell-Yan dimuon events. Studies of flow and correlations with identified particles in the unique forward acceptance region covered by LHCb are also planned.

Substantial development of the heavy ion program in LHCb is also expected during the future LHC runs, taking profit of the currently ongoing detector upgrade [34] which includes a new vertex detector with improved granularity. The proposed plans for future heavy ion running, discussed in detail in Reference [35], foresee an increase in integrated luminosity for p-Pb and Pb-Pb collisions by more than an order of magnitude during the LHC Runs 3 and 4 (2021-2029). This will open novel possibilities, as precision studies using Drell-Yan events and correlations in heavy flavour production. A proposal for a second detector upgrade for the LHC Run 5 (starting 2031) has been put forward [36, 37]. The granularity of such detector, conceived to take profit

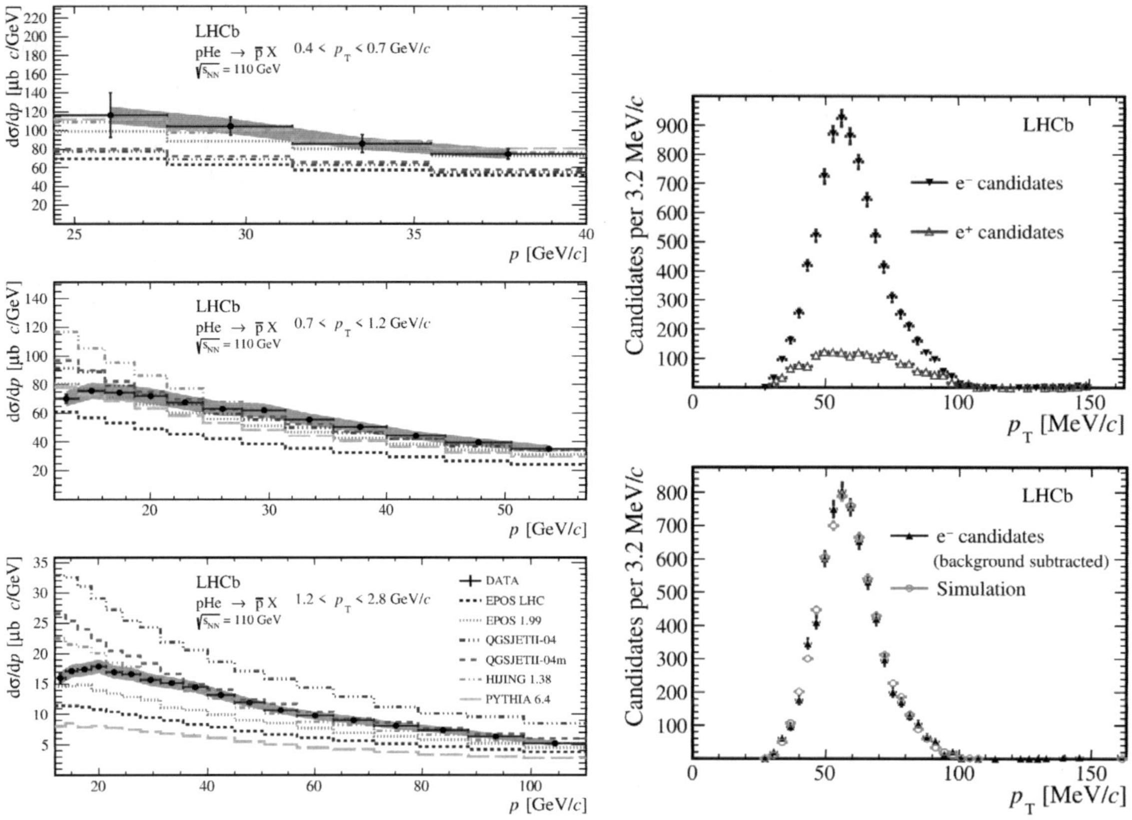

Figure 10. On the left plot, result for antiproton production in p-He collisions at 110 GeV. The differential production cross-section $d\sigma/dp$ is shown as a function of the momentum p for different ranges of $p_{\rm T}$ and is compared with several generators included in the CRMC package [32]. The normalization is obtained from the p-e^- elastic scattering events from the same sample, illustrated in the right plots. The upper plot shows the $p_{\rm T}$ distribution for single scattered electron candidates with negative and positive charge, the latter sample being used to subtract the background from hadronic collisions. After subtraction, the distribution is found to agree well with the expected one from simulated p-e^- scattering, as shown in the lower plot.

of the full potential of LHC luminosity in p-p collisions, would make it possible to reconstruct even the most central Pb-Pb collisions.

The fixed-target program will also be considerably developed, thanks to the installation of a new target device [38]. This will consist of a storage cell containing the injected gas into a 20 cm long region located just upstream the LHCb vertex detector. An increase in fixed-target luminosity by up to two orders of magnitude with the same injected gas flow is anticipated. Furthermore, it will be possible to inject more gas species, notably hydrogen, providing p-p reference for the other fixed-target samples. The physics potential of such program, discussed in more detail in Reference [39], covers a wealth of measurements that won't be possible at any other facility in the coming years, including heavy flavour and Drell-Yan production giving access to nuclear PDFs at large x, measurements of quarkonia suppression in different systems at $\sqrt{s_{\rm NN}} \sim 100$ GeV, and studies of particle production of great interest for cosmic ray physics.

References

[1] Alves Jr A A *et al.* (LHCb collaboration) 2008 *JINST* **3** S08005
[2] Barschel C 2014 Ph.D. thesis RWTH Aachen U. CERN-THESIS-2013-301
[3] Khachatryan V *et al.* (CMS collaboration) 2010 *JHEP* **09** 091 (*Preprint* 1009.4122)
[4] Aaij R *et al.* (LHCb collaboration) 2017 *Phys. Lett.* **B774** 159 (*Preprint* 1706.07122)
[5] Aaij R *et al.* (LHCb collaboration) 2018 *JHEP* **11** 194 (*Preprint* 1810.07655)
[6] Shao H S 2016 *Comput. Phys. Commun.* **198** 238–259 (*Preprint* 1507.03435)
[7] Lansberg J P and Shao H S 2017 *Eur. Phys. J.* **C77** 1 (*Preprint* 1610.05382)
[8] Arleo F and Peigné S 2013 *JHEP* **03** 122 (*Preprint* 1212.0434)
[9] Duclou B, Lappi T and Mntysaari H 2016 *Phys. Rev.* **D94** 074031 (*Preprint* 1605.05680)
[10] Arleo F and Peign S 2017 *Phys. Rev.* **D95** 011502 (*Preprint* 1512.01794)
[11] Aaij R *et al.* (LHCb collaboration) 2016 *JHEP* **03** 133 (*Preprint* 1601.07878)
[12] Abelev B B *et al.* (ALICE collaboration) 2014 *JHEP* **12** 073 (*Preprint* 1405.3796)
[13] Adam J *et al.* (ALICE collaboration) 2016 *JHEP* **06** 050 (*Preprint* 1603.02816)
[14] Aaboud M *et al.* (ATLAS collaboration) 2018 *Eur. Phys. J.* **C78** 171 (*Preprint* 1709.03089)
[15] Sirunyan A M *et al.* (CMS collaboration) 2019 *Phys. Lett.* **B790** 509–532 (*Preprint* 1805.02248)
[16] Ferreiro E G and Lansberg J P 2018 *JHEP* **10** 094 [Erratum: JHEP03,063(2019)] (*Preprint* 1804.04474)
[17] Sirunyan A M *et al.* (CMS collaboration) 2019 *Phys. Lett.* **B790** 270–293 (*Preprint* 1805.09215)
[18] Aaij R *et al.* (LHCb collaboration) 2017 *JHEP* **10** 090 (*Preprint* 1707.02750)
[19] Aaij R *et al.* (LHCb collaboration) 2019 *JHEP* **02** 102 (*Preprint* 1809.01404)
[20] Acharya S *et al.* (ALICE collaboration) 2018 *JHEP* **04** 108 (*Preprint* 1712.09581)
[21] Aaij R *et al.* (LHCb collaboration) 2019 *Phys. Rev.* **D99** 052011 (*Preprint* 1902.05599)
[22] LHCb collaboration 2018 LHCb–CONF–2018–003
[23] Klein S R, Nystrand J, Seger J, Gorbunov Y and Butterworth J 2017 *Comput. Phys. Commun.* **212** 258–268 (*Preprint* 1607.03838)
[24] Aaij R *et al.* (LHCb collaboration) 2018 *Phys. Rev. Lett.* **122** 132002 (*Preprint* 1810.07907)
[25] Pumplin J, Lai H L and Tung W K 2007 *Phys. Rev.* **D75** 054029 (*Preprint* hep-ph/0701220)
[26] Brodsky S J, Kusina A, Lyonnet F, Schienbein I, Spiesberger H and Vogt R 2015 *Adv. High Energy Phys.* **2015** 231547 (*Preprint* 1504.06287)
[27] Maltoni F *et al.* 2006 *Phys. Lett.* **B638** 202–208 (*Preprint* hep-ph/0601203)
[28] Aaij R *et al.* (LHCb collaboration) 2018 *Phys. Rev. Lett.* **121** 222001 (*Preprint* 1808.06127)
[29] Kounine A 2012 *Int. J. Mod. Phys.* **E21** 1230005
[30] Reinert A and Winkler M W 2018 *JCAP* **1801** 055 (*Preprint* 1712.00002)
[31] Korsmeier M, Donato F and Di Mauro M 2018 *Phys. Rev. D* **97**(10) 103019 (*Preprint* 1802.03030)
[32] Pierog T, Baus C and Ulrich R CRMC (Cosmic Ray Monte Carlo package) web.ikp.kit.edu/rulrich/crmc.html
[33] Da Silva C (LHCb collaboration) 2018 Contribution to the Quark Matter 2018 conference URL `https://indico.cern.ch/event/656452/contributions/2859624`
[34] LHCb collaboration 2012
[35] Citron Z *et al.* 2018 *HL/HE-LHC Workshop Report* (*Preprint* 1812.06772)
[36] LHCb collaboration 2017 CERN–LHCC–2017–003
[37] LHCb collaboration 2018 CERN–LHCC–2018–027 LHCb–PUB–2018–009 (*Preprint* 1808.08865)
[38] Di Nezza P *et al.* 2018 Tech. Rep. CERN-PBC-Notes-2018-007
[39] Graziani G *et al.* 2018 Tech. Rep. LHCb-PUB-2018-015

IOP Conf. Series: Journal of Physics: Conf. Series **1271** (2019) 012009 doi:10.1088/1742-6596/1271/1/012009

Tests of Lepton Flavour Universality at LHCb

Katharina Müller on behalf of the LHCb collaboration
Winterthurerstr. 190, CH-8057 Zurich, Switzerland
E-mail: kmueller@physik.uzh.ch

Abstract. In the Standard Model of particle physics the three charged leptons are identical copies of each other, apart from mass differences, and the electroweak coupling of the gauge bosons to leptons is independent of the lepton flavour. This prediction is called lepton flavour universality (LFU) and is well tested. In tree level decays, any violation of LFU would be a clear sign of physics beyond the Standard Model. Experimental tests of LFU in semileptonic decays of b-hadrons or rare b decays are highly sensitive to New Physics particles which preferentially couple to the 2nd and 3rd generations of leptons. Recent results from LHCb on lepton flavour universality in $b \to c\ell\nu$ transitions and rare $b \to s\ell\ell$ decays are discussed. The results are based on $3\,\text{fb}^{-1}$ of proton-proton collisions collected at centre of mass energies $\sqrt{s} = 7\,\text{TeV}$ and $8\,\text{TeV}$.

1. Introduction

In the Standard Model of particle physics (SM), the electroweak gauge bosons Z and $W^{\pm}$ have identical couplings to all three lepton flavours. This means that branching fractions of decays involving different lepton families do not depend on lepton flavour but differ only by phase space and helicity-suppressed contributions. The prediction is called lepton flavour universality (LFU) and is well tested in e.g. decays of tau leptons, light mesons, as well as the gauge bosons. Any experimental evidence of Lepton Flavour Non-Universality would be a clear sign of physics beyond the SM (BSM).

Semileptonic decays of heavy hadrons are an excellent laboratory to test LFU as all three generations can be accessed. Many models extending the SM contain additional interactions that could violate LFU. These are for example BSM theories involving leptoquarks [1, 2] or Z' [3, 4] particles. Processes with third generation of quarks and leptons (B and tau) are well suited to search for LFU violation since many BSM theories with LFU violation predict stronger couplings to the third generation and experimental results have lower precision. One such example is an extended Higgs Sector, which could have a large effect on semitauonic decay rates through the coupling to new charged Higgs scalars [5].

These proceedings summarise recent measurements of these decays by the LHCb collaboration. Feynman diagrams for the SM processes are shown in figure 1. All measurements are based on $3\,\text{fb}^{-1}$ of proton-proton (pp) collisions collected at the LHC collider at centre of mass energies $\sqrt{s} = 7$ TeV and 8 TeV.

The remainder of this paper is organised as follows: section 2 discusses results from $b \to c\ell\nu$ decays with emphasis on the measurements of decays with a D or J/ψ meson in the final state, while section 3 presents results from $b \to s\ell^+\ell^-$ decays and section 4 concludes the paper.

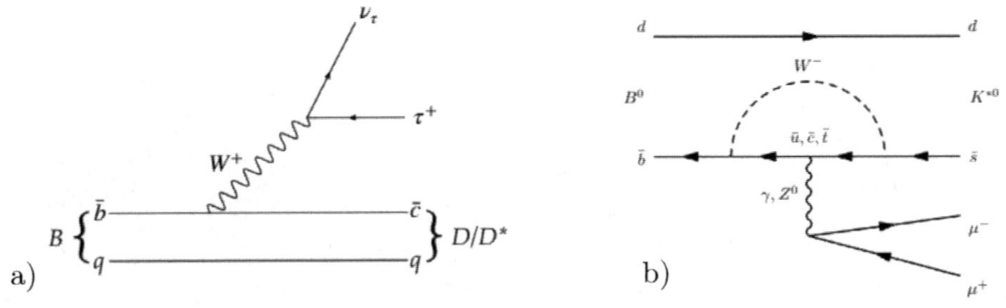

Figure 1. Feynman diagrams in the SM for a semileptonic $b \to c\ell\nu$ (a) and the rare electroweak penguin $b \to s\ell\ell$ (b) decays.

2. Lepton Flavour Universality in $b \to c\ell\nu$ decays

Charged current (semileptonic decays) tree-level $b \to c\ell\nu$ decays have a branching fraction of a few percent and are precisely predicted in the SM. The strongest evidence for a possible violation of LFU is currently seen in measurements of the branching fractions of semileptonic decays involving a tau lepton, in particular the measurement of the observables

$$R(D) = \frac{\mathcal{B}(B^0 \to D^+\tau^-\overline{\nu}_\tau)}{\mathcal{B}(B^0 \to D^+\mu^-\overline{\nu}_\mu)} \text{ [6, 7]} \quad \text{and} \quad R(D^*) = \frac{\mathcal{B}(B^0 \to D^{*+}\tau^-\overline{\nu}_\tau)}{\mathcal{B}(B^0 \to D^{*+}\mu^-\overline{\nu}_\mu)} \text{ [6, 7, 8].} \quad (1)$$

The observed enhancement can be explained in many extensions to the SM, which preferentially couple to third generation leptons.

2.1. Measurements of $R(D^*)$

A precise measurement of a B decay to tau leptons is experimentally challenging at a hadron collider due to the large background from partially reconstructed B-decays with similar topologies. Moreover, the signal decay kinematics can not be fully constrained because of the presence of neutrinos in the final state. Therefore the ratio of $R(D^*)$ is measured using different tau lepton decays to provide independent measurements. LHCb used the muonic $\tau \to \mu\nu_\tau\overline{\nu}_\mu$ as well as the hadronic $\tau \to \pi\pi\pi\nu_\tau$ decay channels.

The muonic decay of the tau has the advantage that both the signal and normalisation channel in $R(D^*)$ contain the same visible final state, with the difference of having one and three neutrinos in the final states, respectively.

The signal and normalisation channels are separated using a fit to three kinematic variables sensitive to the mass difference between the muon and the tau and the presence of additional neutrinos: E^*_μ the energy of the muon in the B^0 rest frame, $m^2_{miss} = (p(B^0) - p(D^*) - p(\mu))^2$, the squared missing mass, and $q^2 = (p(B^0) - p(D^*))^2$ the squared four-momentum transfer to the lepton system. The results of the fit are shown in figure 2. The yields extracted by the fit are used to make a measurement of

$$R(D^*) = 0.336 \pm 0.027(\text{stat}) \pm 0.030(\text{syst}) \text{ [8].}$$

The result was the first measurement of b-hadron decays to tau leptons at a hadron collider and deviates by $2.1\,\sigma$ from the SM (0.252 ± 0.003) prediction [9]. Systematic uncertainties are dominated by the size of the simulated and control samples.

IOP Conf. Series: Journal of Physics: Conf. Series **1271** (2019) 012009 doi:10.1088/1742-6596/1271/1/012009

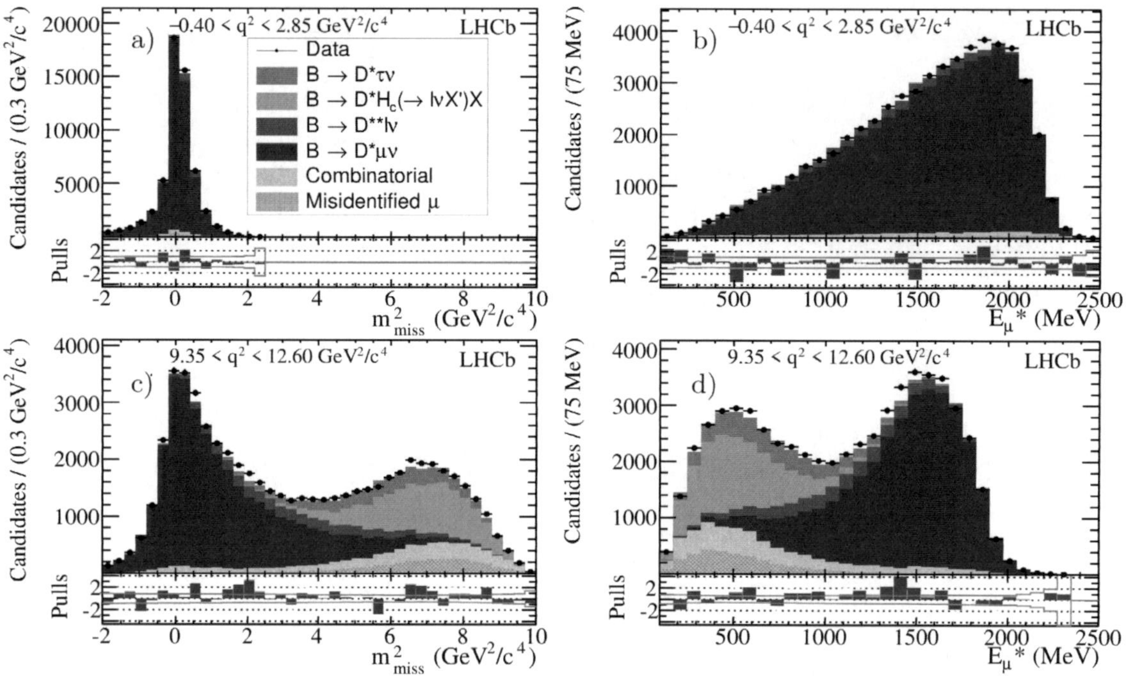

Figure 2. Distributions of $m^2_{\rm miss}$ (a and c) and E^*_μ (b and d) with the fit projections overlaid for the lowest (a and b) and the highest (c and d) q^2 bin (from [8]). The result is based on $3\,{\rm fb}^{-1}$ of pp collisions at $\sqrt{s} = 7$ TeV and 8 TeV.

A more recent measurement of $R(D^*)$ was performed using the hadronic decay of the $\tau^- \to \pi^+\pi^-\pi^-\nu_\tau$ with the D^* meson reconstructed through the decay $D^* \to D^-(\to K^+\pi^-)\pi^-$.

In this analysis the ratio

$$K(D^*) = \frac{\mathcal{B}(B \to D^*\tau\overline{\nu}_\tau)}{\mathcal{B}(B \to D^*3\pi)} \tag{2}$$

is measured. Both decays have the same visible final state which leads to a large cancellation of various experimental uncertainties. Subsequently, $R(D^*)$ is determined as

$$R(D^*) = K(D^*) \times \frac{\mathcal{B}(B \to D^*3\pi)}{\mathcal{B}(B \to D^*\mu\overline{\nu}_\mu)}, \tag{3}$$

with the branching ratios from external inputs [10]. The main background contribution is due to decays of B-hadrons to $D^*3\pi X$, where X represents unreconstructed particles. This decay is about 100 times more abundant. A good suppression is obtained by requiring the tau decay vertex being downstream of the B decay vertex with a $4\,\sigma$ significance as shown in figure 3. The remaining background comes mainly from double-charmed B decays (eg $B \to D^{*+}D^{*-}X$). A Boosted Decision Tree is used to discriminate the signal from doubly-charmed background. The signal yield is obtained from a three-dimensional extended maximum likelihood fit to the BDT output, the tau decay time, and q^2. The results of the fit are shown in figure 4. The yield of the normalisation mode is determined by fitting the invariant mass of the $D3\pi$ system.

The measured value of $K(D^*)$ is found to be

$$K(D^*) = 1.97 \pm 0.13({\rm stat}) \pm 0.18({\rm syst})\ [11]$$

IOP Conf. Series: Journal of Physics: Conf. Series **1271** (2019) 012009 doi:10.1088/1742-6596/1271/1/012009

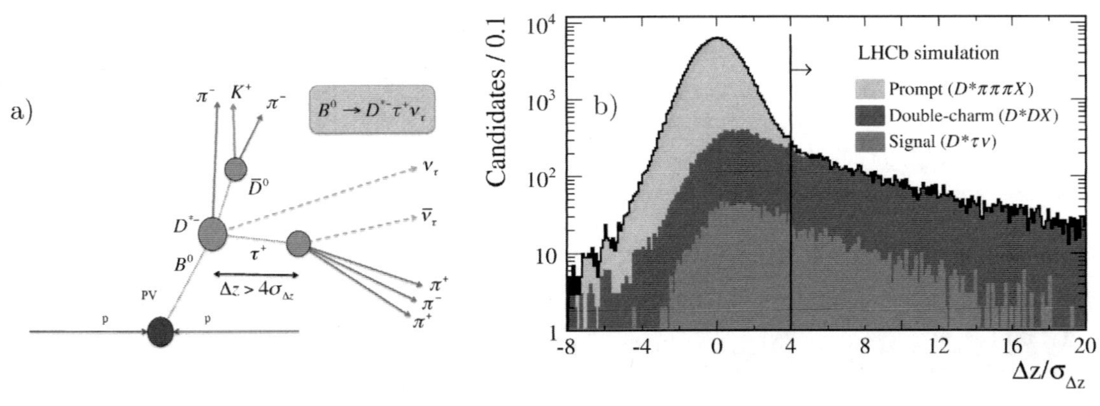

Figure 3. a) Topology of the signal decay. A requirement on the distance between the three-pion and the B^0 vertices along the beam direction to be greater than four times its uncertainty is applied. b) Distribution of the distance between the B^0 vertex and the three-pion vertex along the beam direction, divided by its uncertainty, obtained using simulation. The vertical line shows the 4σ requirement used in the analysis to reject the prompt background component (from [11]).

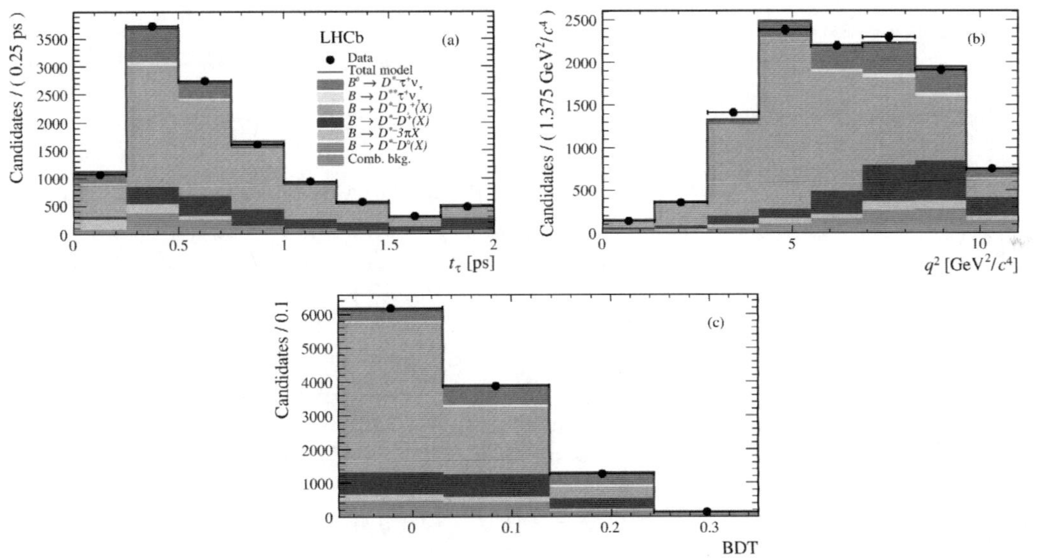

Figure 4. Projections of the three-dimensional fit on the three-pion decay time (a), q^2 (b) and BDT output (c) distributions. The fit components are described in the legend (from [11]).

and $R(D^*)$ is calculated to be

$$R(D^*) = 0.291 \pm 0.019(\text{stat}) \pm 0.026(\text{syst}) \pm 0.013(\text{ext}) \text{ [11]},$$

where the third uncertainty comes from uncertainties on the external input. This measurement is one of the most precise and compatible with the SM within 1σ [9].

Figure 5 shows the combination from the Heavy Flavour Averaging Group (HFLAV)

IOP Conf. Series: Journal of Physics: Conf. Series **1271** (2019) 012009 doi:10.1088/1742-6596/1271/1/012009

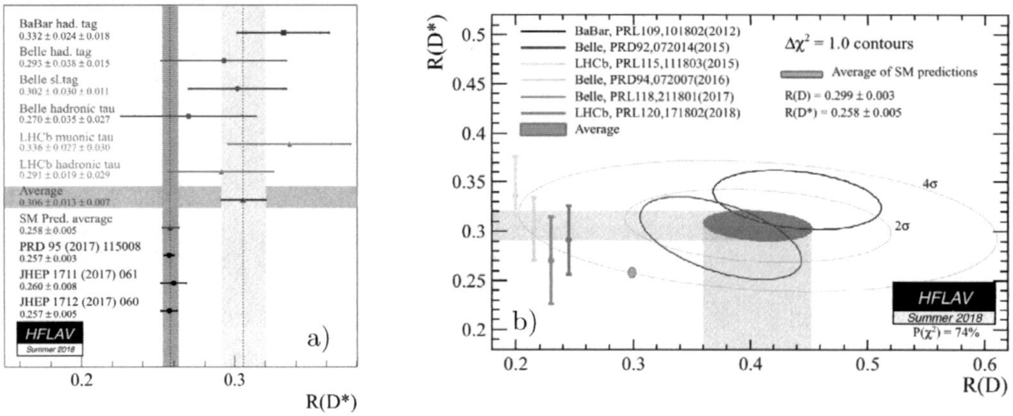

Figure 5. a) Comparison of $R(D^*)$ with the previous results from the B-factories and with SM expectations. b) Summary of the measurements of $R(D)$ and $R(D^*)$ (from [12]).

for $R(D^*)$ versus $R(D)$ along with the SM prediction and individual measurements by the experiments BaBar, Belle and LHCb [12]. All six $R(D^*)$ measurements lie above the SM predictions, the overall combination of the $R(D^*)$ and $R(D)$ measurements yield a $3.8\,\sigma$ tension with the SM.

2.2. Measurement of $R(J/\psi)$
The measurement of $R(J/\psi) = \mathcal{B}(B_c \to J/\psi\tau\overline{\nu}_\tau)/\mathcal{B}(B_c \to J/\psi\mu\overline{\nu}_\mu)$ probes similar physics as $R(D^*)$ but with a different spectator quark. The tau is reconstructed in the muonic and the J/ψ in the di-muon decay channel. The analysis strategy is very similar to the $R(D^*)$ analysis in the muonic channel with the B_c decay time as an additional discriminating variable. A binned maximum likelihood fit to m^2_{miss}, the B_c decay time and the quantity Z, where Z represents eight bins in $(E^*_\mu, q^2)^1$, is performed to determine the signal and normalisation yield. The fit result to the three quantities is shown in figure6. The measured value of

$$R(J/\psi) = 0.71 \pm 0.17(\text{stat}) \pm 0.18(\text{syst})\ [13]$$

is compatible with the SM prediction (0.25-0.28) [14, 15, 16, 17] at the level of $2\,\sigma$. The unknown form factors used in the generation of the simulated samples gives rise to the largest systematic uncertainty.

3. Lepton Flavour Universality in $b \to s\ell^+\ell^-$ decays

A very clean test for New Physics (NP) can be performed by taking ratios of $b \to s\ell^+\ell^-$ decays to different lepton species. Since these decays are not allowed at tree level the branching fractions are highly sensitive to NP effects. Comparisons of decays with different leptons in the final state allow to probe NP models that involve LFU violation among different generations. Currently, $b \to s\ell^+\ell^-$ decays with electrons and muons in the final state are accessible to LHCb. If the momentum transfer of the dilepton system is sufficiently above the dilepton mass, uncertainties

[1] The values 0-3 of Z correspond to $q^2 < 7.15\,\text{GeV}^2/c^4$ and E^*_μ divided in bins with thresholds at [0.68, 1.15, 1.64] GeV. The values 4-7 correspond to bins with the same E^*_μ ranges, and $q^2 \geq 7.15\,\text{GeV}^2/c^4$.

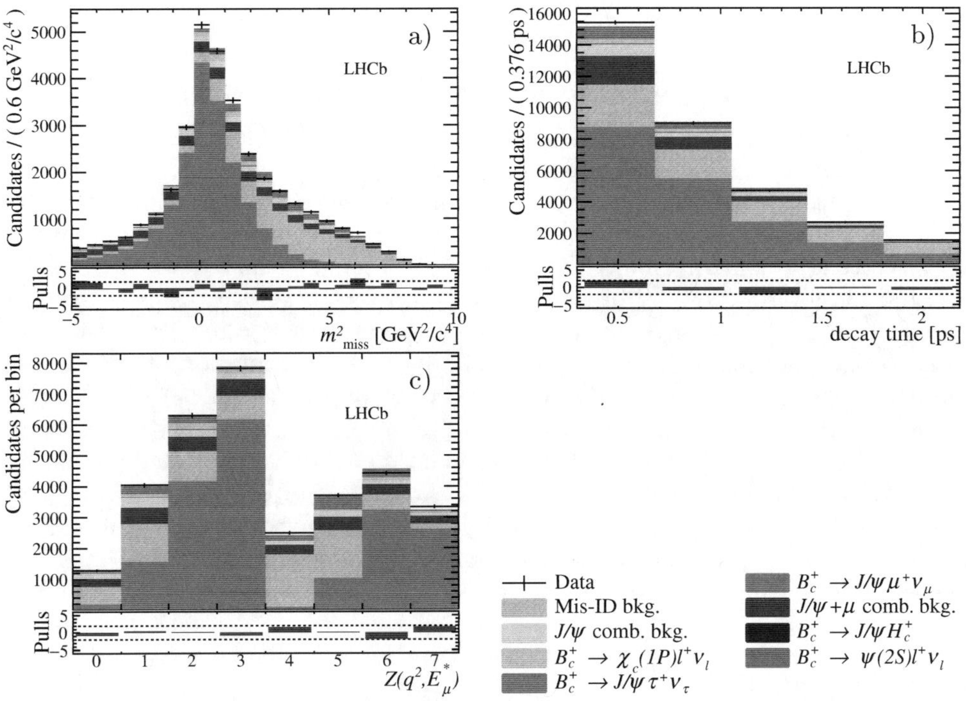

Figure 6. Measurement of $R(J/\psi)$: a) distributions of m^2_{miss}, b) decay time, and c) Z of the signal data, overlaid with projections of the fit model with all normalisation and shape parameters at their best-fit values. Below each panel differences between the data and fit are shown, normalised by the Poisson uncertainty in the data; the dashed lines are at the values ± 2 (from [13]). The result is based on $3\,\text{fb}^{-1}$ of pp collisions at $\sqrt{s} = 7$ TeV and 8 TeV.

in the hadronic form factors cancel to a very good approximation, leading to very precise SM predictions

$$R(K^{(*)}) = \frac{\mathcal{B}(B \to K^{(*)}\mu\mu)}{\mathcal{B}(B \to K^{(*)}ee)} = 1 \pm \mathcal{O}(10^{-3}) \pm \mathcal{O}(10^{-2}) \text{ [18]}, \tag{4}$$

where the first correction accounts for the difference of the lepton masses and the second for QED corrections. In recent years, the interest in LFU tests in $b \to s\ell^+\ell^-$ has increased, mainly driven by two measurements from the LHCb: the ratio $B^+ \to K^+\mu^+\mu^-$ to $B^+ \to K^+e^+e^-$[19] and more recently the ratio $B^0 \to K^*\mu^+\mu^-$ to $B^0 \to K^*e^+e^-$[20], presented below. The LFU testing ratio $R(K^{(*0)})$ is defined as

$$R(K^{(*)}) = \frac{\int \frac{d\Gamma(B\to K^{(*)}\mu^+\mu^-)}{dq^2} dq^2}{\int \frac{d\Gamma(B\to K^{(*)}e^+e^-)}{dq^2} dq^2}. \tag{5}$$

Here, the differential decay rate is measured in a certain q^2 range, where q^2 corresponds to the invariant mass squared of the di-lepton pair. The q^2 ranges corresponding to the J/ψ and $\psi(2S)$ are excluded from the analysis. To cancel experimental uncertainties in the absolute efficiencies of the measurements, the ratio $R(K^{(*)})$ is measured as double ratio, normalising the non-resonant signal mode to the resonant J/ψ mode:

$$R(K^{(*)}) = \frac{\mathcal{B}(B \to K^* \mu^+ \mu^-)}{\mathcal{B}(B \to K^* J/\psi(\to \mu^+ \mu^-))} \times \frac{\mathcal{B}(B \to K^* J/\psi(\to e^+ e^-))}{\mathcal{B}(B \to K^* e^+ e^-)}. \tag{6}$$

The candidates for the normalisation channel $B \to K^* J/\psi(\to \ell^+ \ell^-)$ are selected using similar criteria to that of the non-resonant channel.

The main challenge of the measurement is the reconstruction of the electrons in the LHCb detector. While muons can be reconstructed with high efficiency and low misidentification probability, electrons are highly affected by bremsstrahlung. This results in a degraded B momentum and mass resolution. A bremsstrahlung recovery algorithm is used to improve the electron momentum reconstruction: clusters not associated with charged tracks are searched in the electromagnetic calorimeter in a region defined by the extrapolation of the electron track upstream of the magnet and added to the measured electron momentum. Moreover, the trigger thresholds on electrons are higher than those on muons due to the higher occupancy of the calorimeters which results in a lower efficiency. The distribution of q^2 as a function of the four-body invariant mass for the B^0 candidates is shown in figure 7 (a and b) for both muon and electron final states. In each plot, the contributions due to the charmonium resonances are visible as bands at the J/ψ and $\psi(2S)$ masses. In case of electrons the worse mass resolution due to bremsstrahlung is prominently visible: the reconstructed mass is smeared and generally shifted to lower values.

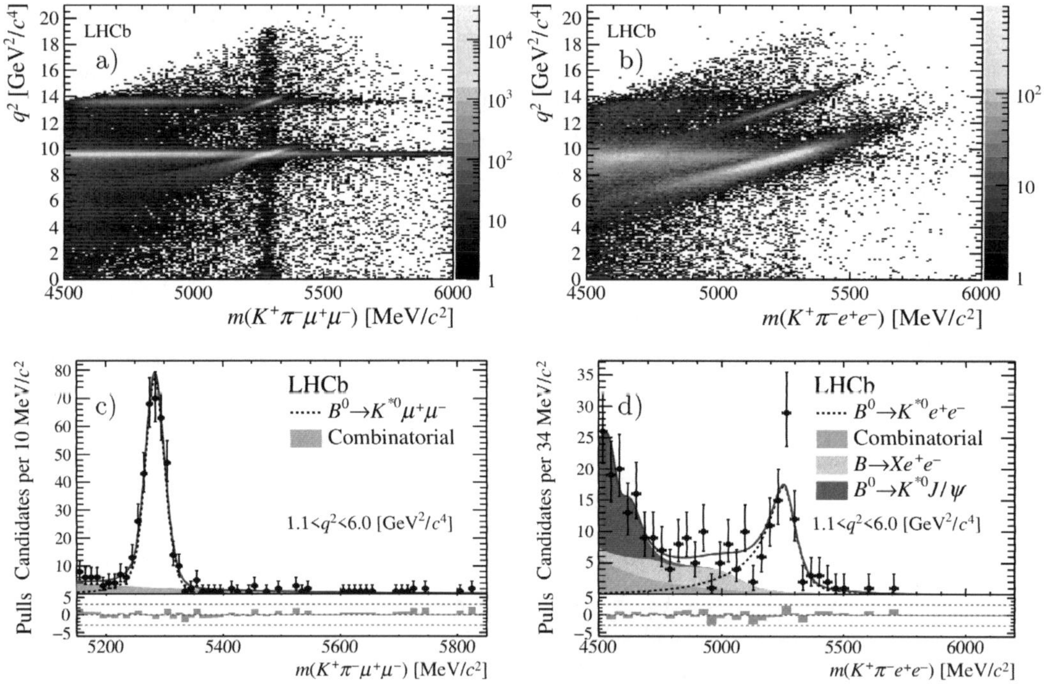

Figure 7. Number of candidates for $B^0 \to K^0 \ell^+ \ell^-$ final states with (a) muons and (b) electrons as a function of the dilepton invariant mass squared, q^2, and the four-body invariant mass of the B^0. Fit to the $m(K^+\pi - \mu^+\pi)$ invariant mass of $B^0 \to K^{*0}\mu^+\mu^-$ (c) and $B^0 \to K^{*0}\mathrm{e}^+\mathrm{e}^-$ (d) in the central-q^2 region. The dashed line is the signal PDF, the shaded shapes are the background PDFs and the solid line is the total PDF. The fit residuals normalised to the data uncertainty are shown at the bottom of each distribution (from [20]).

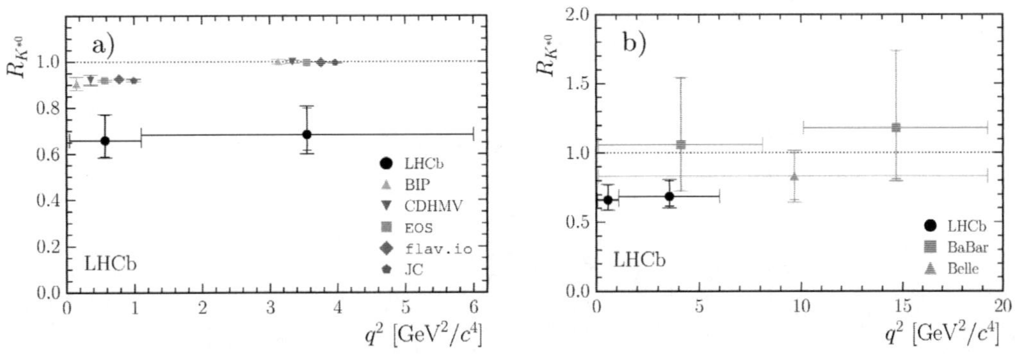

Figure 8. a) Comparison of the LHCb $R(K^*)$ measurements with the SM theoretical predictions. The predictions are displaced horizontally for presentation. b) Comparison of the LHCb $R(K^*)$ measurements with previous experimental results from the B factories [21, 22]. In the case of the B factories the specific vetoes for charmonium resonances are not represented. (from [20]). The result is based on $3\,\mathrm{fb}^{-1}$ of pp collisions at $\sqrt{s} = 7$ TeV and 8 TeV.

The measurement of $R(K^{*0})$ is performed in two q^2 bins that are sensitive to different physics contributions: the first one, $0.045 < q^2 < 1.1\,\mathrm{GeV}^2/c^4$, is dominated by the photon pole while for the central bin, $1.1 < q^2 < 6\,\mathrm{GeV}^2/c^4$, the contribution from the C_9 Wilson coefficient becomes more important.

Signal yields are extracted using an extended maximum likelihood fit to the reconstructed B mass, to the J/ψ and the non-resonant channel in the two q^2 regions; the fits in the central q^2 region are shown in figure 7 (c and d). While the muon sample has only a small background contribution from combinatorial background, several background contributions need to be considered for the electron channel. The ratio $R(K^*)$ is measured to be

$$0.66^{+0.11}_{-0.07}(\mathrm{stat}) \pm 0.03(\mathrm{syst}) \quad \text{for} \quad 0.045 < q^2 < 1.1\,\mathrm{GeV}^2/c^4,$$

$$0.69^{+0.11}_{-0.07}(\mathrm{stat}) \pm 0.05(\mathrm{syst}) \quad \text{for} \quad 1.1 < q^2 < 6.0\,\mathrm{GeV}^2/c^4,$$

where the precision of the measurements are limited by the size of the electron sample. The results are compatible with the SM expectations at the level of 2.1-2.3 σ (low bin) and 2.4-2.5 σ (central bin) [20]. Figure 8 shows the comparison with previous $R(K^*)$ measurements from the B-factories and with theoretical expectations.

The results show the same trend as a previous measurement on $R(K)$ as shown in figure 9 where the LHCb result is displayed together with results from Belle and Babar. In this analysis $R(K)$ as measured by LHCb is about 2.6 σ below the SM [19].

4. Summary and prospects

Three measurements that test LFU using semi-leptonic decays to tau leptons have been presented. They consist of the first measurement of $R(D^*)$ at a hadron collider in the leptonic decay channel of the tau, the most accurate single measurement of $R(D^*)$ using the hadronic decay channel of the tau and the first measurement of $R(J/\psi)$. The tension with the precise SM prediction is about 3.8 σ.

LFU was also tested in rare $b \to s\ell^+\ell^-$ decays by measuring the ratio $R(h)$ of branching fractions of $B \to h\mu^+\mu^-$ and $B \to he^+e^-$ with h either a K^+ or K^*. In both channels the data

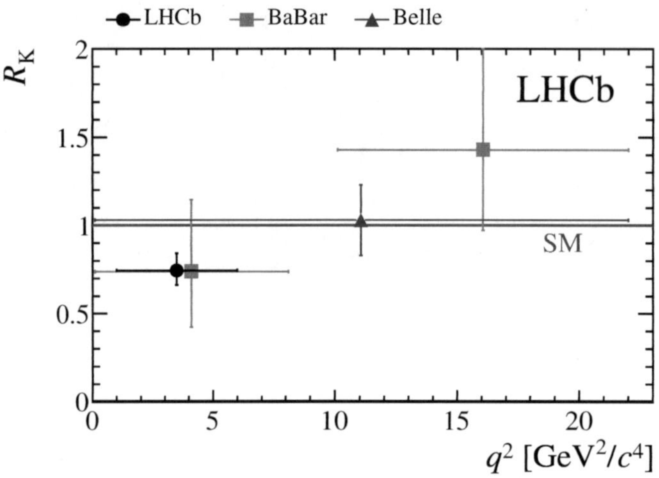

Figure 9. Summary of results on $R(K)$ determined by LHCb experiment together with the results from the BaBar[21] and Belle[22] experiments. The SM prediction is also shown as a continuous function of q^2. The theoretical uncertainty on $R(K)$ is expected to be $\mathcal{O}(10^3)$ (from [19]).

are below the SM prediction; $R(K^*)$ deviates from the SM by 2.1-2.3 σ (low bin) and 2.4-2.5 σ (central bin), while $R(K)$ is about 2.6 σ below the SM.

Many more additional channels are presently analysed, such as the $b \to c\ell\nu$ transitions $R(\Lambda^*)$, $R(D_s)$ or $R(D_s^*)$ or the $b \to s\ell^+\ell^-$ transitions $R(K\pi\pi)$, $R(pK)$, $R(\Phi)$, $R(\Lambda)$ and many more.

All these measurements are based on the Run 1 data collected in 2011 and 2012 with an integrated luminosity of 3 fb^{-1}. The LHCb experiment has collected about 9 fb^{-1} of data that is currently being analyzed. Updates to the analyses presented here are currently ongoing and will provide more accurate measurements. All the systematic uncertainties will decrease with larger simulation and control data samples. Also, the Belle 2 experiment has started to take data and will be able to provide an important cross check of the measurements from the LHCb collaboration.

References

[1] Bečirević D, Fajfer S, Košnik N and Sumensari O 2016 Leptoquark model to explain the B-physics anomalies, R_K and R_D *Phys. Rev.* D**94**(11):115021

[2] Crivellin A, Müller D and Ota T 2017 Simultaneous explanation of R(D$^{(*)}$) and $b \to s\mu^+ \mu$: the last scalar leptoquarks *JHEP* **09**:040

[3] Altmannshofer W, Gori S, Profumo S and Queiroz F S 2016 Explaining dark matter and B decay anomalies with an $L_\mu - L_\tau$ model *JHEP* **12**:106

[4] Buras A J and Girrbach J 2013 Left-handed Z' and Z FCNC quark couplings facing new $b \to s\mu^+\mu^-$ data *JHEP* **12**:009

[5] Tanaka M 1995 Charged higgs effects on exclusive semitauonic B decays *Z. Phys.* C**67**:321326

[6] Lees J P *et al* [BaBar Collaboration] 2013 Measurement of an excess of $\bar{B} \to D^{(*)}\tau^-\bar{\nu}_\tau$ decays and implications for charged higgs bosons *Phys. Rev.* D**88**(7):072012

[7] Huschle M *et al* [Belle Collaboration] 2015 Measurement of the branching ratio of $\bar{B} \to D^{(*)}\tau^-\bar{\nu}_\tau$ relative to $\bar{B} \to D^{(*)}\ell^-\bar{\nu}_\ell$ decays with hadronic tagging at Belle *Phys. Rev.* D**92**(7):072014

[8] Aaij R *et al* [LHCb Collaboration] 2015 Measurement of the ratio of branching fractions $\mathcal{B}(\bar{B}^0 \to D^{*+}\tau^-\bar{\nu}_\tau)/\mathcal{B}(\bar{B}^0 \to D^{*+}\mu^-\bar{\nu}_\mu)$ *Phys. Rev. Lett.* **115**(11):111803 [Erratum: *Phys. Rev. Lett.* **115**(15), 159901(2015)]

[9] Fajfer S, Kamenik J F and Nisandzic I 2012 On the $B \to D^* \tau \bar{\nu}_\tau$ sensitivity to new physics *Phys. Rev.* D**85**:094025
[10] Tanabashi M *et al* [Particle Data Group] 2018 Review of particle physics *Phys. Rev.* D**98**(3):030001
[11] Aaij R *et al* [LHCb Collaboration] 2018 Test of lepton flavor universality by the measurement of the $B^0 \to D^{*-} \tau^+ \nu_\tau$ branching fraction using three-prong τ decays *Phys. Rev.* D**97**(7):072013
[12] Amhis Y *et al* [Heavy Flavor Averaging Group] 2017 Averages of b-hadron, c-hadron, and τ-lepton properties as of summer 2016 *Eur. Phys. J.* C**77**:895 updated results and plots available at https://hflav.web.cern.chhttps://hflav.web.cern.ch
[13] Aaij R *et al* [LHCb Collaboration] 2018 Measurement of the ratio of branching fractions $\mathcal{B}(B_c^+ \to J/\psi \tau^+ \nu_\tau)/\mathcal{B}(B_c^+ \to J/\psi \mu^+ \nu_\mu)$ *Phys. Rev. Lett.* **120**(12):121801
[14] Anisimov A Y, Narodetsky I M, Semay C and Silvestre-Brac B 1999 The B_c meson lifetime in the light front constituent quark model *Phys. Lett.* B**452**:129136
[15] Kiselev V V 2002 Exclusive decays and lifetime of B_c meson in QCD sum rules *Preprint* hep-ph/0211021
[16] Ivanov M A, Korner J G and Santorelli P 2006 Exclusive semileptonic and nonleptonic decays of the B_c meson *Phys. Rev.* D**73**:054024
[17] Hernandez E, Nieves J and Verde-Velasco J M 2006 Study of exclusive semileptonic and non-leptonic decays of B_c - in a nonrelativistic quark model *Phys. Rev.* D**74**:074008
[18] Bordone M, Isidori G, and Pattori A 2016 On the standard model predictions for R_K and R_{K^*} *Eur. Phys. J.* C**76**(8):440
[19] Aaij R *et al* [LHCb Collaboration] 2014 Test of lepton universality using $B^+ \to K^+ \ell^+ \ell^-$ decays *Phys. Rev. Lett.* **113**:151601
[20] Aaij R *et al* [LHCb Collaboration] 2017 Test of lepton universality with $B^0 \to K^{*0} \ell^+ \ell^-$ decays *JHEP* **08**:055
[21] Lees J P *et al* [BaBar Collaboration] 2012 Measurement of branching fractions and rate asymmetries in the rare decays $B \to K^{(*)} l^+ l^-$ *Phys. Rev.* D**86**:032012
[22] Wei J T *et al* [Belle Collaboration] 2009 Measurement of the differential branching fraction and forward-backward asymmetry for $B \to K^{(*)} \ell^+ \ell^-$ *Phys. Rev. Lett.* **103**:171801

IOP Conf. Series: Journal of Physics: Conf. Series **1271** (2019) 012010 doi:10.1088/1742-6596/1271/1/012010

LHCb Upgrades

Olaf Steinkamp ***on behalf of the LHCb Collaboration***
Physik-Institut der Universität Zürich, Winterthurerstrasse 190, 8057 Zürich, Switzerland
E-mail: olafs@physik.uzh.ch

Abstract.
During the LHC long shutdown 2, in 2019/2020, the LHCb collaboration is going to perform a major upgrade of the experiment. The upgraded detector is designed to operate at a five times higher instantaneous luminosity than in Run II and can be read out at the full bunch-crossing frequency of the LHC, abolishing the need for a hardware trigger. This talk will present a short overview of the upgrade activities and highlight the physics reach of the upgraded experiment for selected physics topics. Plans for future upgrades will be briefly outlined.

1. Introduction

One of the main goals of the LHCb experiment [1] is to search for signatures of physics "Beyond the Standard Model" (BSM) in the mixing and decays of particles containing a beauty (b) or charm (c) quark. Most BSM physics models predict the existance of hitherto undiscovered heavy particles that could appear in loop-mediated transitions and lead to sizeable changes in the values of observables with respect to Standard-Model predictions. LHCb is aiming at uncovering such deviations by confronting Standard Model predictions for such observables with precision measurements.

In Runs I and II of the LHC, LHCb has collected pp collisions corresponding to a total integrated luminosity of almost $10\,\text{fb}^{-1}$ at collision energies of 7, 8 and 13 TeV. Using the full Run I and Run II data sets, most measurements will still be limited by statistical uncertainties. As statistical uncertainties scale with the square root of the number of analyzed events, another $30\,\text{fb}^{-1}$ will need to be collected to reduce uncertainties by a factor of two. Under Run-II data taking conditions, this would require about fifteen years of data taking. Therefore, the LHCb collaboration decided to comprehensively upgrade the experiment during the LHC long shutdown, LS2, in 2019/2020.

Throughout Run II, LHCb operated at a leveled instantaneous luminosity of $4 \times 10^{-32}\,\text{cm}^{-2}\text{s}^{-1}$, corresponding to an average pile-up of about 1.1 visible pp interactions[1] per bunch crossing. The leveling of the luminosity was achieved by regularly adjusting a small displacement of the two proton beams in the LHCb collision point [2]. The main reason for selecting the chosen operation point has been the requirement for many physics analyses to assign the displaced decay vertex of a beauty or charm hadron to the pp collision vertex at which this hadron was produced. Operating at low pile-up largely avoids the risk of assigning the decay vertex to a wrong pp vertex.

[1] Interactions that create particles inside the acceptance of the LHCb detector.

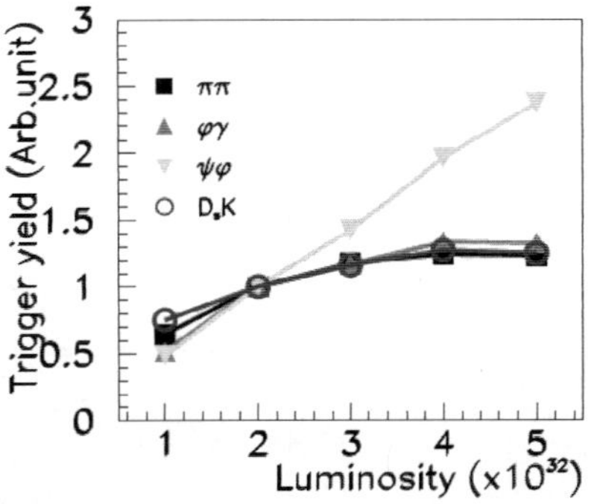

Figure 1. Expected trigger yields for different decays of b mesons, as a function of the instantaneous luminosity (from [5]).

Using the experience gained during Runs I and II, pattern recognition algorithms have been developed that will allow to cope with the pile-up corresponding to a five times higher instantaneous luminosity of $2 \times 10^{-33}\,\mathrm{cm}^{-2}\mathrm{s}^{-1}$. It has been shown that the fraction of mismatched secondary vertices can be kept at the level of a percent at this luminosity. To deal with the higher particle densities, however, detectors with finer granularity and better radiation hardness will be required, in particular in the tracking system.

Another limitation of the original setup is due to the trigger scheme. Only limited detector information from the calorimeters and the muon stations was read out at the LHC bunch crossing rate of 40 MHz. This information was employed in a hardware-based Level 0 (L0) trigger to reduce the rate of candidate events to 1 MHz, the maximum rate at which the full detector could be read out. For events without muons in the final state, the L0 trigger relied on the total energy measured in the electromagnetic calorimeter. Increasing the instantaneous luminosity would require to increase the energy threshold to keep the output rate at the required level. At luminosities exceeding $2 \times 10^{-32}\,\mathrm{cm}^{-2}\mathrm{s}^{-1}$, the threshold starts to cut significantly into the signal efficiency, thereby canceling out any potential gain due to the increased collision rate, as illustrated in Fig. 1. In the upgraded detector, the full detector can therefore be read out at the LHC bunch crossing rate of 40 MHz, removing the need for an L0 trigger. The data are fed directly into a software-based trigger, allowing for more complex and flexible trigger decisions. This new trigger scheme requires the replacement of most of the front-end electronics and of the data acquisition system.

Plans for a long-term upgrade programm, comprising an Upgrade Ib during the long shutdown LS3 in 2024/2025 and an Upgrade II during LS4 around 2030, are taking shape [3, 4]. The goal for Upgrade II is to operate the experiment at an instantaneous luminosity of up to $1 - 2 \times 10^{-34}\,\mathrm{cm}^{-2}\mathrm{s}^{-1}$, corresponding to an increase by another factor of five to ten with respect to Upgrade I. Plans for Upgrade Ib comprise a consolidation of the Upgrade I detector, but no increase in instantaneous luminosity.

In this contribution, the detector for Upgrade I will be discussed, challenges and ideas for Upgrades Ib and II will be outlined and the physics reach of the upgrades will be briefly illustrated for a few selected key analyses. An interesting opportunity in LHCb is to collect data in fixed-target collisions of protons or heavy ions on gas atoms injected into the LHC vacuum. A proposed upgrade of this so-called SMOG system is also discussed.

IOP Conf. Series: Journal of Physics: Conf. Series **1271** (2019) 012010 doi:10.1088/1742-6596/1271/1/012010

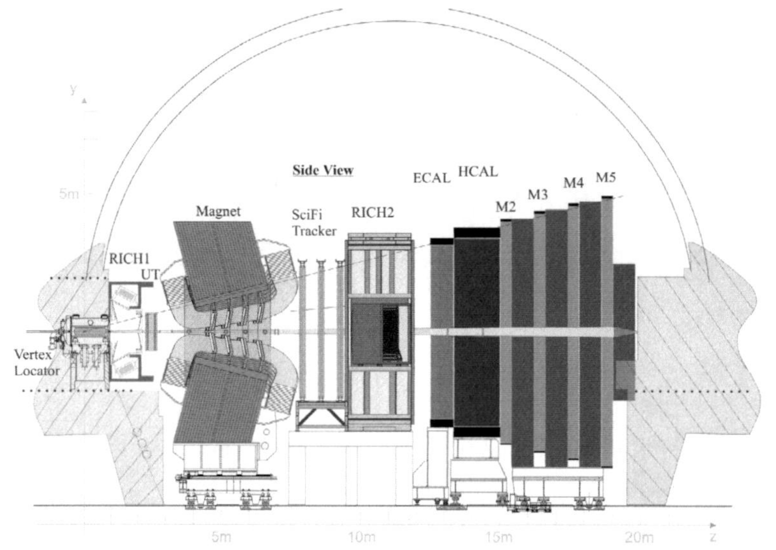

Figure 2. Layout of the LHCb Upgrade I detector (from [8]).

2. The detector for Upgrade I

The LHCb detector for Upgrade I has been described in a series of Technical Design Reports [6, 7, 8, 9]. The layout of the detector, illustrated in Fig. 2, follows closely that of the original detector [1] that was operated very successfully [10] during LHC Runs I and II. The trajectories and momenta of charged particles are measured in a tracking system consisting of a vertex detector (VELO), a planar tracking station (UT) upstream of a dipole magnet and three planar tracking stations downstream of the magnet (SciFi). The tracking system is complemented by two Ring-Imaging CHerenkov detectors (RICH1,2) for particle identification, by electromagnetic and hadronic calorimeters (ECAL,HCAL) and by a muon system (M2-M5). Large parts of the original detector and the entire readout system are replaced in order to adapt to the higher instantaneous luminosity and the 40 MHz readout scheme.

2.1. Vertex detector (VELO)

The original VELO consisted of 21 planar tracking stations arranged over a length of 1 m around the *pp* interaction region [11]. Each station consisted of a double layer of silicon micro-strip sensors, one measuring the azimuthal angle and one measuring the radial distance from the beam axis. Strip pitches ranged from $40\,\mu$m in the region close to the beam axis to $100\,\mu$m in the outer region of the sensor. To obtain the best possible vertex resolution, the detector was installed inside the LHC vacuum vessel, the sensors being separated from the LHC beams only by a $300\,\mu$m thin aluminium foil. During data taking, the active area of the sensors approached the beam axis to 8.2 mm. The in total 170 thousand readout strips of the detector were read out by the custom-developed Beetle ASIC [12], which allowed a maximum readout rate of 1.1 MHz.

The upgraded VELO [13] is going to consist of 26 planar stations of hybrid silicon pixel detectors. Its in total 41 million pixels are $55 \times 55\,\mu$m in size and are read out by the VeloPix ASIC [14], a custom development based on the family of MediPix and TimePix chips [15]. The detector is going to approach the LHC beams even closer than in the original VELO, the active area starting at only 5.1 mm from the beam axis. The thickness of the aluminium foil separating the detectors from the beam is going to be reduced to $200 - 250\,\mu$m. As illustrated in Fig. 3, the combination of less material and smaller extrapolation distance is expected to result in a further improvement of the impact parameter resolution compared to the original detector.

IOP Conf. Series: Journal of Physics: Conf. Series **1271** (2019) 012010 doi:10.1088/1742-6596/1271/1/012010

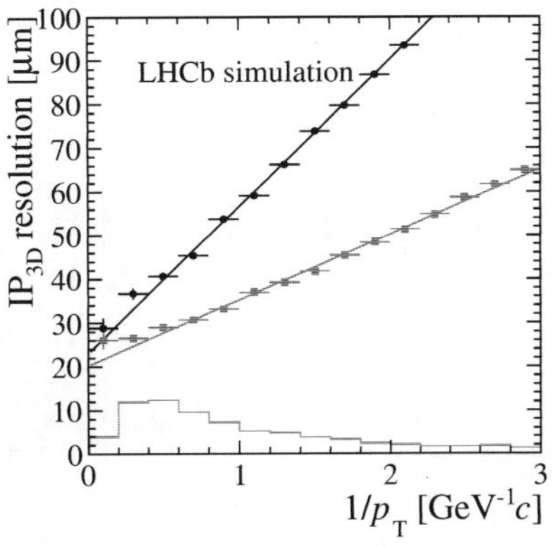

Figure 3. Expected impact parameter resolution as a function of the inverse transverse momentum ($1/p_T$) of the track for (black circles) the original and (red squares) the Upgrade I VELO; the light grey histogram shows the $1/p_T$ distribution for particles produced in decays of b hadrons (from [6]).

2.2. Upstream Tracker (UT)

In the original LHCb detector, the tracking station upstream of the LHCb dipole magnet (Tracker Turicensis, TT) consisted of four planar layers of silicon microstrip detectors with a total of 143 thousand strips of 40, 30, 20 or 10 cm in length and with a pitch of 183 μm [16]. Shorter strips were employed in the inner region of each detection layer, where particle densities are highest. Longer strips were used in the outer regions of the detector. The strips were read out by the Beetle ASIC that was also employed in the orginal VELO.

The new Upstream Tracking station (UT) [17] has again four detection layers using silicon microstrip detectors. Silicon sensors with strips of 10 cm in length and a pitch of 190 μm are used in the outer regions of each detection layer; sensors with strips of 10 cm in length and a pitch of 95 μm are used in an intermediate region and sensors with strips of 5 cm in length and a pitch of 95 μm are used in the innermost region. The sensors surrounding the LHC beam pipe have a cutout in the form of a quarter circle to improve the acceptance of the detector at small polar angles. Sensors for the inner region are produced in n-in-p technology for improved radiation hardness. The 537 thousand strips of the UT are going to be read out by a new custom-developed front-end ASIC, the SALT [18], compatible with 40 MHz readout.

The importance of the UT for pattern recognition is illustrated in Fig. 4, which shows the expected number of wrongly reconstructed track candidates ("ghosts") per event, with and without using information from the UT.

2.3. Downstream Tracker (SciFi)

In the original LHCb detector, two detector technologies were employed in the three planar tracking stations downstream of the dipole magnet: a silicon micro-strip detector with strips of 198 μm pitch (Inner Tracker, IT) in the inner region of each station and a straw drift-tube detector (Outer Tracker, OT) with drift tubes of 5 mm in diameter in the outer region [19]. Both detectors had four detection layers per station, for a total of 130 thousand readout channels in the IT and 54 thousand readout channels in the OT.

The upgraded LHCb detector is also going to have three downstream tracking stations with four planar detection layers each. A single detector technology, scintillating fibres, is going to be employed to cover the entire surface of each detetion layer [20]. Fibres have a diameter of 250 μm and are 2.5 m long, covering half the height of the detection layer. They are going to

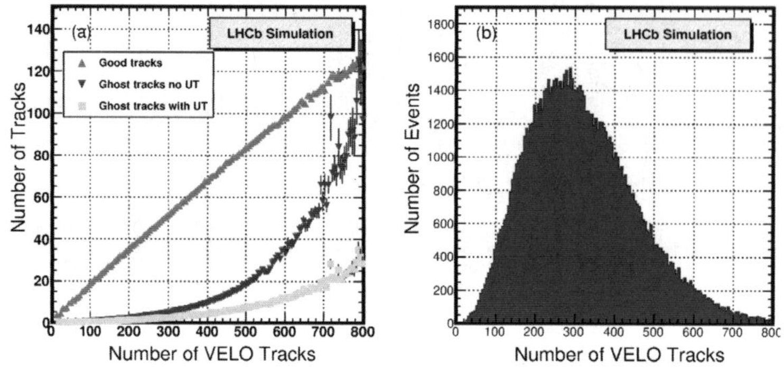

Figure 4. (Left) Expected numbers of correctly and wrongly ("ghosts") reconstructed track candidates as a function of the number of track segments in the VELO; (right) distribution of the number of track segments in the VELO (from [8]).

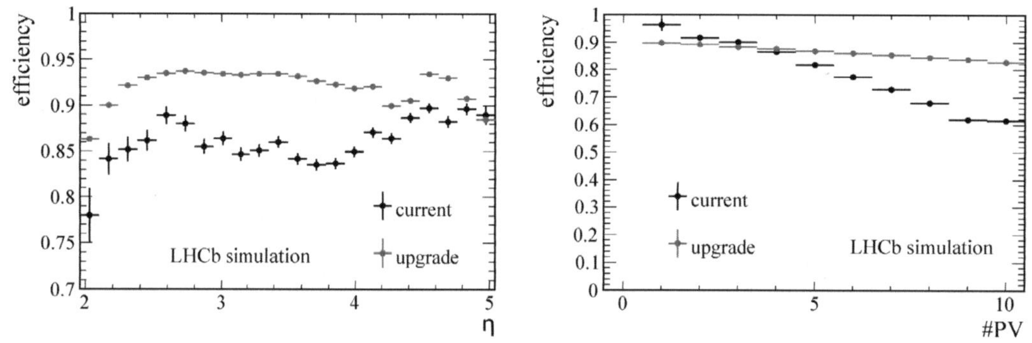

Figure 5. Expected track reconstruction efficiency as a function of (left) the pseudorapidity, η, of the track and (right) the number of reconstructed pp interaction vertices (from [8]).

be read out by silicon photomultipliers (SiPM) at the top and bottom of each layer. The other end of the fibres is clad with a thin mirror to improve light collection efficiency. The SiPM are going to be operated at -40°C to mitigate the effects of radiation damage.

The expected performance of the tracking system in terms of track reconstruction efficiency is illustrated in Fig. 5, as a function of the pseudorapidity, η, of the track and as a function of the number of reconstructed pp interaction vertices (PV). While the efficiency is expected to deteriorate a little for events with a small number of PV, due to the less fine granularity in the inner region of the detection layers, it is significantly more robust for events with larger numbers of PV. Integrated over the expected distribution in the number of PV, the efficiency improves over the entire range in η.

2.4. Ring Imaging Cherenkov detectors (RICH)

Both the original and the upgraded LHCb detector comprise two RICH detectors [21]: the first of these (RICH1) is located upstream of the LHCb dipole magnet and covers the full acceptance of the experiment. It uses gaseous C_4F_{10} as a radiator and provides efficienct kaon/pion separation up to about 50 GeV/c in momentum. The second RICH detector (RICH2) is located downstream of the magnet, uses gaseous CF_4 as a radiator and provides efficienct kaon/pion separation up to about 100 GeV/c in momentum. Exploiting the fact that high-momentum particles tend to be produced at smaller polar angles, RICH2 covers only the forward part of the accpetance. In the original RICH detectors, Cherenkov photons were detected in Hybrid Photo Diodes (HPD) [22] with a granularity of $2.5 \times 2.5\,\mathrm{mm}^2$.

The upgraded RICH detectors [23] are going to occupy the same locations and employ the same gas radiators as the original detectors. The system of light-collecting mirrors for RICH1, upstream of the magnet, is going to be replaced with a new optical design to deal with the increased particle density and keep the expected occupancy in the photon detection plane at an acceptable level. In both RICH detectors, Multi-anode Photo Multiplier Tubes (MaPMT) with a granularity of $2.9 \times 2.9\,\mathrm{mm}^2$ are going to be employed to detect the Cherenkov photons [24]. The MaPMTs are going to be read out by new front-end electronics compatible with 40 MHz readout.

2.5. Calorimeters (ECAL and HCAL)
The original LHCb detector comprised a calorimeter system [25] consisting of a pre-shower / scintillating pad detector (PS/SPD), an electromagnetic calorimeter (ECAL) using lead-scintillator Shashlik technology and a hadronic calorimeter (HCAL) using iron absorbers and scintillator tiles. The main purpose of the PS/SPD system has been to provide input for the L0 hardware trigger, and it is therefore going to be removed for the LHCb upgrade. The detector modules and photomultipliers of the ECAL and HCAL are going to be kept. Both calorimeters are going to be equipped with new readout electronics, compatible with 40 MHz readout [26].

2.6. Muon system (M2-M5)
The LHCb muon system [27] originally comprised five detector stations. Four of these stations (M2-M5) are located downstream of the calorimeter systems and consist of multi-wire proportional chambers interspersed with iron absorbers. A fifth muon station (M1) was located in front of the calorimeter system. It used wire chambers in the outer regions and Triple-GEM detectors in the region of highest occupancy close to the beam axis. The main purpose of M1 has been to serve in the L0 hardware trigger and it is going to be removed for the upgrade. Stations M2-M5 will be kept for the upgrade. The innermost region of M2 is going to be equipped with new Triple-GEM detectors [28] to cope with the expected increase in particle density and all stations are going to be equipped with new off-detector readout electronics, compatible with 40 MHz readout [29].

3. Upgrades Ib and II
Interest in a second comprehensive upgrade of the LHCb experiment (Upgrade II), envisaged to be installed in the long shutdown LS 4 of the LHC around the year 2030, is rapidly gaining momentum. An Expression of Interest [3] was submitted in 2017, a document discussing the physics case [4] has been published in 2018. The CERN LHCC has encouraged the collaboration to proceed to a Framework TDR.

The goal of Upgrade II is to be able to operate at an instantaneous luminosity of $1 - 2 \times 10^{-34}\,\mathrm{cm}^{-2}\mathrm{s}^{-1}$ [30] and to collect an integrated luminosity of 300 fb^{-1}. A luminosity of $1.5 \times 10^{-34}\,\mathrm{cm}^{-2}\mathrm{s}^{-1}$ corresponds to an average pile-up of 42 visible pp interactions per bunch crossing and more than 2000 charged particles per bunch crossing produced on average inside the LHCb acceptance. Operating a precision experiment at these conditions will put extreme requirements on detectors, readout and reconstruction algorithms. One crucial aspect for a large part of the LHCb physics programme is to maintain an efficient and clean association of beauty and charm decay vertices to the corresponding proton-proton collision vertex. Simulation studies show that excellent timing resolution for charged particles is going to play a crucial role in this. It is expected that a timing resolution of the order of 50 ps per track will allow to keep the fraction of misassociated vertices at the level of a few percent.

IOP Conf. Series: Journal of Physics: Conf. Series **1271** (2019) 012010 doi:10.1088/1742-6596/1271/1/012010

Table 1. Expected sensitivities in selected key analyses (reproduced from Ref. [4]).

Observable	Current LHCb	LHCb 2025	Belle II	Upgrade II	ATLAS & CMS
EW Penguins					
R_K $(1 < q^2 < 6\,\mathrm{GeV}^2c^4)$	0.1 [274]	0.025	0.036	0.007	–
R_{K^*} $(1 < q^2 < 6\,\mathrm{GeV}^2c^4)$	0.1 [275]	0.031	0.032	0.008	–
R_ϕ, R_{pK}, R_π	–	0.08, 0.06, 0.18	–	0.02, 0.02, 0.05	–
CKM tests					
γ, with $B_s^0 \to D_s^+ K^-$	$(^{+17}_{-22})^\circ$ [136]	4°	–	1°	–
γ, all modes	$(^{+5.0}_{-5.8})^\circ$ [167]	1.5°	1.5°	0.35°	–
$\sin 2\beta$, with $B^0 \to J/\psi K_S^0$	0.04 [609]	0.011	0.005	0.003	–
ϕ_s, with $B_s^0 \to J/\psi\phi$	49 mrad [44]	14 mrad	–	4 mrad	22 mrad [610]
ϕ_s, with $B_s^0 \to D_s^+ D_s^-$	170 mrad [49]	35 mrad	–	9 mrad	–
$\phi_s^{s\bar{s}s}$, with $B_s^0 \to \phi\phi$	154 mrad [94]	39 mrad	–	11 mrad	Under study [611]
a_{sl}^s	33×10^{-4} [211]	10×10^{-4}	–	3×10^{-4}	–
$\lvert V_{ub}\rvert/\lvert V_{cb}\rvert$	6% [201]	3%	1%	1%	–
$\boldsymbol{B_s^0, B^0 \to \mu^+\mu^-}$					
$\mathcal{B}(B^0 \to \mu^+\mu^-)/\mathcal{B}(B_s^0 \to \mu^+\mu^-)$	90% [264]	34%	–	10%	21% [612]
$\tau_{B_s^0 \to \mu^+\mu^-}$	22% [264]	8%	–	2%	–
$S_{\mu\mu}$	–	–	–	0.2	–
$\boldsymbol{b \to c\ell^-\bar{\nu}_l}$ **LUV studies**					
$R(D^*)$	0.026 [215, 217]	0.0072	0.005	0.002	–
$R(J/\psi)$	0.24 [220]	0.071	–	0.02	–
Charm					
$\Delta A_{CP}(KK - \pi\pi)$	8.5×10^{-4} [613]	1.7×10^{-4}	5.4×10^{-4}	3.0×10^{-5}	–
A_Γ $(\approx x \sin\phi)$	2.8×10^{-4} [240]	4.3×10^{-5}	3.5×10^{-4}	1.0×10^{-5}	–
$x \sin\phi$ from $D^0 \to K^+\pi^-$	13×10^{-4} [228]	3.2×10^{-4}	4.6×10^{-4}	8.0×10^{-5}	–
$x \sin\phi$ from multibody decays	–	$(K3\pi)$ 4.0×10^{-5}	$(K_S^0\pi\pi)$ 1.2×10^{-4}	$(K3\pi)$ 8.0×10^{-6}	–

Besides timing resolution, detectors with finer spatial granularity will be required to cope with the increased particle densities. Developments under consideration comprise silicon pixel detectors with timing resolution, such as Low-Gain Avalanche Devices (LGAD) [31], for the vertex locator; a new silicon-pixel tracker, possibly using Depleted monolithic CMOS (DCMOS, HV-CMOS) sensors [32], for the inner part of the tracking stations downstream of the LHCb magnet; an ECAL with finer segmentation and employing materials with smaller Moliere radius in the inner regions of the detector; and detectors with finer segmentation, such as μ-RWELL [33], for the inner regions of the muon stations.

The long shutdown LS 3, during which the HL-LHC and the ATLAS and CMS phase 2 upgrades are scheduled to be installed, falls in between the two major LHCb upgrades. The LHCb collaboration envisages to make use of LS 3 for a consolidation of the Upgrade I detector, dubbed Upgrade Ib. One aspect of this consolidation is the installation of a silicon-pixel inner tracker in the tracking stations downstream of the magnet. Such a detector, smaller in size but using the same technology as that foreseen for Upgrade II, could help to speed up track reconstruction algorithms and at the same time gain valuable experience for Upgrade II. Other additions to the detector are considered that have the potential to extend the physics reach of LHCb: a new set of tracking stations located along the inner sides of the LHCb dipole magnet could extend the acceptance of the tracking system for low-momentum tracks. A TORCH ("Time Of interally Reflected CHerenkov light") detector [34], installed in between RICH2 and the calorimeters could extend the particle identification capabilities for hadrons below $10\,\mathrm{GeV}/c$ in momentum and has the potential to provide excellent timing resolution for charged particles.

4. Physics reach for the LHCb Upgrades

The latest studies of the expected physics reach for the LHCb Upgrades can be found in Ref. [4]. The results of these studies for selected key analyses are reproduced here in Tab. 1.

As an example, Fig. 6 illustrates the current and expected precision on the CKM

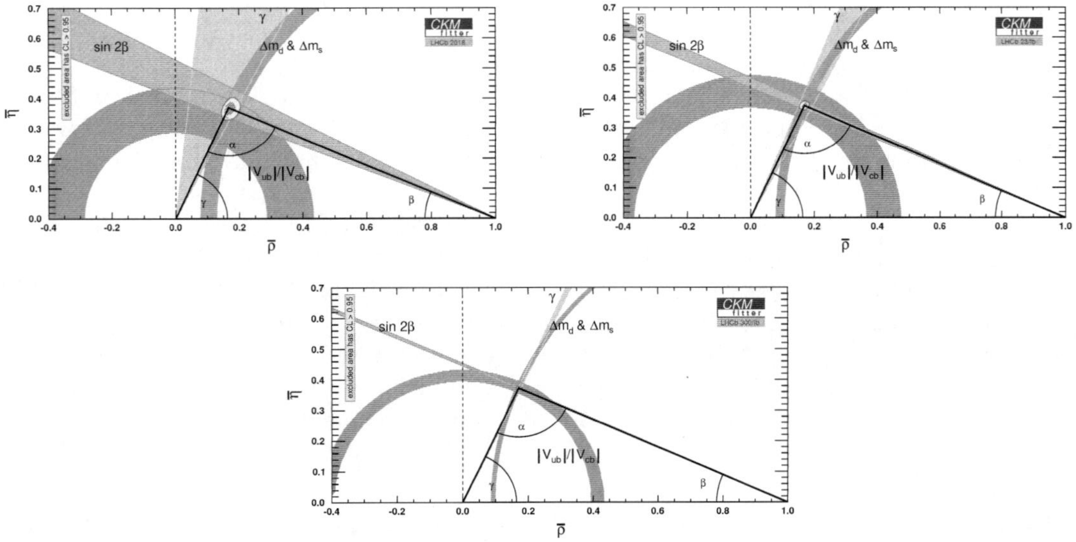

Figure 6. Constraints on the CKM unitarity triangle, (top left) reconstructed using LHCb measurements published by summer 2018 and expected precision for integrated luminosities of (top right) 23 fb^{-1} expected by 2025 and (bottom) 300 fb^{-1} expected for Upgrade II (from [4]).

Unitarity Triangle using LHCb measurements alone. LHCb measurements testing lepton-flavour universality (LFU) by comparing the decay rates of $B^+ \to K^+\mu^+\mu^-$ and $B^+ \to K^+e^+e^-$ [35], of $B^0 \to K^{*0}\mu^+\mu^-$ and $B^0 \to K^{*0}e^+e^-$ [36] and of $B^0 \to D^{*-}\tau^+\nu_\tau$ and $B^0 \to D^{*-}\mu^+\nu_\mu$ [38, 39] receive a lot of attention in the community[2]. Improvements in precision by a factor of three to four are expected in such tests by 2025 and improvements by more than an order of magnitude are expected for the full statistics of Upgrade II. The measurement of the branching fraction of the very rare decay $B^0_s \to \mu^+\mu^-$ and its first observation were highlights of Run I of the LHC [40, 41]. The branching fraction for the related decay $B^0 \to \mu^+\mu^-$ is predicted to be an order of magnitude smaller in the Standard Model. A measurement of the ratio of the two branching fractions is expected to be possible with a precision of 34% by 2025 and with a precision of 10% using the full statistics of LHCb Upgrade II. Significant improvements are also predicted in the field of charm physics.

5. Fixed-target collisions in LHCb (SMOG)

The System for Measuring Overlap with Gas (SMOG) allows to inject small amounts of a nobel gas into the LHC vacuum vessel in the region of the LHCb VELO, increasing the gas pressure in this region from about 10^{-9} mbar to 10^{-7} mbar. Its original purpose is to measure the trajectories of charged particles produced in the interaction of beam particles colliding on the gas atoms and to reconstruct the profiles of the circulating beams for precise luminosity determinations. However, collisions of circulating protons or heavy ions on gas atoms also allow for a unique fixed-target physics programme at the LHC. Since 2015, data have been collected for protons colliding on Helium, Neon and Argon atoms at beam energies of 2.5, 4.0 and 6.5 GeV as well as collisions of Lead ions on Neon atoms. The largest sample, collected in p-Ne collisions at 2.5 GeV in 2017, corresponds to an integrated luminosity of about 100 nb^{-1}. Two measurements

[2] A new LHCb result on the search for LFU in $B^+ \to K^+\mu^+\mu^-$ and $B^+ \to K^+e^+e^-$ has been presented [37] recently.

have been published so far, one on the production of hidden and open charm in p-He and p-Ar collisions [44] and one on the production of antiprotons in p-He collisions [43]. The latter measurement provides valuable input for cosmic-ray physics.

An upgrade of the system has been proposed for installation during long shutdown LS 2. The physics opportunities are discussed in Ref. [45]. A technical proposal [46] has been approved by the LHCb collaboration and is going to be presented to the LHC machine committee in January 2019. It comprises a 20 cm long storage cell to be installed just upstream of the VELO. This would allow for up to 100 times larger target densities, resulting in much larger data samples. For example, the measurements published in Ref. [44] are based on samples of 400 reconstructed J/ψ candidates and 2000 reconstructed D^0 candidates; with the propsed SMOG2 setup, yields of the order of 15 million J/ψ and 150 million D^0 mesons can be expected. Other gas species, such as Hydrogen and Deuterium, Nitrogen or Oxygen, could also be considered. The density of the injected gas could be controlled with better precision, allowing for a more accurate determination of the collected luminosity and therefore more precise measurements of absolute cross sections.

6. Summary

The LHCb experiment has operated extremely well during LHC Runs I and II, yielding a wealth of exciting results. Even using the full accumulated data, corresponding to an integrated luminosity of almost $10\,\mathrm{fb}^{-1}$, most measurements are expected to be limited by statistical uncertainties. The LHCb collaboration has therefore decided for a comprehensive upgrade (Upgrade I) of the experiment already during the long shutdown LS 2 in 2019/2020. A further comprehensive upgrade (Upgrade II) is envisaged for LS4 around the year 2030. Upgrade I should allow to collect $23\,\mathrm{fb}^{-1}$ by 2025 and $50\,\mathrm{fb}^{-1}$ by 2030, while the ultimate goal of Upgrade II is to collect $300\,\mathrm{fb}^{-1}$, allowing for tests of the Standard Model in a wide range of flavour measurements and with unprecedented precision. Measuring the collisions of proton and ion beams on a variety of gas atoms at rest gives unique opportunities in fixed-target physics, with interesting links to the cosmic-ray community.

References

[1] A. A. Alves, Jr. *et al.* [LHCb Collaboration], "The LHCb Detector at the LHC," JINST **3** (2008) S08005.
[2] F. Follin and D. Jacquet, "Implementation and experience with luminosity levelling with offset beam," arXiv:1410.3667.
[3] R. Aaij *et al.* [LHCb Collaboration], "Expression of Interest for a Phase-II LHCb Upgrade," CERN-LHCC-2017-003.
[4] R. Aaij *et al.* [LHCb Collaboration], "Physics case for an LHCb Upgrade II," CERN-LHCC-2018-027.
[5] LHCb Collaboration, "Letter of Intent for the LHCb Upgrade," CERN-LHCC-2011-001.
[6] LHCb Collaboration, "LHCb VELO Upgrade Technical Design Report," CERN-LHCC-2013-021.
[7] LHCb Collaboration, "LHCb PID Upgrade Technical Design Report," CERN-LHCC-2013-022.
[8] LHCb Collaboration, "LHCb Tracker Upgrade Technical Design Report," CERN-LHCC-2014-001.
[9] LHCb Collaboration, "LHCb Trigger and Online Upgrade Technical Design Report," CERN-LHCC-2014-016.
[10] R. Aaij *et al.* [LHCb Collaboration], "LHCb Detector Performance," Int. J. Mod. Phys. A **30** (2015) 07, 1530022 [arXiv:1412.6352 [hep-ex]].
[11] R. Aaij *et al.*, "Performance of the LHCb Vertex Locator," JINST **9** (2014) P09007 [arXiv:1405.7808 [physics.ins-det]].
[12] S. Löchner and M. Schmelling, "The Beetle Reference Manual - chip version 1.3, 1.4 and 1.5," CERN-LHCb-2005-105.
[13] E. Lemos Cid *et al.* [LHCb Collaboration], "The LHCb Vertex Locator Upgrade," PoS Vertex **2017** (2018) 002.
[14] T. Poikela *et al.*, "The VeloPix ASIC," JINST **12** (2017) no.01, C01070.
[15] R. Ballabriga, M. Campbell and X. Llopart, "Asic developments for radiation imaging applications: The medipix and timepix family," Nucl. Instrum. Meth. A **878** (2018) 10.
[16] C. Abellan Beteta *et al.*, "Monitoring radiation damage in the LHCb Tracker Turicensis," [arXiv:1809.05063 [physics.ins-det]].

[17] O. Steinkamp [LHCb Upstream Tracker Group], "The upstream tracker for the LHCb upgrade," Nucl. Instrum. Meth. A **831** (2016) 367.

[18] S. Bugiel *et al.*, "SALT, a dedicated readout chip for high precision tracking silicon strip detectors at the LHCb Upgrade," JINST **11** (2016) no.02, C02028. doi:10.1088/1748-0221/11/02/C02028

[19] P. d'Argent *et al.* [LHCb Outer Tracker Group], "Improved performance of the LHCb Outer Tracker in LHC Run 2," JINST **12** (2017) no.11, P11016 [arXiv:1708.00819 [physics.ins-det]].

[20] P. Hopchev [LHCb SciFi Tracker Collaboration], "SciFi: A large Scintillating Fibre Tracker for LHCb," arXiv:1710.08325 [physics.ins-det].

[21] M. Adinolfi *et al.* [LHCb RICH Group], "Performance of the LHCb RICH detector at the LHC," Eur. Phys. J. C **73** (2013) 2431 [arXiv:1211.6759 [physics.ins-det]].

[22] L. Carson, "Hybrid Photon Detectors for the LHCb RICH: Performance and operational experience," Nucl. Instrum. Meth. A **604** (2009) 202.

[23] S. Easo [LHCb RICH Collaboration], "Overview of LHCb-RICH upgrade," Nucl. Instrum. Meth. A **876** (2017) 160.

[24] L. Cassina [LHCb RICH Collaboration], "Photodetectors and front-end electronics for the LHCb RICH upgrade," Nucl. Instrum. Meth. A **876** (2017) 217

[25] M. Chefdeville [LHCb Collaboration], "Operation and performance of the LHCb calorimeters," JINST **13** (2018) no.03, C03021.

[26] Y. Guz, "The LHCb Calorimeter system: design, performance and upgrade," JINST **12** (2017) no.07, C07024.

[27] A. A. Alves, Jr. *et al.*, "Performance of the LHCb muon system," JINST **8** (2013) P02022 doi:10.1088/1748-0221/8/02/P02022 [arXiv:1211.1346 [physics.ins-det]].

[28] D. Pinci and M. Santimaria, "Study of the GEM chamber for the upgrade of the LHCb muon system," Nuovo Cim. C **39** (2016) no.1, 265.

[29] S. Cadeddu *et al.*, "The nSYNC ASIC for the new readout electronics of the LHCb Muon Detector Upgrade," Nucl. Instrum. Meth. (2018).

[30] J. Albrecht *et al.*, "Luminosity scenarios for LHCb Upgrade II," LHCb-PUB-2019-001.

[31] G. Pellegrini *et al.*, "Technology developments and first measurements of Low Gain Avalanche Detectors (LGAD) for high energy physics applications," Nucl. Instrum. Meth. A **765** (2014) 12.

[32] I. Peric, "A novel monolithic pixelated particle detector implemented in high-voltage CMOS technology," Nucl. Instrum. Meth. A **582** (2007) 876.

[33] M. Poli Lener *et al.*, "The μ-RWELL: A compact, spark protected, single amplification-stage MPGD," Nucl. Instrum. Meth. A **824** (2016) 565.

[34] M. Charles, R. Forty, "TORCH: Time of Flight Identification with Cherenkov Radiation," LHCb-PROC-2010-038.

[35] R. Aaij *et al.* [LHCb Collaboration], "Test of lepton universality using $B^+ \to K^+\ell^+\ell^-$ decays," Phys. Rev. Lett. **113** (2014) 151601 doi:10.1103/PhysRevLett.113.151601 [arXiv:1406.6482 [hep-ex]].

[36] R. Aaij *et al.* [LHCb Collaboration], "Test of lepton universality with $B^0 \to K^{*0}\ell^+\ell^-$ decays," JHEP **1708** (2017) 055 [arXiv:1705.05802 [hep-ex]].

[37] R. Aaij *et al.* [LHCb Collaboration], "Search for lepton-universality violation in $B^+ \to K^+\ell^+\ell^-$ decays," arXiv:1903.09252 [hep-ex].

[38] R. Aaij *et al.* [LHCb Collaboration], "Measurement of the ratio of branching fractions $\mathcal{B}(\bar{B}^0 \to D^{*+}\tau^-\bar{\nu}_\tau)/\mathcal{B}(\bar{B}^0 \to D^{*+}\mu^-\bar{\nu}_\mu)$," Phys. Rev. Lett. **115** (2015) no.11, 111803 Erratum: [Phys. Rev. Lett. **115** (2015) no.15, 159901] [arXiv:1506.08614 [hep-ex]].

[39] R. Aaij *et al.* [LHCb Collaboration], "Measurement of the ratio of the $B^0 \to D^{*-}\tau^+\nu_\tau$ and $B^0 \to D^{*-}\mu^+\nu_\mu$ branching fractions using three-prong τ-lepton decays," Phys. Rev. Lett. **120** (2018) no.17, 171802 [arXiv:1708.08856 [hep-ex]].

[40] R. Aaij *et al.* [LHCb Collaboration], "Measurement of the $B_s^0 \to \mu^+\mu^-$ branching fraction and search for $B^0 \to \mu^+\mu^-$ decays at the LHCb experiment," Phys. Rev. Lett. **111** (2013) 101805 [arXiv:1307.5024 [hep-ex]].

[41] V. Khachatryan *et al.* [CMS and LHCb Collaborations], "Observation of the rare $B_s^0 \to \mu^+\mu^-$ decay from the combined analysis of CMS and LHCb data," Nature **522** (2015) 68 [arXiv:1411.4413 [hep-ex]].

[42] C. Barschel, "Absolute luminosity measurement at LHCb with beam-gas imaging," LHCb-TALK-2013-056.

[43] R. Aaij *et al.* [LHCb Collaboration], "Measurement of antiproton production in pHe collisions at $\sqrt{s_{NN}} = 110$ GeV," Phys. Rev. Lett. **121** (2018) no.22, 222001 [arXiv:1808.06127 [hep-ex]].

[44] R. Aaij *et al.* [LHCb Collaboration], "First measurement of charm production in fixed-target configuration at the LHC," Phys. Rev. Lett. **122** (2019) no.13, 132002 [arXiv:1810.07907 [hep-ex]].

[45] A. Bursche *et al.*, "Physics opportunities with the fixed-target program of the LHCb experiment using an unpolarized gas target," LHCb-PUB-2018-015.

[46] P. Di Nezza *et al.*, "The SMOG2 project," CERN-PBC-Notes-2018-007.

IOP Conf. Series: Journal of Physics: Conf. Series **1271** (2019) 012011 doi:10.1088/1742-6596/1271/1/012011

The Belle II Experiment

Steven H. Robertson

Associate Professor and IPP Research Scientist, Department of Physics, McGill University, 3600 University St., Montreal, Quebec H3A2T8, Canada

E-mail: steven.robertson@mcgill.ca

Abstract. The Belle II experiment has recently completed its "Phase 2" commissioning operations run, during which first collisions at the $\Upsilon(4S)$ resonances were recorded. Belle II is the successor of the *BABAR* and Belle B Factory experiments, and is designed to operate at a peak luminosity of 8×10^{35} cm^{-2}s^{-1} with a target data set of 50 ab^{-1}. This large and inclusive data set will permit studies of B meson decays, along with charm, τ and many other physics topics, with unprecedented precision in the clean e^+e^- collider environment. In this presentation I describe the Belle II experiment, its planned physics program and prospects as we head toward the first "physics" run beginning in 2019.

1. Introduction

The Belle II experiment [1] is the successor experiment to the previous generation of asymmetric B factory experiments, *BABAR* and Belle [2], intended to extend the reach of these experiments within the domain of precision flavour measurements by collecting a data sample equivalent to more than 30 times the combined integrated luminosity of these previous experiments. Belle II is a collaboration of over 900 physicists from 26 countries and regions world wide, including more than 260 graduate students. The Belle II detector has been installed in the interaction region of the SuperKEKB e^+e^- collider [3] at the KEK laboratory in Tsukuba, Japan, which is designed to operate at instantaneous luminosities of up to 8×10^{35} cm^{-2}s^{-1} at, and in the vicinity of, the $\Upsilon(4S)$ resonance, at approximately 10.5 GeV centre of mass (CM) energy. Asymmetric beam energies of 7 and 4 GeV for the e^- and e^+ beams, respectively, result in a CM system which is boosted along the beam (z) axis, in order to facilitate separation of B meson decay vertices produced in $\Upsilon(4S) \to B\bar{B}$ interactions. First collisions were recorded by Belle II during the "Phase 2" accelerator commissioning run in early 2018, and "factory mode" physics data taking begins in early 2019. In this presentation, I give an overview of the SuperKEKB accelerator and the Belle II detector, and describe some of the unique features of this experimental environment. I then discuss the Belle II physics program, with a focus on prospects for studies of recent flavour "anomalies." Finally, I briefly present some initial results from the Phase 2 commissioning run and discuss prospects for the 2019 run and beyond.

2. SuperKEKB and the Belle II experiment

The Belle II experiment detector [4] is based on the original Belle detector, with substantial upgrades to all detector subsystems, as illustrated in Figure 1, in order to address the experimental challenges presented by the extremely high luminosity environment provided by SuperKEKB. In particular, the entire inner detector region which provides the tracking system

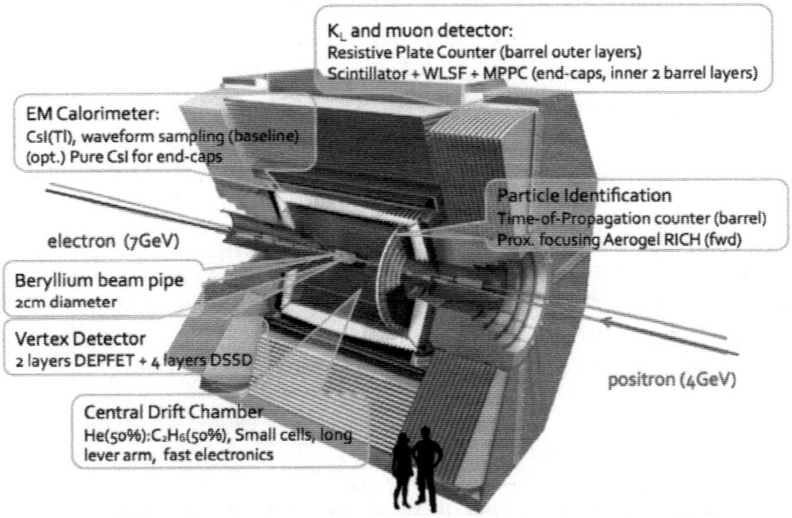

Figure 1. The Belle II detector.

has been replaced. The interaction point (IP) is surrounded by a low-mass beryllium beam pipe with a 2 cm diameter. The small diameter enables two layers of DEPFET silicon pixel detectors to be placed in close proximity to the IP for precise decay vertex reconstruction. These detectors are surrounded by an additional four layers of double-sided silicon strip detectors, with strips running perpendicular on the two sides, to provide additional $x-y$ space points for track and vertex reconstruction. These detectors are designed to have extremely low material content, to minimize the degradation of tracking resolution due to multiple scattering. Additional tracking at radii between 160 and 1130 mm from the beam axis is provided by a large-volume cylindrical drift chamber based on a 50:50 mixture of helium and ethane. A total of 56 layers of tracking cells radially are arranged in nine groups of "superlayers," in which the innermost superlayer has a reduced cell size to help address occupancy issues arising from the high luminosity environment. Superlayers alternate between axial and "stereo" wires, in which wires are offset between 45 and 74 mrad between the two ends, in order to provide z-coördinate position information. The tracking detectors additionally provide particle identification (PID) capability via ionization energy deposition ($\mathrm{d}E/\mathrm{d}x$), and momentum measurement of charged particles as a result of a 1.5 T axial magnetic field provided by a large-bore solenoid positioned outside of the electromagnetic calorimeter. Additional PID, and in particular $K-\pi$ separation, in the central "barrel" region of the detector is provided by a Time of Propagation (TOP) detector based on a novel configuration of a ring-imaging Cherenkov device which exploits the total internal reflection of Cherenkov photons in synthetic quartz bars. This is complemented in the forward endcap region by a aerogel-based proximity focusing ring imaging Cherenkov system, with Cherenkov photons detected by an array of approximately 60000 channels of hybrid APDs. Electrons and photons are reconstructed using energy measurements from a CsI(Tl) crystal calorimeter, reusing crystals and photodetectors from the original Belle detector. New front-end electronics and upgraded feature extraction capability permit a faster and more precise energy determination in the high-luminosity environment of Belle II. The magnetic flux return for the solenoid is instrumented with a combination of RPCs and scintillators read out by MPPCs via wavelength-shifting fibres in order to provide identification of muons and neutral hadrons (K_L) which penetrate the calorimeter.

The SuperKEKB accelerator is a substantially upgraded version of the KEKB collider with design luminosity of 8×10^{35} cm^{-2}s^{-1}, representing a 40-fold increase compared with its predecessor. This luminosity is achieved via a novel low-emittance "nanobeam" scheme exploiting ILC and light-source technologies, combined with a new positron damping ring and positron beam vacuum chamber. At full luminosity, each beam will circulate 2500 bunches, with bunch crossings approximately every 4 ns. Given the cross sections for $e^+e^- \rightarrow f\bar{f}$ (where f represents any fermion) we expect a rate of order 1 kHz of $B\bar{B}$ events recorded, accompanied by several kHz of charm ($c\bar{c}$), light quark $q\bar{q}$ and $\tau^+\tau^-$ events. Additionally, we expect approximately 30 kHz of Bhabha ($e^+e^- \rightarrow e^+e^-$) events within the detector acceptance. This later contribution must be rejected in real time by the first-level event trigger, while the other components are recorded and made available for data analysis via an inclusive trigger strategy. Consequently, the target integrated luminosity of the Belle II collaboration of 50 ab^{-1} implies a data sample containing in excess of 50 billion $B\bar{B}$ pairs, with similar size samples of τ and charm events.

3. The Belle II physics program

The previous generation of B Factory experiments, Belle and BABAR, operated for approximately a decade beginning in 1999. These experiments collectively integrated data sets in excess of 1 ab^{-1} of data at the $\Upsilon(4S)$ resonance, as well as substantial data samples at higher and lower mass Υ resonances as well as off-peak. The primary objective of these experiments was the confirmation of the Kobayashi-Maskawa mechanism for CP violation within the Standard Model (SM) via the CKM matrix. While many of these measurements remain statistically limited, the target data sets for these experiments were chosen so as to provide sufficient measurement precision to indisputably validate or refute the SM picture of the origin of CP violation within the CKM matrix. With this picture now firmly established experimentally, the scientific objectives of the next generation of flavour physics experiments has evolved: the Belle II experiment is designed to look for deviations from SM expectations indicative of new physics contributions across a very broad class of flavour observables. This scientific program requires a much higher degree of precision, and hence a much larger data sample, than previous B Factory experiments. In addition to CP violation measurements, potential flavour observables include branching fractions, kinematic distributions, and angular distributions and asymmetries in B, charm and τ decays.

Observation of new physics would require one or more measurements to deviate significantly from theoretical predictions within the context of the SM. This can occur in one of two ways: either precise measurements of specific flavour observables can be shown to deviate from (equally precise) theoretical predictions; or very rare or forbidden processes can be observed when SM predictions are well below experimental sensitivity. CP violation studies, as illustrated by the ubiquitous "unitarity triangle" representations [5], are an excellent example of the first of these ways, while studies of electroweak flavour changing neutral current (FCNC) B decay processes exemplify the second. In both cases, however, a key feature is that the relevant decay processes can occur via loop-level processes which can potentially contain high-mass non-SM particles, and the presence of these non-SM contributions can measurably alter various experimental observables. Moreover, the existence of a particular non-SM particle or interaction would be expected to contribute in predictable ways across a variety of related B meson decay modes. Consequently, these measurements have the potential to not only discover new physics phenomena arising from mass scales in excess of those directly accessible at the LHC, but also to elucidate the nature of this high-mass new physics via the pattern of deviations observed across a broad class of measurements.

While Belle II shares many of the same physics objectives as the LHCb experiment [6], there exists a high degree of complementarity between these two experiments. As a hadron collider experiment, LHCb benefits from a large $b\bar{b}$ production cross section and highly boosted topologies, with hadronization into a variety of B_d, B_s mesons and b-baryons. However, the high background environment imposes limitations on the accessibility of some decay modes for analysis, in particular inclusive processes, some decays with neutral particles and decays with missing energy. In contrast, Belle II exploits the extremely clean e^+e^- environment and exclusive $\Upsilon(4S) \to B_d\bar{B}_d$ production to access a broader range of B_d final states, but this production mechanism has a relatively low cross section and is limited to B_d meson decays (although B_s mesons can be produced through dedicated running at the $\Upsilon(5S)$ resonance). This complementarity is exemplified by the recent measurements of anomalies in $B \to K^*\ell^+\ell^-$ (where $\ell = e, \mu$) by LHCb [7, 8]. Although these modes have been studied previously at B factory experiments, the large sample of B_d mesons available at LHCb have resulted in significantly improved precision of determinations of the lepton universality ratio, $R_{K^{(*)}}(q^2) \equiv \mathcal{B}(B \to K^{(*)}\mu^+\mu^-)/\mathcal{B}(B \to K^{(*)}e^+e^-)$, as a function of q^2, resulting in a deviation from SM expectations with a combined significance of approximately 4σ. As highly suppressed SM FCNC processes, these modes are excellent places to look for new physics, so such a discrepancy is tantalizing. Belle II will also be able to study these modes, in both charged and neutral B decays. With the much higher luminosity compared to the previous B factory experiments, it is anticipated that Belle II will have comparable precision to the current LHCb measurements with an integrated luminosity of about 5 ab^{-1}, or about 10% of the full dataset. Moreover, this measurement will use a distinct methodology compared to LHCb and hence provide an important validation and crosscheck of the LHCb results. The two experiments will then collect data concurrently following the LHC LS2 shutdown, providing further precision to these measurements over the coming decade.

However, it is likely that any new physics effect which may contribute to $R_{K^{(*)}}(q^2)$ would also result in potentially observable impacts on other related B decay processes which proceed via similar one-loop FCNC decays. In particular, with the excellent $K-\pi$ separation in Belle II, exclusive $b \to d\ell^+\ell^-$ decays will be accessible, while the inclusive trigger and hermetic detector will permit measurements of the inclusive processes $B \to X_{s,d}\ell^+\ell^-$ (where $X_{s,d}$ represents a hadronic system containing an s or d quark). The self consistency of these measurements will provide further validation of any evidence of non-SM contributions to these decays. Many new physics scenarios in which lepton universality is violated between the two light charged lepton generations also predict similar effects for the third generation τ leptons. Consequently, $B \to K\tau^+\tau^-$ is also very much of interest. A recent BABAR measurement [9] has demonstrated the potential for this decay at B factories, and Belle II is expected to be able to extend limits on this process to close to the SM range. Similarly, lepton non-universality in the charged lepton sector could potentially also impact the corresponding neutral-lepton decays $B \to K^*\nu\bar{\nu}$. In contrast to $B \to K^*\ell^+\ell^-$, the neutrino modes proceed in the SM only via FCNC processes in which the $\nu\bar{\nu}$ couples to Z^0 bosons rather than photons or $W^\pm$ bosons. Consequently, the presence or absence of a deviation from SM predictions in these modes would help elucidate the nature of any non-SM effect seen in $B \to K^*\ell^+\ell^-$. Finally, lepton non-universality in new physics models is often accompanied by violations of lepton flavour. As such, it would also be interesting to search for decays of the type $B \to K^*\ell^\pm\ell'^\mp$, where $\ell' \neq \ell$ could be any of e, μ or τ. While the light-lepton LFV modes $B \to K^*e^\pm\mu^\mp$ are accessible at LHCb, Belle II possesses a unique capability, discussed below, for studying decays with missing energy such as $B \to K^*\ell^\pm\tau^\mp$. Clearly, the strength of the Belle II physics program lies in breadth, and the ability to test new physics scenarios with high precision across a large number of decay modes.

3.1. Processes with missing energy

As a consequence of the precisely known e^+e^- initial state and the hermeticity of the detector, Belle II possesses a unique ability to reconstruct final states with one or more undetected particles, including neutrinos or possible non-interacting new physics particles. Many interesting

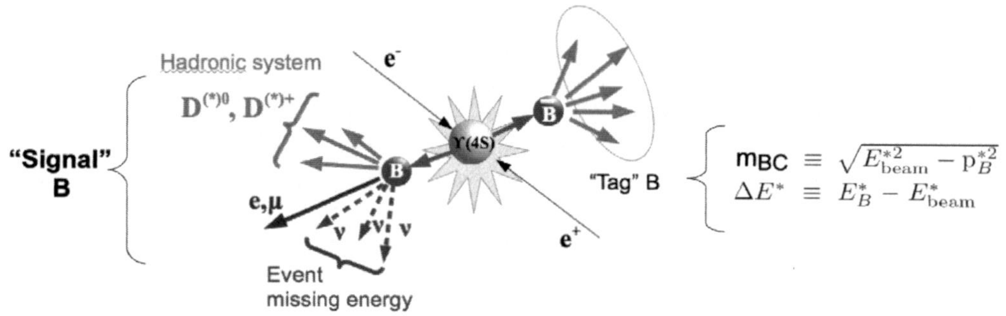

Figure 2. Hadronic B tag reconstruction methodology as applied to a $B \to D^{(*)}\tau\nu$ signal event produced in $\Upsilon(4S) \to B\bar{B}$.

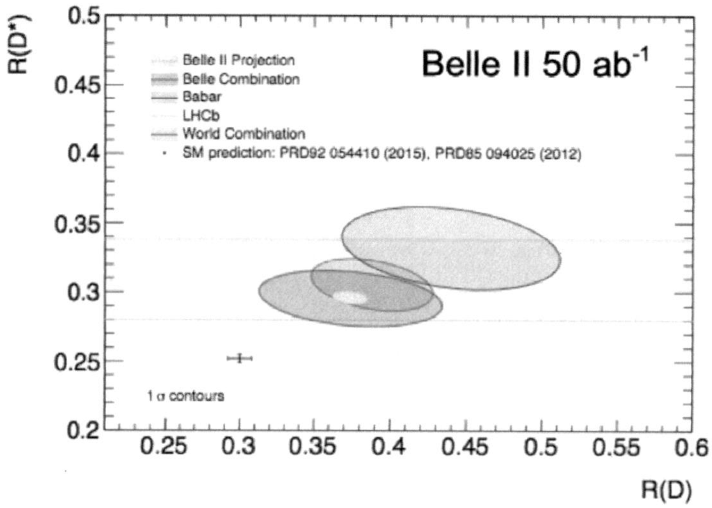

Figure 3. Expected Belle II sensitivity to $R(D^{(*)})$ with the full 50ab ^{1} data sample.

B meson decay processes fall into this category, including $B \to D^{(*)}\tau\nu$ and the related tree-level leptonic decays $B^{+} \to \tau^{+}\nu$ and $B^{+} \to \mu^{+}\nu$, as well as the flavour changing neutral current decays $B \to K^{(*)}\nu\bar{\nu}$, $B \to K^{(*)}\tau^{+}\tau^{-}$ and the lepton flavour violating decays $B \to K^{(*)}\tau^{\pm}\ell^{\mp}$ and $B^{0} \to \tau^{\pm}\ell^{\mp}$ ($\ell = e, \mu$). Common features of these decays is that they contain multiple neutrinos in the final state, either produced directly in the B decay or via the secondary decays of τ leptons. Because B meson production occurs exclusively via $e^{+}e^{-} \to \Upsilon(4S) \to B\bar{B}$, reconstructed particles, and also the missing energy 4-vector, can be associated with a signal decay mode by process of elimination; the accompanying "tag B" is exclusively reconstructed into one of many possible hadronic decay modes (see Figure 2), utilizing kinematic constraints to ensure that all of the tag B daughters are correctly identified. Having done this, all remaining reconstructed particles in the event, as well as any missing energy, can then be attributed to the "signal B". In this way, Belle II is able to access many missing-energy modes which would otherwise be intractable due to the lack of a reconstructable signature. An important example is the decay $B \to D^{(*)}\tau\nu$, which has shown indications of tension with respect to SM predictions in the ratio $R(D^{(*)}) \equiv \mathcal{B}(B \to D^{(*)}\tau\nu)/\mathcal{B}(B \to D^{(*)}\ell\nu)$ ($\ell = e, \mu$) [10]. The expected precision

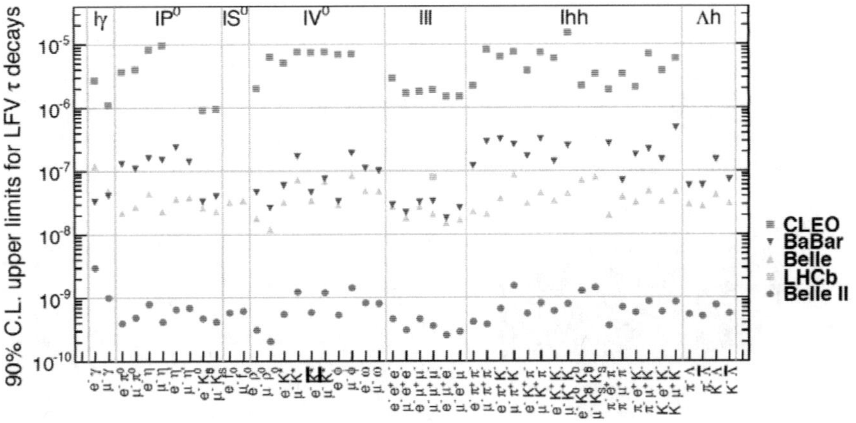

Figure 4. Expected Belle II sensitivity to lepton flavour violating τ decay modes.

of these measurements with the full Belle II dataset is shown in Figure 3, along with current measurements from other experiments.

4. Rare and forbidden decays

Decays which are extremely rare or entirely forbidden within the SM would provide definitive evidence of new physics if observed. This is true not only for B mesons, but also for rare charm and τ lepton decays. The high continuum production cross section for $c\bar{c}$ and $\tau^+\tau^-$ at the $\Upsilon(4S)$ CM energy, combined with the inclusive trigger strategy of Belle II, means that of order 50 billion of each of these types of events are expected to be recorded over the lifetime of the experiment, providing extremely stringent probes for new physics. A particularly interesting class of such decays are LFV, or neutrino-less, τ decays, the archetype of which is the process $\tau \to \mu\gamma$. Because these processes lack the neutrinos that are usually present in τ decays, these searches have very low backgrounds and possess a distinctive signature of an invariant mass which peaks at the τ lepton mass. Approximately 50 distinct τ decay modes have been studied at previous B factory experiments, resulting in branching fraction limits which in some cases approach 1×10^{-8}. Simulation studies for Belle II have indicated that the vastly increased data statistics will enable these sensitivities to be lowered by 1- 2 orders of magnitude, typically to the level of a few$\times 10^{-9}$, as shown in Figure 4.

5. Phase 2 performance

First collisions of the SuperKEKB B factory took place in early 2018 and the accelerator subsequently achieved a peak instantaneous luminosity of 5.5×10^{33} cm^{2}s^{-1} during the Phase 2 commissioning run. As the primary purpose of this run was the optimization of accelerator performance, the Belle II experiment was installed in the beam line without the silicon vertex detector installed, in order to prevent possible damage to this device. However, in spite of this limitation, the experiment successfully recorded collisions data corresponding to 472 pb^{-1} of integrated luminosity, for a total of roughly one million B meson decays. These data have been used to optimize detector performance and to validate the reconstruction of various physics objects within the detector. Examples of performance during this run are shown in Figure 5, including (top left) the dE/dx for reconstructed tracks in the drift chamber as a function of momentum, and (top right) the $\gamma\gamma$ invariant mass of clusters reconstructed in the calorimeter, showing a clear π^0 mass peak. The lower two plots illustrate fits to a pion (left) and kaon (right)

IOP Conf. Series: Journal of Physics: Conf. Series **1271** (2019) 012011 doi:10.1088/1742-6596/1271/1/012011

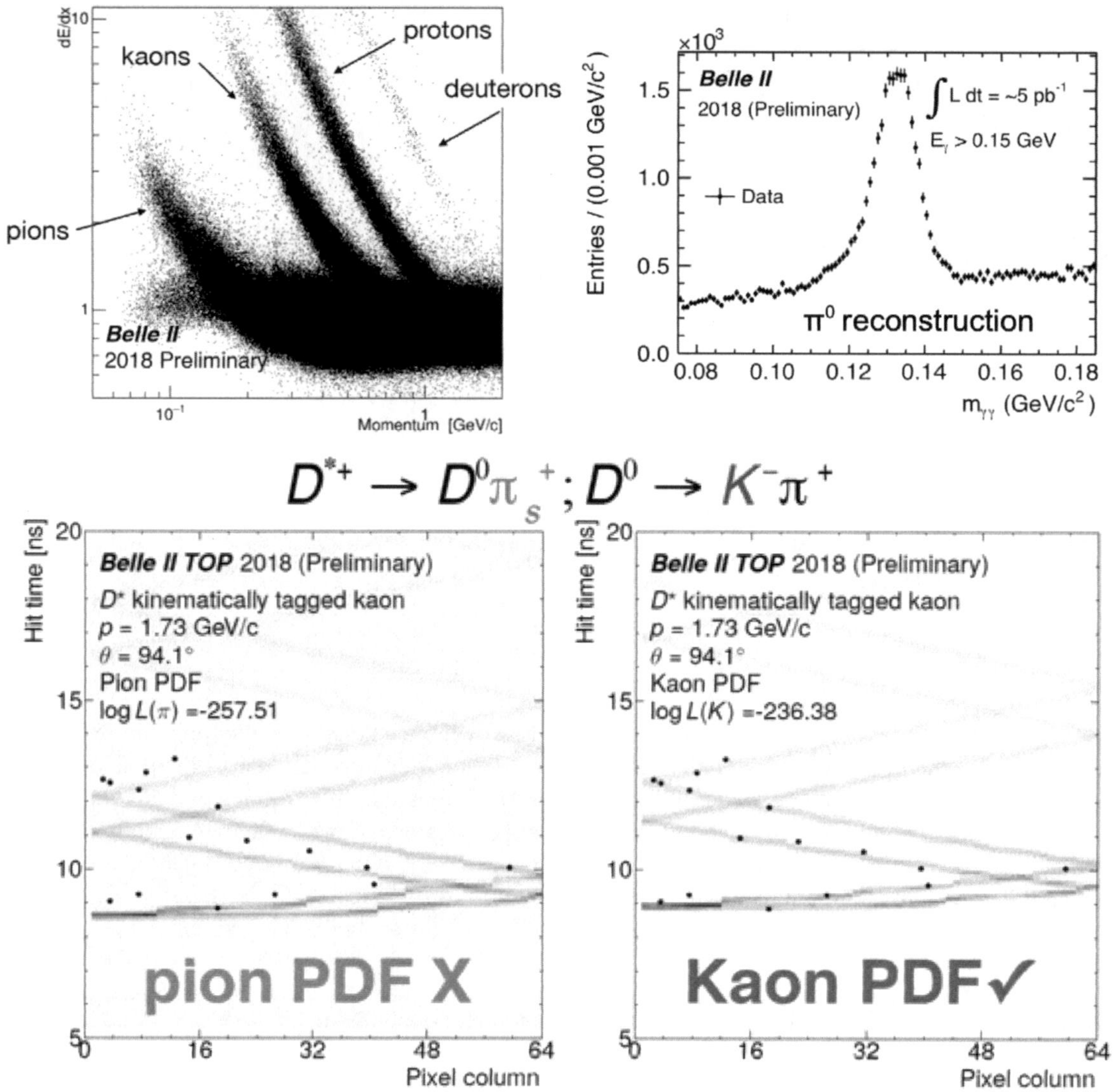

Figure 5. Belle II detector performance during Phase 2 commissioning. See text for description.

hypothesis for Cherenkov photons produced in the TOP by a charged track identified as a kaon via tagging of a soft pion in D^{*+} decays.

In addition to detector performance studies, initial physics studies have been performed on this dataset in order to begin the process of "re-discovering" known physics processes. In particular, topological differences between $B\bar{B}$ events produced at the $\Upsilon(4S)$ and "continuum" $q\bar{q}$ production have been used to demonstrate (see Figure 6, left) that $B\bar{B}$ events are present in the dataset at the expected rate, confirming the the SuperKEKB accelerator was operating at the peak of the $\Upsilon(4S)$ resonance. Exclusive reconstruction of hadronic B decay modes has also shown clear evidence of B mesons (Figure 6, right).

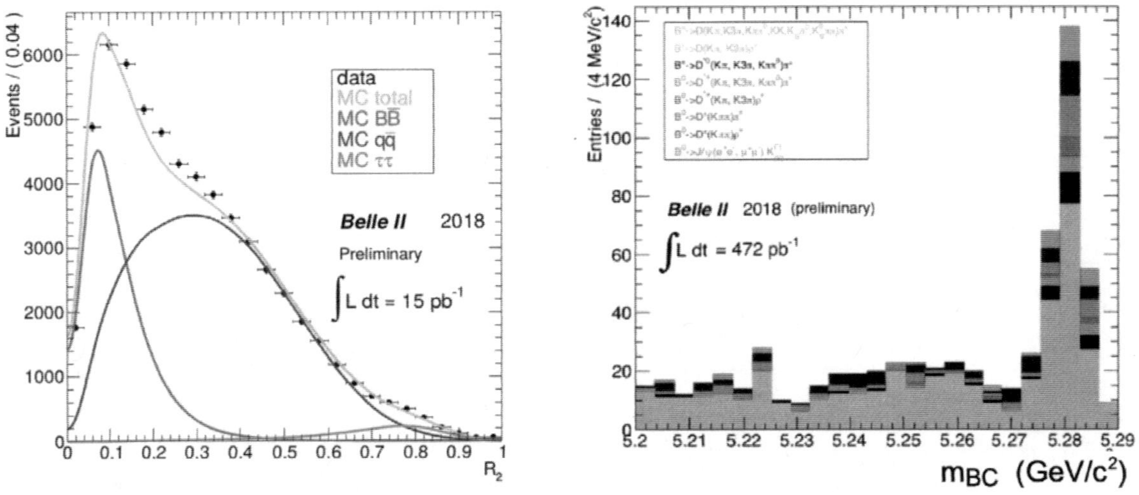

Figure 6. Evidence of $B\bar{B}$ events in Belle II Phase 2 data from event shapes (left) and exclusive hadronic B decays (right).

6. Future prospects

Belle II is currently preparing for "Phase 3" operations, with high-luminosity factory-mode data taking, beginning in March 2019. Beam background studies from Phase 2 have indicated that the vertex detectors can be safely operated, and they have now been installed in the interaction region, completing the Belle II detector. Approximately seven months of operations are foreseen for 2019, with instantaneous luminosity initially comparable to *BABAR* or Belle, but increasing rapidly in the following years to the design value, with a competitive $\Upsilon(4S)$ dataset anticipated by early in the coming decade.

References

[1] E. Kou *et al.* [Belle-II Collaboration], arXiv:1808.10567 [hep-ex].
[2] A. J. Bevan *et al.* [BaBar and Belle Collaborations], Eur. Phys. J. C **74**, 3026 (2014) doi:10.1140/epjc/s10052-014-3026-9 [arXiv:1406.6311 [hep-ex]].
[3] K. Akai *et al.* [SuperKEKB Collaboration], Nucl. Instrum. Meth. A **907**, 188 (2018) doi:10.1016/j.nima.2018.08.017 [arXiv:1809.01958 [physics.acc-ph]].
[4] T. Abe *et al.* [Belle-II Collaboration], arXiv:1011.0352 [physics.ins-det].
[5] See for example http://www.slac.stanford.edu/xorg/hflav/triangle/summer2018/index.shtml
[6] R. Aaij *et al.* [LHCb Collaboration], arXiv:1808.08865.
[7] R. Aaij *et al.* [LHCb Collaboration], JHEP **1708**, 055 (2017) doi:10.1007/JHEP08(2017)055 [arXiv:1705.05802 [hep-ex]].
[8] C. Bobeth, G. Hiller and G. Piranishvili, JHEP **0712**, 040 (2007) doi:10.1088/1126-6708/2007/12/040 [arXiv:0709.4174 [hep-ph]].
[9] J. P. Lees *et al.* [BaBar Collaboration], Phys. Rev. Lett. **118**, no. 3, 031802 (2017) doi:10.1103/PhysRevLett.118.031802 [arXiv:1605.09637 [hep-ex]].
[10] https://hflav-eos.web.cern.ch/hflav-eos/semi/summer18/RDRDs.html

IOP Conf. Series: Journal of Physics: Conf. Series **1271** (2019) 012012 doi:10.1088/1742-6596/1271/1/012012

Strongly coupled $\Upsilon(1S)$ suppression in $\sqrt{s}_{NN} = 2.76$ TeV Pb+Pb collisions

N N Barnard and W A Horowitz

Department of Physics, University of Cape Town, Private Bag X3, Rondebosch 7701, South Africa

E-mail: brnnad007@myuct.ac.za

Abstract. In a revolutionary paper, Matsui and Satz proposed using the suppression of quarkonia as a smoking gun signature of deconfinement in relativistic heavy ion collisions. The stunning success of using strong-coupling, AdS/CFT techniques to predict the viscosity to entropy density ratio extracted from RHIC and LHC heavy ion collision data using sophisticated 3+1D relativistic viscous hydrodynamics has prompted further investigation into the physics of a strongly-coupled plasma. We compute, for the first time, the suppression of bottomonia in a strongly coupled QGP, and compare the results to those from a weakly coupled QGP and to data. The complex binding energies which inform the thermal width and hence the R_{AA} of $\Upsilon(1S)$ are determined using imaginary time techniques.

1. Introduction

The relativistic heavy-ion collisions at the Large Hadron Collider (LHC) and the Relativistic Heavy Ion Collider (RHIC) are sufficiently energetic for hadrons to transition into a new phase of colored matter, known as the quark-gluon plasma (QGP) [1]. Quarkonia may theoretically exist in conjunction with the QGP at $T > T_c$, where T_c is the critical temperature required for QGP formation, due to its small binding radii relative to the screening radius. At some T, the screening radius becomes smaller than the typical quarkonia radii, leading to their dissolution. In addition, excited states of quarkonia dissociate before the ground state [2]. The suppression of the bound states of quarkonia in heavy-ion collisions is hence considered a valuable indicator of the formation of QGP, and the comparison of this suppression to that in minimum bias $p+p$ collisions where no QGP is formed, is a useful probe of the QGP's properties.

Potential models can be used to describe the interaction of the quark and antiquark in the $q\bar{q}$ pair to calculate the suppression of quarkonia production in heavy-ion collisions [3]. This potential at finite temperature contains not only a standard real Debye-screened term, but also an imaginary part which gives the thermal width of the state, and hence its suppression [4]. One of the first to show this was [5], which made use of perturbative methods to find the static potential of quarkonia at finite temperature. They concluded that the thermal width of the state increases with T, suggesting that at high T the dissociation due to the effect described by the imaginary part of the potential occurs before color screening can even come into effect.

The complex-valued potential was explored further using non-perturbative lattice QCD by [4], among others, allowing for the study of strongly-coupled quarkonia as well. An important consideration in finding heavy quarkonium suppression is the velocity of the $q\bar{q}$ in relation to the

IOP Conf. Series: Journal of Physics: Conf. Series **1271** (2019) 012012 doi:10.1088/1742-6596/1271/1/012012

surrounding QGP, however, while perturbative and lattice QCD calculations generally consider the $q\bar{q}$ meson to be at rest in the medium.

The suppression of quarkonia moving at velocity in a QGP hence requires holographic techniques such as the Anti-de Sitter/Conformal Field Theory (AdS/CFT) correspondence. The potential for static quarkonia at finite temperature in $\mathcal{N} = 4$ super Yang-Mills theory was calculated by [6, 7, 8], among others. Liu, Rajagopal and Wiedemann (LRW) [9] were the first to present a quantitative description from AdS/CFT of the consequences of velocity on the screening length of charmonium, suggesting that for strongly coupled J/ψ, velocity could result in a significant additional source of suppression at high transverse momentum p_T in the form of a decrease in dissociation energy with increased velocity. Since then, many have performed similar investigations, and while the aforementioned are limited in their scope of application, it is interesting to note that [10] in particular concludes that the effect of velocity may not be as consequential as postulated in LRW.

On the other hand, from pQCD, [11] found a potential for weakly coupled $q\bar{q}$ states which is dependent on velocity and shows that the dissociation energy increases with quarkonia velocity.

We would ultimately like to investigate the consequences of these different velocity dependence pictures from AdS/CFT compared to pQCD. Here we have a more modest goal: to compare the suppression of quarkonia at rest with respect to the QGP for pQCD vs. AdS/CFT potentials.

We consider the case of ground state bottomonia at rest with respect to an isotropic quark-gluon plasma. We follow the methodology outlined in Krouppa et al. [12], with a number of improvements. Given a potential, we evolve a random wave function through imaginary time; after a sufficiently long evolution, only the ground state wave function remains. This ground state wave function then determines the ground state binding energy. We independently confirmed these binding energies by an application of the complex Ritz variational method [13]. Finally, following Krouppa, we used the complex binding energies in a quarkonia suppression model to compute R_{AA}.

2. Methodology

2.1. Potential Models

The potential model presented here for weakly coupled quarkonia is taken from [12] and is complex-valued. Both the real and the imaginary parts of the potential were found using leading-order perturbative calculations. We plot the real part and imaginary part of the potential $V(r)$ as a function of quark separation r in Fig. 1a and Fig. 1b, respectively.

We modeled the strongly coupled quarkonia at rest in a QGP with the potential given in Albacete et al. [6]. In that work, the authors derive the potential in $\mathcal{N} = 4$ super Yang-Mills at finite temperature using AdS/CFT. Fig. 2a and 2b show the real and imaginary parts of $V_s(r)$ as a function of quark separation r, taking the 't Hooft coupling $\lambda = 10$. See [14] for full functional forms of the potential models.

Note that the real parts of the pQCD and AdS/CFT potentials, shown in Fig. 1a and 2a, respectively, are similar in form, but the imaginary parts from pQCD and AdS/CFT, shown in Fig. 1b and 2b, respectively, differ greatly: the imaginary part of the pQCD potential saturates as a function of r whereas that of the AdS/CFT potential diverges.

2.2. Numerical Integration of TDSE

The methodology used here follows that of [12, 15], adapted to the special case of an isotropic plasma, with various modifications of the discretization (see [14] for details). In order to compute the ground state wave function, and hence the binding energy, we need to solve the non-relativistic, time dependent Schrödinger Equation (nRTDSE):

$$i\partial_t \Psi(r,t) = H\Psi(r,t). \tag{1}$$

IOP Conf. Series: Journal of Physics: Conf. Series **1271** (2019) 012012 doi:10.1088/1742-6596/1271/1/012012

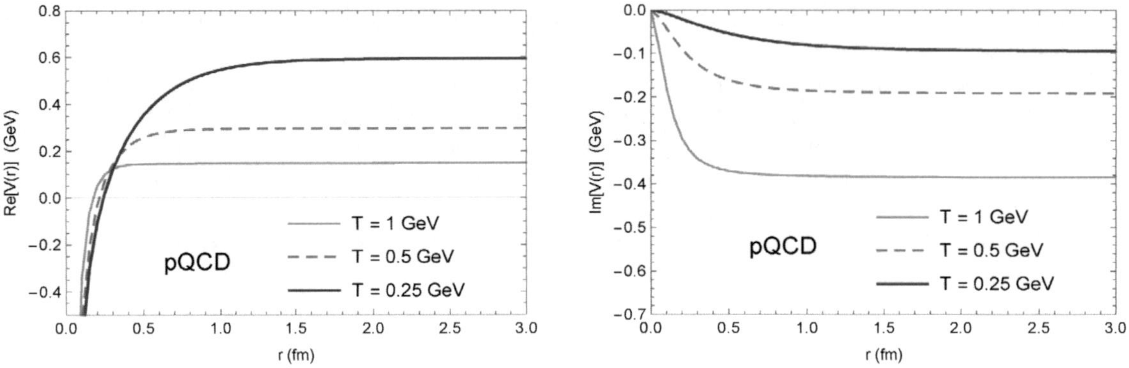

(a) (b)

Figure 1. Plot of the (a) real part of the potential $\Re[V(r)]$, and the (b) imaginary part of the potential $\Im[V(r)]$, as a function of the distance r between the quark and anti-quark in the $b\bar{b}$, for various T in an isotropic plasma.

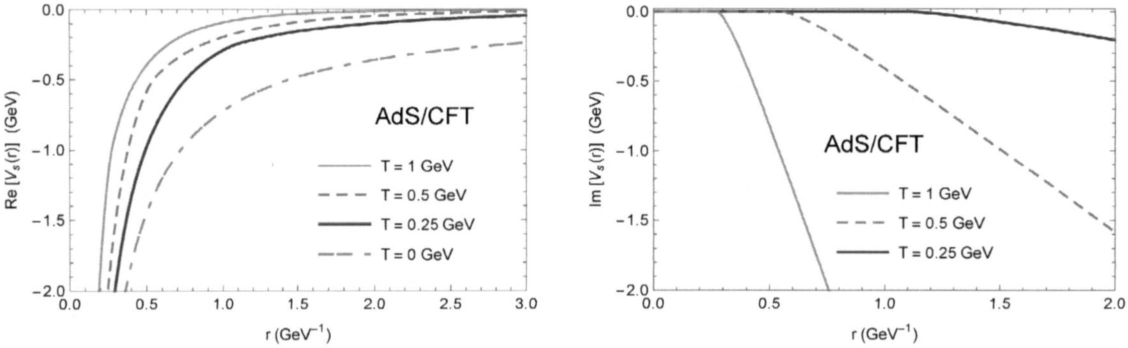

(a) (b)

Figure 2. The (a) real part of the strongly coupled potential $\Re[V_s(r)]$ and the (b) imaginary part of the potential $\Im[V_s(r)]$ as a function of the distance r between the quark and anti-quark in the $b\bar{b}$, for various temperatures T in an isotropic plasma.

Performing a Wick rotation to an imaginary time $\tau \equiv it$, Eq. (1) has the general solution

$$\Psi(r,\tau) = \sum_{n=0}^{\infty} c_n \psi_n(r) e^{-E_n \tau}. \tag{2}$$

Since $E_n > E_0$ for all $n > 0$, one can evolve forward in imaginary time such that all the higher order wave functions are suppressed and only the ground state wave function remains:

$$\lim_{\tau\to\infty} \Psi(r,t) \to c_0 \psi_0(r) e^{-E_0 \tau}, \tag{3}$$

where $\psi_0(r)$ is the ground state wave function and E_0 the ground state energy. The binding energy of the state can then be found from

$$E_{\text{bind}} \equiv E_0 - \Re[V(|r| \to \infty)], \tag{4}$$

where the ground state energy E_0 can be found from the ground state wave function,

$$E_0 = \frac{\int r^2\, dr\, \psi_0(r)^*\, H\, \psi_0(r)}{\int r^2\, dr\, |\psi_0|^2}. \tag{5}$$

2.3. Suppression

We would like to make quantitative predictions for the suppression of bottomonia in heavy ion collisions and compare to measured data. The nuclear modification factor R_{AA} is calculated following [12]:

$$R_{AA}(p_T, y, \mathbf{x}_\perp, b) = e^{-\zeta(p_T, y, \mathbf{x}_\perp, b)}, \tag{6}$$

$$\zeta \equiv \Theta(\tau_f - \tau_{\text{form}}) \int_{\max(\tau_{\text{form}}, \tau_0)}^{\tau_f} d\tau\, \Gamma(\tau, \mathbf{x}_\perp, \varsigma = y),$$

where the thermal width $\Gamma(\tau, \mathbf{x}_\perp, \varsigma)$ is given as

$$\Gamma(\tau, \mathbf{x}_\perp, \varsigma) = \begin{cases} -2\Im[E_{bind}] & \Re[E_{bind}] < 0 \\ \gamma_{\text{dis}} & \Re[E_{bind}] \geq 0. \end{cases} \tag{7}$$

We take $\gamma_{\text{dis}} = 10$ GeV, as was done in [12]. Furthermore, b is the impact parameter, and y the rapidity, taken to be zero. The formation time τ_{form} is calculated using

$$\tau_{\text{form}} = E_T\, \tau^0_{\text{form}} / m_Q \tag{8}$$

where $\tau^0_{\text{form}} = 0.2$ fm is taken for the initial formation time of the state [12]. Lastly, the final time τ_f is taken as the time at which the temperature T of the QGP drops below the critical temperature T_c.

We use the optical limit of the Glauber model [16] to describe the background in the case of $\sqrt{s_{\text{NN}}} = 2.76$ TeV Pb+Pb collisions. Taking a weighted average over the region with limits $\mathbf{x}_\perp = [-10, 10]$ fm, we have

$$R_{AA}(p_T, b) = \frac{\int d^2\mathbf{x}_\perp d\phi\, T_{\text{AA}}(\mathbf{x}_\perp, b)\, R_{AA}(p_T, y, \mathbf{x}_\perp, b)}{2\pi N_{\text{coll}}}, \tag{9}$$

where $T_{\text{AA}}(\mathbf{x}_\perp, b)$ is the nuclear overlap function and $N_{\text{coll}} \equiv \int d^2\mathbf{x}_\perp\, T_{\text{AA}}(\mathbf{x}_\perp, b)$ is the number of binary nucleon-nucleon collisions in the region. We set a central temperature $T_0 = 522$ MeV and initial formation time $\tau_0 = 0.3$ fm, as is done in [12].

3. Results

3.1. Binding Energies

Fig. 3a is a plot of the real part of the binding energy of $\Upsilon(1S)$ from the pQCD potential, and strongly coupled potential, as a function of temperature. Similarly, Fig. 3b gives the imaginary part of the binding energies for all cases mentioned above. The binding energy results for weakly coupled bottomonium from [15] are labeled "pQCD (KRS)" and are included for comparison.

For the AdS/CFT results, we show the binding energy, both for the case where the coupling constant is $\lambda = 10$ (labeled as $\alpha_s = 0.27$) and where $\lambda = 5.5$, in an attempt to at least partially map out some of the systematic theoretical uncertainties associated with the use of the AdS/CFT correspondence – this is discussed in Section 3.2.

Both the binding energy results presented for the pQCD potential and the AdS/CFT potential taking $\lambda = 10$ were independently confirmed using a complex variational method, further explained in [14].

The binding energy found from our adapted methodology for the pQCD potential differs quantitatively from that presented in [15], which was used in Krouppa et al. [12] to calculate suppression. In the case of $\Upsilon(1S)$, this difference does not change the qualitative behavior of the quarkonia, since both results suggest that the quarkonia remain bound up to at least $T = 3\,T_c$.

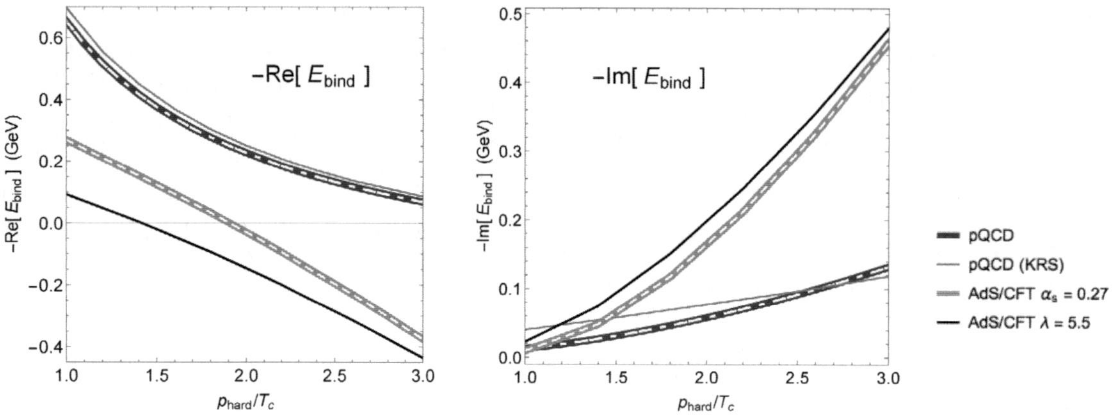

(a) (b)

Figure 3. The (a) negative real part of $E_{\rm bind}$ and (b) negative imaginary part of $E_{\rm bind}$ for Υ(1S). The blue, green, and black curves give the results for weakly coupled and strongly coupled ($\lambda = 10$ and $\lambda = 5.5$) Υ(1S), respectively. The dashed white curves are from the independent evaluation using a complex variational method. The results from KRS [15], which should be identical to the blue curves, are given as solid red for comparison.

However, we see in the results that follow that the small quantitative differences in the derived binding energies lead to a significant quantitative difference in the predicted suppression.

The imaginary part of the binding energies from AdS/CFT are notably larger than those of weakly coupled quarkonia, and rise more steeply. This result is not surprising as the AdS/CFT potential has a divergent imaginary part, compared to the saturation of the imaginary part of the pQCD potential.

Unlike in the case of the weakly coupled quarkonia where the Υ(1S) remains bound for the temperature range considered, the strongly coupled Υ(1S) dissociates at $\sim 1.9\,T_c$. The comparatively larger imaginary part of the binding energy up to the temperature at which the bottomonium dissociates implies a much larger thermal width at higher T, and hence a larger suppression.

3.2. Suppression

Fig. 4a gives the nuclear modification factor R_{AA} for each of the sets of binding energies shown in Fig. 3a and 3b as a function of the number of participating nucleons $N_{\rm part}$. To this end, the $R_{AA}(p_T, b)$ from Eq. (9) is averaged over the transverse momentum range $0 \leq p_T \leq 40$ with a weighting of E^{-4} [12].

Fig. 4b shows $R_{AA}(p_T)$, where all centrality classes are included, weighed by the number of binary nucleon-nucleon collisions $N_{\rm coll}$. Suppression results for mid-rapidity ($|y| < 2.4$) Pb+Pb collisions at $\sqrt{s_{\rm NN}} = 2.76$ TeV from the CMS Collaboration [17] are included in Fig. 4a and 4b for comparison.

We show in Fig. 4a and 4b two predictions for the suppression of strongly coupled bottomonia. Since we used a potential derived in AdS-space dual to maximally supersymmetric Yang-Mills theory, there is no single obvious map between the parameters of QCD and of $\mathcal{N} = 4$ SYM. For the $\alpha_s = 0.27$ curve, we took $\lambda_{SYM} = 10$ and $T_{SYM} = T_{QCD}$, where $\lambda_{SYM} = 10 = \sqrt{4\pi\alpha_s N_c}$ (and thus $\alpha_s = 0.27$ for $N_c = 3$) is approximately the value of the QCD running coupling constant evaluated at the first Matsubara frequency of the plasma. For the $\lambda = 5.5$ curve, the coupling constant was set by a comparison to the $q\bar{q}$ potential from lattice and $T_{SYM} = T_{QCD}/3^{-1/4}$ is a result of assuming the entropies of the QCD and SYM plasmas are the same [18].

IOP Conf. Series: Journal of Physics: Conf. Series **1271** (2019) 012012 doi:10.1088/1742-6596/1271/1/012012

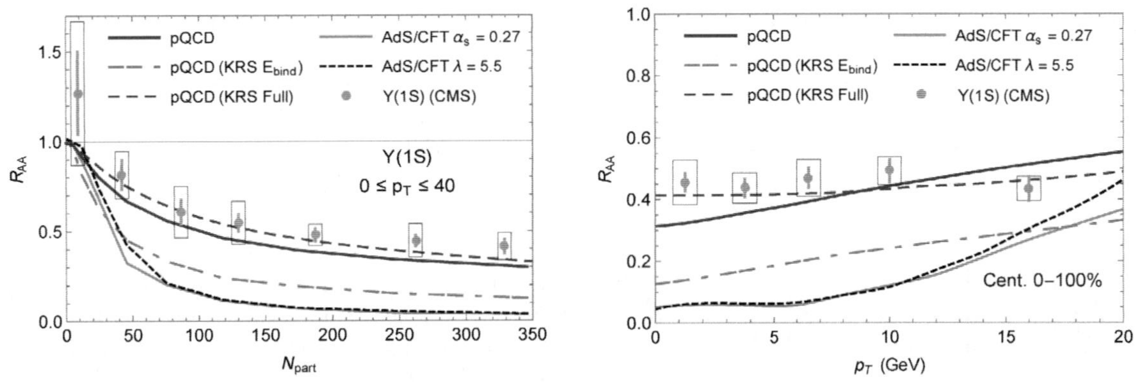

(a) (b)

Figure 4. (a) Nuclear modification factor R_{AA} as a function of the number of participating nucleons N_{part} for $0 \leq p_T \leq 40$. (b) Nuclear modification factor R_{AA} as a function of transverse momentum p_T for combined centrality classes. In both subfigures, the thick solid blue line gives our results for weakly coupled $\Upsilon(1S)$, and the dashed-dotted red line that calculated for the binding energy from KRS [15] using our suppression model. The R_{AA} presented in KRS [12] as calculated using their suppression model is given in dashed purple. The solid green and dotted black lines give the results for strongly coupled $\Upsilon(1S)$ with coupling constants $\alpha_s = 0.27$ (and $T_{SYM} = T_{QCD}$) and $\lambda = 5.5$ (and $T_{SYM} = T_{QCD}/3^{1/4}$), respectively. Data from CMS [17] is included in orange.

We show in Fig. 4a and 4b three predictions for weakly coupled bottomonia: 1) the suppression using the binding energies we compute from the potential in [15] run through our medium background, 2) the suppression using the binding energies computed in [15] run through our medium background, and 3) the suppression quoted in [12] in which they run the binding energies computed in [15] through their background.

4. Discussion and Outlook

We computed for the first time the suppression of bottomonia in an isotropic strongly coupled QGP and compared the results to those from a weakly coupled QGP and to data from the CMS Collaboration [17].

The non-relativistic, time dependent, radially symmetric Schrödinger Equation was solved numerically in order to find the ground state wave functions for two potential models: one from pQCD and one from AdS/CFT. The numerical evaluation of the imaginary time Schrödinger Equation was performed by evolving forward in imaginary time until all higher order wave functions were sufficiently suppressed. The potential used for weakly coupled quarkonia was taken from [15], in which the potential came from leading-order pQCD with various corrections. The strongly coupled quarkonia potential was taken from [6], who obtained their potential from AdS/CFT.

The ground state wave functions obtained were then used to find the (complex) ground state energies for $\Upsilon(1S)$. These ground state energies were then independently confirmed using a complex variational technique [13]. Our binding energies for the weakly coupled potential in [15] differed somewhat from those found in [15], likely due to extending the physical region under consideration and from a possibly more careful treatment of the potential.

Our first results for $\Upsilon(1S)$ strongly coupled to a strongly coupled plasma show binding energies with much larger imaginary parts than those found from the pQCD potential, as well as real parts that become positive within the T_c to $3\,T_c$ range considered. Thus, for the potential

IOP Conf. Series: Journal of Physics: Conf. Series **1271** (2019) 012012 doi:10.1088/1742-6596/1271/1/012012

models considered here, a strongly coupled $\Upsilon(1S)$ interacting with a strongly coupled plasma melts at a *lower temperature* than a weakly coupled $\Upsilon(1S)$ interacting with a weakly coupled plasma. The $\Upsilon(1S)$ hence appears more strongly bound at weak coupling than at strong coupling, which is surprising.

Since the weak coupling bottomonia become more strongly bound as the coupling is increased and the strong coupling bottomonia become less strongly bound as the coupling decreases, that the weak coupling bottomonia is more strongly bound at weak coupling than at strong coupling suggests some non-monotonic behavior of the binding energies at the threshold between the weak and strong coupling regimes. This non-monotonic behavior possibly stems from deriving the potential at weak coupling in QCD whereas the strong coupling potential was derived in the slightly different theory, $\mathcal{N} = 4$ SYM; it would be interesting to compare binding energies from the quarkonium potential at weak and strong coupling consistently within $\mathcal{N} = 4$ SYM.

We then input the complex ground state binding energies we found using the imaginary time techniques into an implementation of the suppression model described in [12] to determine the $\Upsilon(1S)$ nuclear modification factor R_{AA} as a function of the number of participating nucleons, N_{part}, and of transverse momentum, p_T, respectively. The difference in binding energies for the two coupling scenarios is echoed in the R_{AA} results: from the larger imaginary parts of the strongly coupled binding energies, we see a significantly larger suppression for strongly coupled $\Upsilon(1S)$ than for weakly coupled $\Upsilon(1S)$. Quantitatively, our full model—comprised of the potential, the resulting quarkonia binding energies, and the translation to R_{AA}—significantly overpredicts the suppression of strongly coupled bottomonia compared to data. At the same time, our predictions for weakly coupled bottomonia are consistent with data.

We note that our model for the medium is significantly less sophisticated compared to that used in [12]: our background is an optical Glauber model as opposed to the 3+1D viscous anisotropic hydrodynamics in that work. Our medium incorporates only Bjorken expansion, whereas the background in [12] includes transverse expansion and entropy production. Therefore the plasma in [12] cools faster than ours, leading to our model showing more dissociation for the same binding energies. The extent of the sensitivity of R_{AA} to the background used is surprisingly large. With the only difference being the background geometry used, we ran the binding energies from [15] through our suppression model and found an R_{AA} a factor of two smaller than that shown in [12].

In contrast to the favorable comparison between the pQCD-based results of [12] and the CMS data [17], if we assume our weak coupling binding energies are more accurate than those of [15], then computing R_{AA} with the more sophisticated background from [12] would likely yield a significant underprediction of the suppression of bottomonia.

At strong coupling, with a potential derived from AdS/CFT as described in [6], it seems unlikely that the use of a more sophisticated background would reduce the suppression of bottomonia enough that the predicted R_{AA} would be consistent with data; however, the differences from using a more sophisticated background, suppression model, and velocity dependent potential may ultimately be sufficient for future strongly coupled quarkonia predictions to be consistent with current data.

We note that our suppression calculations do not consider feed-down from higher excited states $\Upsilon(\text{nS})$, $n > 1$ for either the pQCD or AdS/CFT potentials. Considering feed-down, however, would only serve to suppress the R_{AA} further. Hence, our qualitative conclusions about strongly-coupled quarkonia would remain unchanged should higher excited states be included.

We leave the implementation of more advanced calculations of quarkonia suppression—including better modeling of the medium background, more accurate initial quarkonia production, a more realistic dissociation model, and the use of velocity dependent potentials—and a more thorough investigation of systematic theoretical uncertainties in quarkonia R_{AA} for future work.

Acknowledgments
The authors wish to thank the South African National Research Foundation and the SA-CERN Collaboration for their financial support.

References

[1] Gyulassy M and McLerran L 2005 *Nucl. Phys.* A **750** 30–63
[2] Karsch F, Kharzeev D and Satz H 2006 *Phys. Lett.* B **637** 75–80
[3] Matsui T and Satz H 1986 *Phys. Lett.* B **178** 416–22
[4] Rothkopf A, Hatsuda T and Sasaki S 2012 *Phys. Rev. Lett.* **108** 162001
[5] Laine M, Philipsen O, Romatschke P and Tassler M 2007 *J. High Energy Phys.* JHEP03(2007)054
[6] Albacete J L, Kovchegov Y V and Taliotis A 2008 *Phys. Rev.* D **78** 115007
[7] Noronha J and Dumitru A 2009 *Phys. Rev. Lett.* **103** 152304
[8] Ewers C, Kaczmarek O and Samberg A 2016 arXiv:1605.07181
[9] Liu H, Rajagopal K and Weidemann U A 2007 *Phys. Rev. Lett.* **98** 182301
[10] Finazzo S I and Noronha J 2015 *J. High Energy Phys.* JHEP01(2015)051
[11] Escobedo M A, Giannuzzi F, Mannarelli M and Soto J 2013 *Phys. Rev.* D **87** 114005
[12] Krouppa B, Ryblewski R and Strickland M 2015 *Phys. Rev.* C **92** 061901
[13] Kraft D and Plessas W 2016 *J. Phys.: Conf. Ser.* **738** 012029
[14] Barnard N N and Horowitz W A 2017 arXiv:1706.09217
[15] Margotta M, McCarty K, McGahan C, Strickland M and Yager-Elorriaga D 2011 *Phys. Rev.* D **83** 105019
[16] Miller M L, Reygers K, Sanders S J and Steinberg P 2007 *Ann. Rev. Nucl. Part. Sci.* **57** 205–43
[17] Khachatryan V *et al* (CMS) 2017 *Phys. Lett.* B **770** 357–9
[18] Gubser S S 2008 *Nucl. Phys.* **B790** 175–199

BSM constraints from model-independent measurements: A Contur Update

Jon Butterworth

Department of Physics & Astronomy, UCL, London, Gower St., London WC1E 6BT, UK

E-mail: j.butterworth@ucl.ac.uk

Abstract. Particle-level measurements, especially of differential cross-sections, made in fiducial regions of phase-space have a high degree of model-independence and can therefore be used to give information about a wide variety of Beyond the Standard Model (BSM) physics implemented in Monte Carlo generators, using a broad range of final states. The Contur package is used to make such comparisons. We summarise a snapshot of current results for a number of BSM scenarios; a UV complete model in which the global Baryon-number minus Lepton-number symmetry is gauged; several Dark Matter simplified models, and two generic light scalar models.

1. Introduction

The Contur programme was presented in Ref. [1]. It makes use of the well-developed chain of phenomenological and experimental software tools which allows a new model to be coded in FeynRules[2], exported in a standard format [3] intelligible to full final-state event generators [4] such as Herwig [5], and compared to particle-level collider data stored in HEPDATA [6] using the analyses encoded in Rivet [7]. Essentially the idea is to look at the predicted contributions from a BSM model to the fiducial phase space of measurements which have been shown to be in good agreement with the Standard Model (SM), and see if the additional contributions would have been visible in the measurement. If so, that model (or those parameters of that model) are said to be disfavoured by the data, at some confidence level. More details are given in [1]. In each case, the Herwig event generator [5, 8] is used to generate inclusively all signatures involving the new particle content of the model. In this presentation we give an update of some recent results.

2. Gauged B-L model with heavy neutrinos

There is significant interest in extensions to the SM in which the global symmetry behind the conservation of $B - L$ (Baryon number minus lepton number) is gauged, giving an additional $U(1)_{B-L}$ symmetry and an associated new gauge boson. One such model was discussed in [9], in which the additional $U(1)_{B-L}$ gauge symmetry is spontaneously broken by an extra singlet Higgs. To make the model anomaly-free it also incorporates three generations of neutral leptons sterile under the SM gauge interactions, thereby enabling the Seesaw mechanism of light neutrino mass generation.

In [10], the potential signatures from a range of model parameters and processes were considered. While the heavy neutrinos may give rare and distinct signals (as discussed in [9]),

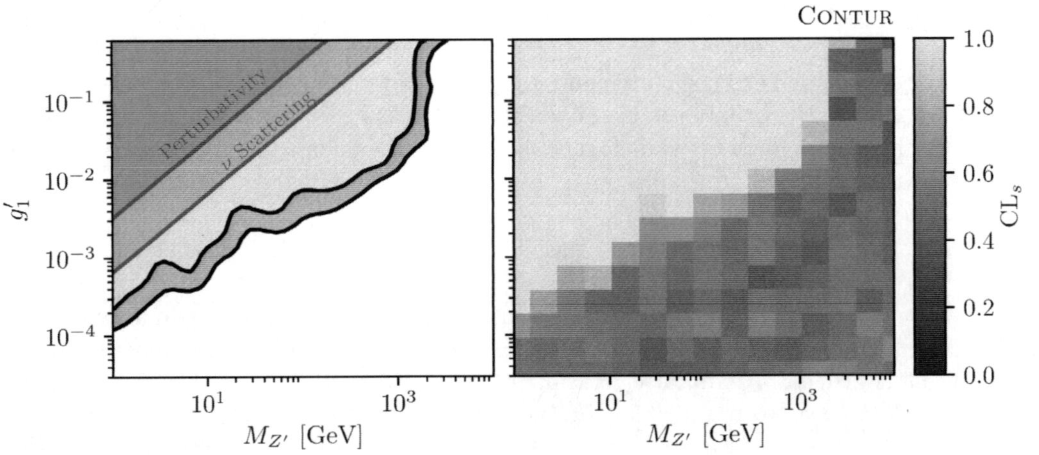

Figure 1. Sensitivity of LHC measurements to the BSM contribution from a gauged B-L model in the $M_{Z'}$ vs g'_1 plane. Left, 95% (yellow) and 68% (green) excluded contours. Right, underlying heatmap of exclusion at each scanned parameter space point. $\sin\alpha = 0.2, M_{h_2} = 200$ GeV; theory bounds and constraints from M_W and from neutrino scattering cross sections are also shown.

they play no significant role in the signatures addessed by CONTUR. The important parameters are the mass of the new gauge boson $M_{Z'}$ and its coupling, g'_1, to the SM, and the mass of the new Higgs, M_{h_2}, and its mixing, $\sin\alpha$, with the SM Higgs. If the Higgs sector is decoupled ($\sin\alpha = 0$), the model reduces a rather standard Z' model, and the majority of the parameter space reachable by CONTUR is already excluded by dedicated searches. More interesting is the case where $\sin\alpha$ is non-zero. An example, for $M_{h_2} = 200$ GeV and $\sin\alpha = 0.2$, is shown in Figure 1. The CONTUR analysis of ATLAS, CMS and LHCb data disfavours a substantial region of the plane.

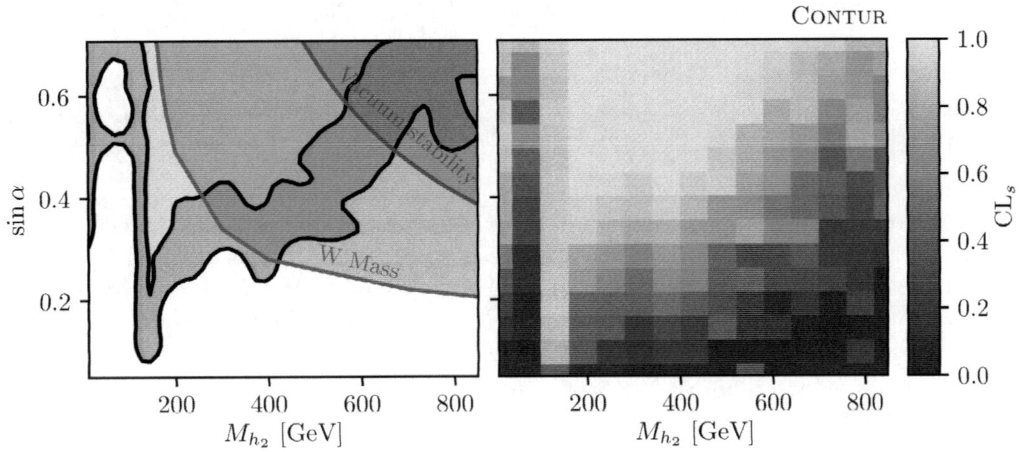

Figure 2. Sensitivity of LHC measurements to the BSM contribution from a gauged B-L model in the M_{h_2} vs $\sin\alpha$ plane. Legend as Figure 1, $M_{Z'} = 35$ GeV, $\sin\alpha = 0.2$. The constraint from measurements of the W mass is also shown.

In regions where signatures involving direct production of the Z' are not visible, either because

IOP Conf. Series: Journal of Physics: Conf. Series **1271** (2019) 012013 doi:10.1088/1742-6596/1271/1/012013

it is too massive or g_1' is too low, it is useful to scan the M_{h_2} vs $\sin\alpha$ plane to see where signatures involving the BSM Higgs might play a role. An example is given in Figure 2. The constraints from W mass measurements (where the additional Higgs can contribute to loop corrections) are stringent. The CONTUR analysis extends them slightly, and there is lower sensitivity beyond this, apparent in the heatmap, showing that future measurements should be able to probe lower $\sin\alpha$ values.

3. Dark Matter

A Dark Matter (DM) 'simplified model' was considered in the first CONTUR paper [1]. In this model, DM is a Majorana fermion which couples to a mediating spin-1 boson via an axial-vector current with strength g_{DM}. The boson in turn couples to first generation SM quarks with a vector coupling of strength g_q. An update of the results for $g_{\text{DM}} = 1, g_q = 0.25$ is shown in Figure 3. The data which have been added to HEPDATA and Rivet since 2016 lead to improved sensitivity, although the lack of the published 8 TeV and 13 TeV dijet data still weakens the analysis compared to searches with those datasets. It is hoped that these data will soon be available in Rivet. The diagonal structure along the line $M_{Z'} \approx 2M_{\text{DM}}$ is caused by the reduced sensitivity when the Z' decays to DM rather than to jets.

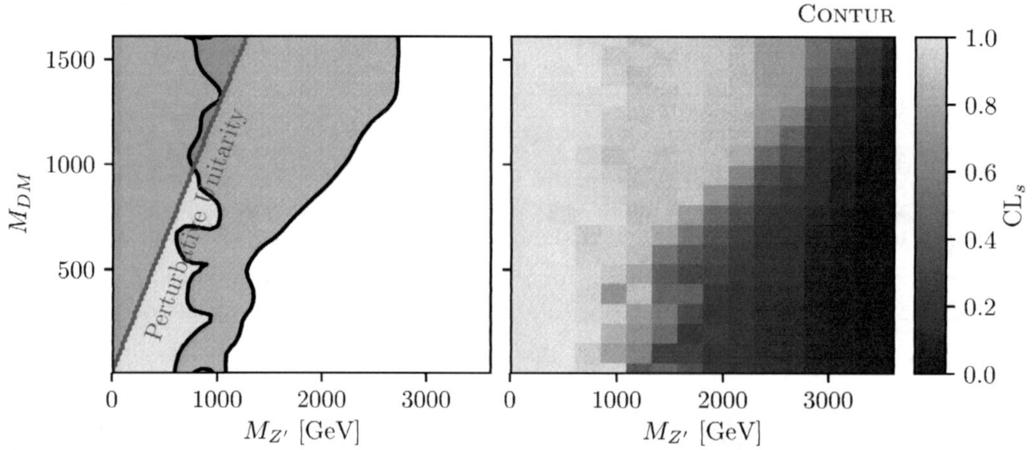

Figure 3. Sensitivity of LHC measurements to a Majorana DM candidate coupling to a spin-1 mediator Z', which in turn couples to first generation quarks with coupling strength $g_q = 0.25$.

A simple modification of this model such that the Z' couples to all three flavours of quarks has also been studied. The main impact of this change is to open up decays to top quark pairs once $M_{Z'} > 2M_{\text{top}}$. Many top measurements from both ATLAS and CMS are available in Rivet for 8 TeV and 13 TeV data, so the lack of dijet data is less important. The improved sensitivity is clearly visible in Figure 4.

This variant, with the Z' coupling to all three generations, is much closer to the benchmark models studied by the experiments, for example in the summaries shown by CMS at ICHEP 2018 (see Figure 5), and the recent ATLAS compilation [13]. The most similar of these models has the same Z' couplings, the only difference being that the DM candidate is now a Dirac fermion.

Exactly this model was also tested with CONTUR, with the result shown in Figure 6. The CONTUR limits are close to those given by the searches, despite the lack of dijet data and generally lower luminosity used in the measurements so far.

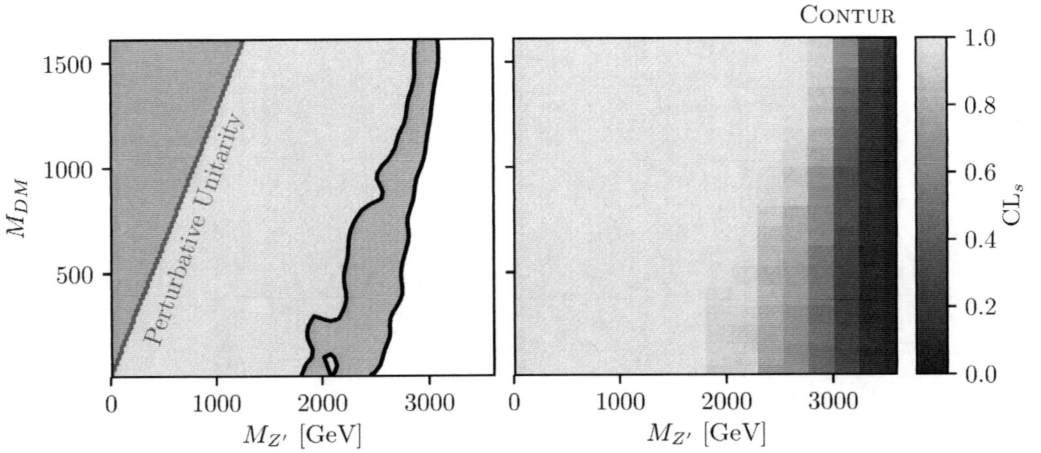

Figure 4. Sensitivity of LHC measurements to a Majorana DM candidate coupling to a spin-1 mediator Z', which in turn couples to all three generations of quarks with coupling strength $g_q = 0.25$.

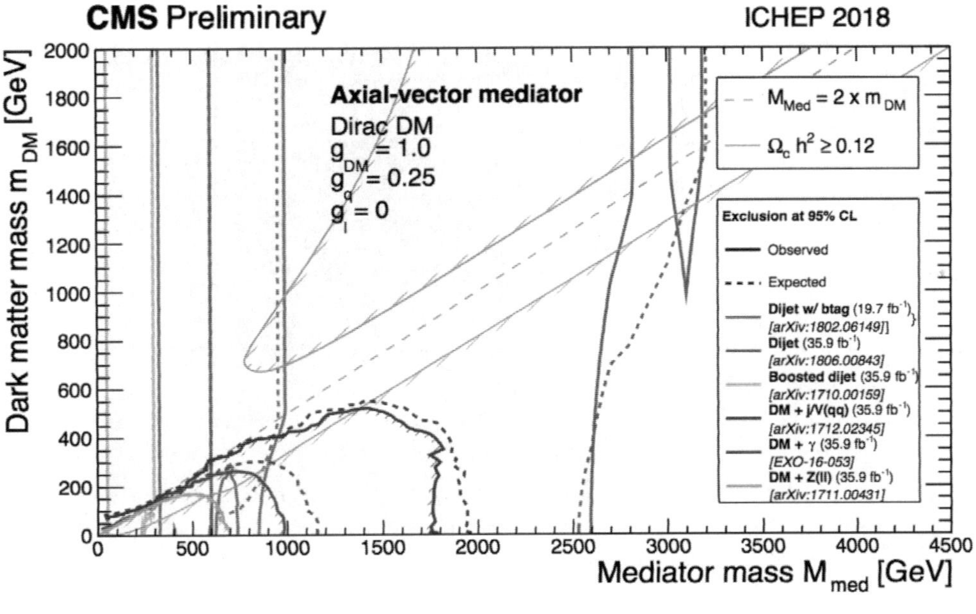

Figure 5. 95% CL observed and expected exclusion regions for dijet searches and missing energy based DM searches from CMS in the lepto-phobic Axial-vector model. Following the recommendation of the LHC DM working group [11, 12] the exclusions are computed for a universal quark coupling of $g_q = 0.25$ and $g_{\rm DM} = 1.0$. It should also be noted that the absolute exclusion of the different searches as well as their relative importance, will strongly depend on the chosen coupling and model scenario. Therefore, the exclusion regions, relic density contours, and unitarity curve shown in this plot are not applicable to other choices of coupling values or model.

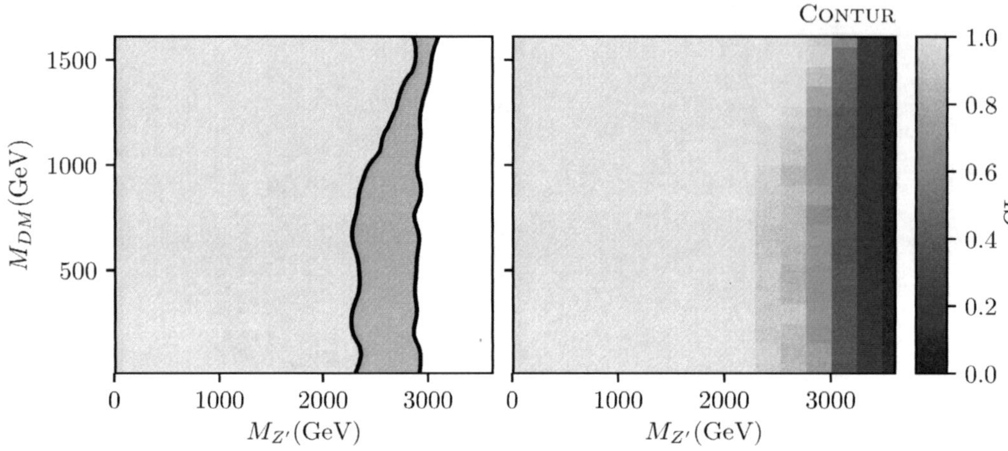

Figure 6. Sensitivity of LHC measurements to a Dirac DM candidate coupling to a spin-1 mediator Z', which in turn couples to all three generations of quarks with coupling strength $g_q = 0.25$.

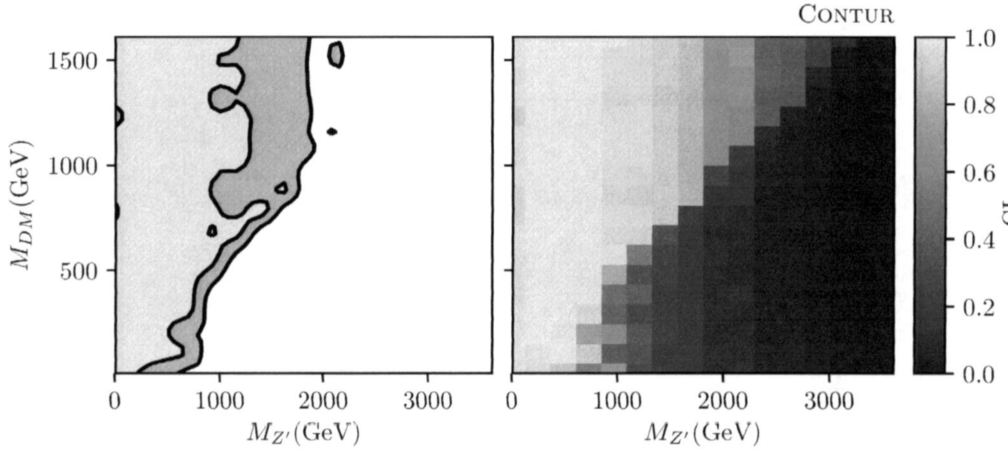

Figure 7. Sensitivity of LHC measurements to a Dirac DM candidate coupling to a spin-1 mediator Z', which in turn couples to all three generations of quarks with coupling strength $g_q = 0.1$ and to leptons with a coupling strength $g_l = 0.01$.

Another variant of the model, studied in [13], reduces g_q to 0.1 but allows a small coupling to leptons, $g_l = 0.01$, opening up dilepton signatures. The CONTUR result for this scenario is shown in Figure 7. The sensitivity is not as strong as that of the searches in [13], because in addition to the lack of 8 and 13 TeV dijets, there is no 13 TeV dilepton measurement yet available in Rivet.

4. Light Scalars

Finally we consider LHC sensitivity to additional light scalar particles. These are a common feature in extensions of the SM, for example appearing in composite Higgs scenarios, or as the radion in models with extra dimensions [14]. Consideration of precision electroweak measurements, collider searches and flavour physics does not completely exclude the existence of light neutral CP-odd or CP-even scalar particles below the mass of the observed Higgs boson[15]. A CP-even scalar can for example be identified as the radion mode present in warped extra-dimension models with bulk gauge fields. A CP-odd scalar is typically a pseudo Nambu Goldstone boson from an approximate global symmetry, just like those appearing in composite Higgs models. The couplings to gauge fields are induced by the many fermion resonances populating the TeV scale (see e.g [16] or also [17]).

As part of the 2017 Les Houches workshop on TeV scale physics [18], a simplified model was used to examine whether measurements at the LHC can give information about such possible particles. Here we present updated results of that study.

The study uses an effective theory approach to describe a scalar with mass M_ϕ interacting with gauge bosons. The effective theory has $SU(2) \times U(1)_Y$ symmetry. When the scalar is light, with a mass below the electroweak scale, we assume that it has large tree-level $SU(2) \times U(1)_Y$ couplings, so that the loop-induced electroweak-breaking contributions are subleading. Under these conditions the interactions of a CP-even and CP-odd scalars with gauge bosons are described by dimension-5 effective Lagrangians. Mixing with the SM Higgs is assumed to be small to ensure that the SM Higgs has SM-like couplings compatible with measurements. A common scale Λ for all couplings was assumed. More details are given in [18].

The measurements of interest are those involving isolated photons, or pairs of photons, in the final state. These have been measured inclusively [19, 20, 21], and in association with jets [22, 23, 24], W or Z bosons[25, 26] (i.e. leptons and/or missing energy). At low M_ϕ and low-ish Λ, one of the most sensitive measurements is the $\gamma + E_T^{\text{miss}}$ measurement from [25]. The Higgs fiducial diphoton measurements [27] are also of interest. These were studied and in principle have some sensitivity – events generated by the models considered do contribute to the fiducial region. However, since the value of M_ϕ considered here lie below the SM Higgs mass, the events which will enter the fiducial phase space of the Higgs measurement will arise from combinatorial backgrounds of pairs of photons, and thus will not exhibit a peak at the Higgs mass. Because of this, they are likely to removed as part of the background fitting and subtraction process in that analysis. We therefore do not include the Higgs cross sections when calculating the sensitivity.

A study carried out around the same time as the Les Houches study [28] using lower energy data as well as LHC data, but not the LHC boson+γ data, obtains similar results, although due to differences in the definition of the couplings a precise comparison has not been carried out.

Figure 8 shows the LHC sensitivity in the Λ vs M_ϕ plane for the CP-even scalar. Figure 9 the equivalent sensitivity for the CP-even scalar.

Dependent on M_ϕ, the Λ values up to 4.5 to 10.5 TeV are excluded, under the assumptions of our procedure, for the CP-even scalar. For the CP-odd scalar, the maximum reach is around 8.5 TeV in Λ.

5. Summary and Future Plans

The Contur approach exposes significant sensitivity to SM extensions in unfolded particle-level measurements. Where dedicated searches for these models have already been performed, see [29, 30], Contur has similar sensitivity if the same data set is used; however, in several cases it lags behind either because the measurements have not yet been made, or because they have not been made available in HEPDATA and Rivet. For BSM scenarios which have not been considered by dedicated searches, Contur provides an efficient way to identify models and regions of parameter space which are disfavoured by existing data, and those which remain of interest. The ability to consider the exact, inclusive phenomenology of a new model means that Contur smoothly transitions between different signals as the parameters of the model change the dominant processes.

At present, Contur does not make use of the full correlation information which is available

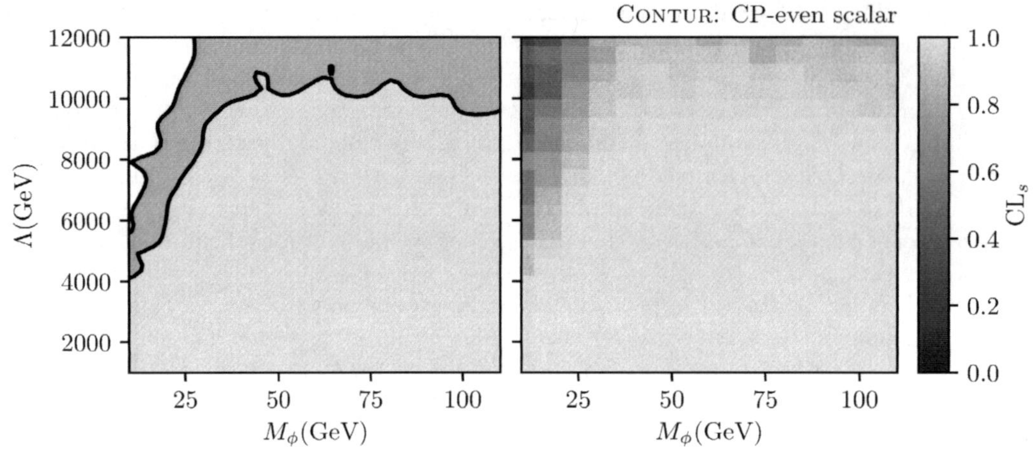

Figure 8. Sensitivity of LHC measurements to a light CP-even scalar particle ϕ decaying to photons.

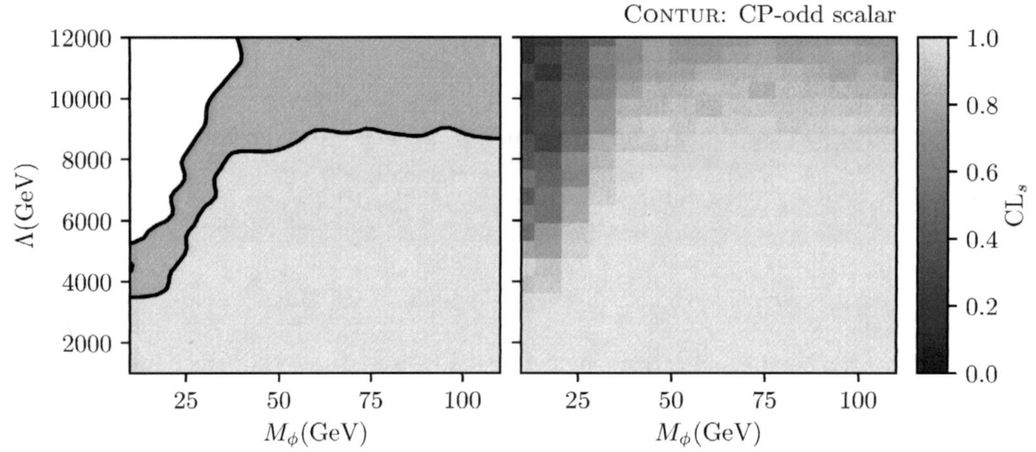

Figure 9. Sensitivity of LHC measurements to a light CP-odd scalar particle ϕ decaying to photons.

for several of the measurements considered. The intention is to do so, which should increase the sensitivity is several cases. Also, in these studies we have taken the data – which are consistent with the SM – to be exactly equal to the SM, looking for whether the BSM contributions would have made a visible difference compared to the experimental uncertainties. In this mode, CONTUR can only ever exclude BSM scenarios, and where the SM theory uncertainty is large, it may in fact overestimate the exclusion. It is however possible to make use of precision final-state SM calculations, which means that in future CONTUR could potentially identify BSM scenarios which describe the data *better* than the SM.

More results and updates are available at https://contur.hepforge.org .

Acknowledgments
This work has received funding from the European Union's Horizon 2020 research and innovation programme as part of the Marie Sklodowska-Curie Innovative Training Network MCnetITN3

(grant agreement no. 722104). I'd like to thank my CONTUR co-authors, especially David Yallup, as well as Peter Richardson who provided impressive support with the Herwig UFO interface. Finally, I would like to thank the conference organisers for the invitation and for hosting a really stimulating meeting, and especially Deepak Kar for his hospitality at University of the Witwatersrand.

References

[1] Butterworth J M, Grellscheid D, Krämer M, Sarrazin B and Yallup D 2017 *JHEP* **03** 078 (*Preprint* 1606.05296)
[2] Alloul A, Christensen N D, Degrande C, Duhr C and Fuks B 2014 *Comput. Phys. Commun.* **185** 2250–2300 (*Preprint* 1310.1921)
[3] Degrande C, Duhr C, Fuks B, Grellscheid D, Mattelaer O and Reiter T 2012 *Comput. Phys. Commun.* **183** 1201–1214 (*Preprint* 1108.2040)
[4] Buckley A *et al.* 2011 *Phys. Rept.* **504** 145–233 (*Preprint* 1101.2599)
[5] Bellm J *et al.* 2016 *Eur. Phys. J.* **C76** 196 (*Preprint* 1512.01178)
[6] Maguire E, Heinrich L and Watt G 2017 *J. Phys. Conf. Ser.* **898** 102006 (*Preprint* 1704.05473)
[7] Buckley A, Butterworth J, Lonnblad L, Grellscheid D, Hoeth H *et al.* 2013 *Comput.Phys.Commun.* **184** 2803–2819 (*Preprint* 1003.0694)
[8] Bellm J *et al.* 2017 (*Preprint* 1705.06919)
[9] Deppisch F F, Liu W and Mitra M 2018 *JHEP* **08** 181 (*Preprint* 1804.04075)
[10] Amrith S, Butterworth J M, Deppisch F F, Liu W, Varma A and Yallup D 2018 LHC Constraints on a $B-L$ Gauge Model using Contur Tech. rep. UCL (*Preprint* 1811.11452)
[11] Busoni G *et al.* 2016 Recommendations on presenting LHC searches for missing transverse energy signals using simplified s-channel models of dark matter Tech. rep. CERN (*Preprint* 1603.04156)
[12] Albert A *et al.* 2017 Recommendations of the LHC Dark Matter Working Group: Comparing LHC searches for heavy mediators of dark matter production in visible and invisible decay channels Tech. rep. CERN (*Preprint* 1703.05703)
[13] Aad G *et al.* (ATLAS Collaboration) 2018 Constraints on mediator-based dark matter models using $\sqrt{s} = 13$ TeV pp collisions at the LHC with the ATLAS detector Tech. Rep. ATLAS-CONF-2018-051 CERN Geneva URL https://cds.cern.ch/record/2646248
[14] Angelescu A, Moreau G and Richard F 2017 *Phys. Rev.* **D96** 015019 (*Preprint* 1702.03984)
[15] Cacciapaglia G, Deandrea A, Gascon-Shotkin S, Le Corre S, Lethuillier M and Tao J 2016 *JHEP* **12** 068 (*Preprint* 1607.08653)
[16] Belyaev A, Cacciapaglia G, Cai H, Ferretti G, Flacke T, Parolini A and Serodio H 2017 *JHEP* **01** 094 (*Preprint* 1610.06591)
[17] Fichet S, von Gersdorff G, Ponton E and Rosenfeld R 2016 *JHEP* **09** 158 (*Preprint* 1607.03125)
[18] Brooijmans G *et al.* 2018 *Les Houches 2017: Physics at TeV Colliders New Physics Working Group Report* (*Preprint* 1803.10379) URL http://lss.fnal.gov/archive/2017/conf/fermilab-conf-17-664-ppd.pdf
[19] Aad G *et al.* (ATLAS) 2013 *JHEP* **01** 086 (*Preprint* 1211.1913)
[20] Aad G *et al.* (ATLAS) 2014 *Phys. Rev.* **D89** 052004 (*Preprint* 1311.1440)
[21] Aad G *et al.* (ATLAS) 2016 *JHEP* **08** 005 (*Preprint* 1605.03495)
[22] Chatrchyan S *et al.* (CMS) 2014 *JHEP* **06** 009 (*Preprint* 1311.6141)
[23] Aad G *et al.* (ATLAS) 2013 *Nucl. Phys.* **B875** 483–535 (*Preprint* 1307.6795)
[24] Aad G *et al.* (ATLAS) 2012 *Phys. Rev.* **D85** 092014 (*Preprint* 1203.3161)
[25] Aad G *et al.* (ATLAS) 2013 *Phys. Rev.* **D87** 112003 [Erratum: Phys. Rev.D91,no.11,119901(2015)] (*Preprint* 1302.1283)
[26] Aad G *et al.* (ATLAS) 2016 *Phys. Rev.* **D93** 112002 (*Preprint* 1604.05232)
[27] Aad G *et al.* (ATLAS) 2014 *JHEP* **09** 112 (*Preprint* 1407.4222)
[28] Mariotti A, Redigolo D, Sala F and Tobioka K 2018 *Phys. Lett.* **B783** 13–18 (*Preprint* 1710.01743)
[29] Whalen K these proceedings
[30] Pandolfi F these proceedings

IOP Conf. Series: Journal of Physics: Conf. Series **1271** (2019) 012014 doi:10.1088/1742-6596/1271/1/012014

Composite Models on a safe road to the Planck scale

Giacomo Cacciapaglia[1,2], **Teng Ma**[3] **and Yongcheng Wu**[4]

[1] Université de Lyon, Lyon, France; Université Lyon 1, Villeurbanne, France
[2] Institut de Physique Nucléaire de Lyon, UMR5822, CNRS/IN2P3, Villeurbanne, France
[3] CAS Key Laboratory of Theoretical Physics, Institute of Theoretical Physics, Chinese Academy of Sciences, Beijing 100190, China
[4] Ottawa-Carleton Institute for Physics, Carleton University, 1125 Colonel By Drive, Ottawa, Ontario K1S 5B6, Canada

E-mail: g.cacciapaglia@ipnl.in2p3.fr

Abstract. We present the first serious attempt to define a model of composite pseudo-Nambu-Goldstone Higgs with partial compositeness for all standard fermions that is valid up to the Planck scale. The main ingredient is the presence of a large multiplicity of fermions in the microscopic gauge-fermion description, which allows us to use large-N_f techniques to show the presence of an interactive UV fixed point for the gauge couplings (UV safety). We also present results for the Dark Matter relic density and direct detection in the example model that is UV-completed.

1. Introduction

The possibility that the electroweak symmetry breaking in the Standard Model (SM) is dynamically generated by a confining and condensing strong interaction is still a valid alternative to the Higgs mechanism in the SM. The latter is described in terms of an elementary scalar field, with all the drawbacks associated to it (hierarchy, non-explanation of the negative mass squared, *etc.*). In dynamical models, the scalar sector is replaced by a new strong sector: this idea is in fact as old as the SM itself [1, 2]. In the original proposal, heavily based on Quantum Chromodynamics (QCD), no light scalar arose in the spectrum, thus this possibility is ruled out by the recent discovery of a 125 GeV Higgs-like boson. On the other hand, a light Higgs boson can still be present arising as a pseudo-Nambu-Goldstone boson (pNGB) of a broad global symmetry breaking pattern [3]. This mechanism, first proposed in the early 80's, has received a revival with the discovery of the holographic principle [4] that links a strongly interacting conformal field theory to extra dimensions. Thus, the Higgs arises as a gauge field in the holographic picture [5], in analogy to models of Gauge-Higgs Unification [6].

Following the holographic approach, most of the recent work has been focusing on effective field theory approaches [7, 8], where little attention is given to the microscopic origin of the confining dynamics. The UV completion of the model is thus assumed to be either a conformal theory of unknown origin, or extra dimensions. An alternative, inspired by QCD, is to describe an underlying microscopic model in terms of gauge and fermion degrees of freedom. By doing so, one first realises that non-minimal cosets in terms of number of pNGBs are the norm. For instance, the minimal allowed coset would be $SU(4)/Sp(4)$ [9], which contains 5 pNGBs organised as a Higgs doublet (4 degrees of freedom) plus a gauge singlet. All the allowed cosets fall in 3 classes: $SU(N)/Sp(N)$, with N even, if the fermions are in a pseudo-real irreducible representation (*irrep*) of the confining gauge force, $SU(N)/SO(N)$ if they are in a real *irrep*, and finally $SU(N) \times SU(N)/SU(N)$ if they are in a complex *irrep*. This observation strongly limits the possible cosets to study.

Another ingredient that came into play, and is important in model building, is the concept of fermion partial compositeness [10]. In this paradigm, the SM fermions acquire their mass by coupling linearly to some fermionic operators in the composite sector. This needs to be compared to the older idea to couple them in bi-linear operators, and it was introduced to give a generic solution to the problem of potentially dangerous flavour changing neutral currents. In the holographic approach partial compositeness materialises in the fact that the SM fermions are the lightest states of a bulk propagating fermion, thus their localisation in the extra dimension determines the degree of compositeness [11, 12]. In underlying gauge-fermion theories, this mechanism requires the presence of four-fermion couplings between one SM fermion and fermions in the strong sector, which have the proper quantum numbers. This calls for the issue of assigning QCD quantum numbers to the strong sector in order to allow couplings to the quarks.

In these proceedings we will focus on models that feature an underlying gauge-fermion description, as described above. In particular, we will focus on two aspects: the possibility to UV complete such models in order to give them a microscopic description valid up to the Planck scale, and the possibility to feature a Dark Matter candidate among the additional pNGBs. As we will see, both features will heavily rely on the underlying gauge-fermion description.

About the first point, attempts to build *"UV completions"* were put forward in Refs [13, 14, 15]: however, they are not genuine UV completions but simple *underlying descriptions*. The main reason is that no attempt is given to explaining the origin of the couplings responsible for top partial compositeness, thus the presence of four-fermion interactions imposes a cut-off to the theory that is very close to the one of the effective description in terms of pNGBs. Nevertheless, such descriptions have the benefit of lattice studies that can shed some light on the low-energy properties of the models [16, 17, 18, 19]. One interesting idea emerged in Ref. [13]: top partial compositeness can be accommodated by adding a second species of fermions, which transform under a different *irrep* than those condensing in the Higgs sector. In this way, QCD interactions are sequestered to the new sector and do not interfere with the electroweak symmetry breaking. [1] Here we will go beyond this approach and try to genuinely define the theory above the validity of the effective pNGB description: our approach will be based on large N_f resummation techniques, see Ref. [21, 22, 23]. In fact, extending partial compositeness to all fermions of the SM necessarily implies the presence of a large number of fermions in the second sector. This naturally drives the theory towards a UV interacting fixed point, which renders the model valid up to arbitrary scales.

The second point is the presence of a Dark Matter candidate. As already mentioned, underlying gauge-fermion descriptions naturally predict the presence of non-minimal cosets, which include additional pNGBs. In the minimal model, $SU(4)/Sp(4)$, the additional singlet has been the first Dark Matter candidate of this kind [24], however it decays via topological anomalies. The most minimal model that features a Dark matter candidate is the minimal coset with complex *irreps*, based on $SU(4) \times SU(4)/SU(4)$ [25]. The properties of the Dark matter candidate in the case of bilinear top mass terms have been studied in Ref. [26]. Other examples include $SU(6)/Sp(6)$ [27] and $SU(6)/SO(6)$ [28]. For completeness, works considering cosets that do not have a simple underlying gauge-fermion description have also been considered in Refs [29, 30, 31, 32].

[1] The only exception is a QCD-like model in Ref. [20].

Table 1. Minimal cosets with a pNGB Higgs doublet arising from an underlying gauge-fermion theory. The fourth column shows the $SU(2)_L$ *irrep*, with the hypercharge as subscript. The last three columns show some properties of the explicit models, with the nomenclature M1-M12 from Ref. [33], and MV being the model from Ref. [20].

ψ irrep	coset	pNGBs	EW charges	models	$\mathcal{G}_{\rm HC}$	Lattice results
pseudo-real	$\frac{SU(4)}{Sp(4)}$	5	$2_{\pm 1/2} \oplus 1_0$	M8-M9	$Sp(4)$, $SO(11)$	$Sp(4)$
real	$\frac{SU(5)}{SO(5)}$	14	$2_{\pm 1/2} \oplus 1_0$ $\oplus 3_{\pm 1} \oplus 3_0$	M1-M7	$SU(4)$, $Sp(4)$, $SO(7)$, $SO(9)$, $SO(10)$	$SU(4)$, $Sp(4)$
complex	$\frac{SU(4)\times SU(4)}{SU(4)}$	15	$2 \times 2_{\pm 1/2} \oplus 3_0$ $\oplus 1_{\pm 1} \oplus 2 \times 1_0$	M10-M12 MV	$SO(10)$, $SU(4)$, $SU(5)$ $SU(3)$	$SU(4)$ $SU(3)$

2. Running safely to the Planck scale

As already mentioned, our analysis is based on the low energy models listed in Ref. [13] (see also Ref. [33]). These models were designed to give underlying fermion-gauge descriptions to the interactions that give rise to the composite pNGB Higgs and top partial compositeness. The QCD interactions are sequestered to a second sector of underlying fermions, χ, that transform under a different *irrep* under the confining Hypercolour (HC) than the underlying fermions, ψ, generating the Higgs. A complete, but brief, list of models is presented in table 1. We also included the model of Ref. [20] where ψ and χ transform under the same *irrep*. The top partners, i.e. the operators that couple to the top fields, arise as bound states of the form $\psi\psi\chi$ or $\psi\chi\chi$, depending on the specific model.

One limitation of these models is the fact that they can accommodate only for the generation of the top mass. The reason is that many more underlying fermions need to be added in order to include operators that can couple to the bottom, and also to leptons and the other two generations. Thus, the model will feature too many fermionic degrees of freedom and asymptotic freedom is lost. In other words, the theory will not confine at low energies. In our work in Ref. [34] we turned this drawback into a benefit: in fact, adding a fermion χ for each SM fermion (i.e., allowing partial compositeness for all SM fermions) increases dramatically the fermionic degrees of freedom to the point that we can employ large N_f techniques to resum their effect on the running of the gauge couplings. It has been observed in the literature that the resummed beta function allows for a zero in the UV, thus there is a strong hint of the presence of a UV interacting fixed point [35]. In turn this implies that our models also feature an interacting UV fixed point, thus they can be trusted up to arbitrarily high scales.

While this picture sounds simple, it is not that easy to realise it in practice. We will present here an explicit example with model M10, motivated by the fact that it has a pNGB Dark Matter candidate. The model is based on the HC group $SO(10)_{\rm HC}$, and the underlying fermions charged under $SU(10)_{\rm HC}$ are listed in table 2. The underlying fermions that characterise M10 (i.e., the low energy part of it) are the ones with ≈ 0 mass. They will make sure that the theory confines and generates a pNGB effective theory below the scale $\Lambda_{\rm HC} \approx 4\pi f$, where f is the decay constant of the composite Higgs. As a reference, we can keep in mind the following scales: $f \approx 1$ TeV (in order to escape bounds from electroweak precision tests) and $\Lambda_{\rm HC} \approx 10$ TeV.

At the scale $\Lambda_{\rm HC}$, we introduce two more χ's, with the quantum numbers that allow for the generation of partial compositeness for the bottom quark and tau lepton (thus completing the third generation). Their presence pushes the theory inside an IR conformal window [36, 37, 38], preferably at strong coupling. The role of this walking region [39] is to generate a large enough

IOP Conf. Series: Journal of Physics: Conf. Series **1271** (2019) 012014 doi:10.1088/1742-6596/1271/1/012014

Table 2. UV-completed model M10 - all fermions are Dirac spinors.

	$SO(10)_{\mathrm{HC}}$	$SU(3)_c$	$SU(2)_L$	$U(1)_Y$	mass
ψ_L	16	1	2	0	
$\psi_R^{u,d}$	16	1	1	$\pm 1/2$	~ 0
χ_u^3	10	3	1	$2/3$	
χ_d^3	10	3	1	$-1/3$	$\sim \Lambda_{\mathrm{HC}}$
χ_l^3		1	1	-1	
$\chi_u^{1,2}$		3	1	$2/3$	
$\chi_d^{1,2}$	10	3	1	$-1/3$	$\Lambda_{\mathrm{FI}} \gg \Lambda_{\mathrm{HC}}$
$\chi_l^{1,2}$		1	1	-1	

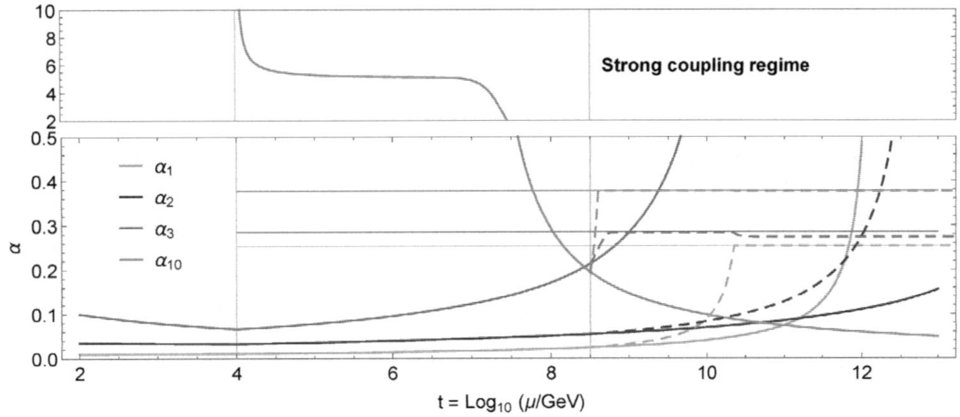

Figure 1. Plot of the renormalisation group running of the gauge couplings, α_i, above the electroweak scale. The continuous lines correspond to one-loop running, while the dashed lines show the large-N_f resummation above $\Lambda_{\mathrm{FI}} = 10^{8.5}$ GeV. The upper panel contains the running of α_{10} at strong coupling. Note that α_2 also runs towards the UV fixed point $\alpha_2 = 0.86$ (not shown in the plot).

split between the scale where the partial compositeness couplings are generated and the scale of compositeness. At a much higher scale, Λ_{FI}, we introduced two more copies of the χ's, in order to generate partial compositeness for all generations. Now, the number of fermions is such that large-N_f resummation can be employed for all gauge couplings. The running, without large-N_f, is shown in figure 1 by the solid lines. The green one corresponds to the confining $SO(10)_{\mathrm{HC}}$, with the strong coupling walking represented as a sketch because a perturbative calculation in that regime is clearly not trustworthy. Lattice studies will therefore be necessary to study this region and the validity and duration of the walking regime [16, 17, 18, 19]. It is clear that all the gauge couplings start growing towards the UV once the third-generation partial compositeness is introduced. This growth will be tamed by large-N_f resummation.

2.1. Large-N_f results

Contrary to calculations already present in the literature [40, 41, 42, 43], where N_f refers to the multiplicity of a single fermion type, in our case we have fermions that have different quantum numbers under the 4 gauge groups of the model. Thus, we need to define a different effective N_f for each group and resum them all at the same order.

For each gauge group, therefore, we define a normalised coupling [35] that takes into account the multiplicity of fermions f charged under it:

$$K_i \equiv N_i T_i \frac{\alpha_i}{\pi} = \frac{\alpha_i}{\pi} \sum_{\mathrm{f}} n_{\mathrm{f}} T(r_{\mathrm{f}})\,; \tag{1}$$

with $i = 1, 2, 3, 10$ labelling the 4 gauge groups. We fix the overall normalisation with the fundamental *irreps*, i.e. $T_1 = T_{10} = 1$ and $T_2 = T_3 = 1/2$. Moreover, for $U(1)$ it suffices to replace in the formula $T(r_{\mathrm{f}}) \to Y_{\mathrm{f}}^2$. The multiplicities N_i, that vary with each group, are all considered formally to be of order "N_f", thus we can resum them at the same time and at the same order. For the UV-complete model M10, we find the following values:

$$N_1 = 93\,, \quad N_2 = 22\,, \quad N_3 = 66\,, \quad N_{10} = 25\,. \tag{2}$$

The resummed evolution equations, keeping all terms in the expansion, in general read

$$\frac{\partial \ln K_i}{\partial \ln \mu} \equiv \beta_i(K_i) = \frac{2K_i}{3}\left[1 + \sum_n \frac{1}{N_f^n} B_i^{(n)}\right]\,, \tag{3}$$

with the first-order terms equal to

$$\frac{B_i^{(1)}}{N_f} = \frac{1}{N_i}\left[c_{i,i} H_1(K_i) + \sum_{j\neq i} c_{i,j} F_1(K_j)\right]\,, \quad \text{for } i = 2, 3, 10\,; \tag{4}$$

and

$$\frac{B_1^{(1)}}{N_f} = \frac{1}{N_1}\left[c_{1,1} F_1(K_1) + \sum_{j\neq 1} c_{1,j} F_1(K_j)\right]\,. \tag{5}$$

The coefficients $c_{i,j}$ can be computed numerically in terms of the quantum numbers of the fermions. For M10, this yields

$$\begin{array}{llll}
c_{1,1} = \frac{1927}{3348}\,, & c_{1,2} = \frac{3}{44}\,, & c_{1,3} = \frac{211}{99}\,, & c_{1,10} = 15\,; \\
c_{2,1} = \frac{1}{186}\,, & c_{2,2} = \frac{3}{2}\,, & c_{2,3} = \frac{2}{11}\,, & c_{2,10} = \frac{9}{5}\,; \\
c_{3,1} = \frac{211}{1116}\,, & c_{3,2} = \frac{9}{44}\,, & c_{3,3} = \frac{8}{3}\,, & c_{3,10} = \frac{54}{5}\,; \\
c_{10,1} = \frac{3}{31}\,, & c_{10,2} = \frac{3}{11}\,, & c_{10,3} = \frac{8}{11}\,, & c_{10,10} = \frac{9}{2}\,.
\end{array} \tag{6}$$

The functions H_1 and F_1 are generated by the resummation of fermion bubbles on the gauge propagators of the two-loop diagrams, and their explicit expression can be found in Ref. [41]. They have a special property: they feature a pole above which the resummation does not converge, and the pole drives the function towards negative values. This implies that near the pole the beta function at leading order will vanish. The poles, therefore, act as an effective barrier that does not allow the gauge couplings to grow beyond such a point in the UV. This

IOP Conf. Series: Journal of Physics: Conf. Series **1271** (2019) 012014 doi:10.1088/1742-6596/1271/1/012014

property is at the origin of the postulated UV fixed point. For the two functions relevant here, the poles stand at:

$$H_1(K^\star) \to -\infty \text{ for } K^\star = 3\,, \qquad F_1(K^\star) \to -\infty \text{ for } K^\star = \frac{15}{2}\,. \tag{7}$$

These barriers will thus drive our model to a complete UV fixed point only if the value of the couplings at the threshold $\Lambda_{\rm Fl}$ are below the pole: the evolution towards the UV will stop at that value where the beta function vanishes and the theory approaches a fixed point. Numerically, in M10 this implies:

$$\begin{aligned} &\alpha_1(\Lambda_{\rm Fl}) < 0.25\,,\ \alpha_2(\Lambda_{\rm Fl}) < 0.86\,, \\ &\alpha_3(\Lambda_{\rm Fl}) < 0.28\,,\ \alpha_{10}(\Lambda_{\rm Fl}) < 0.38\,. \end{aligned} \tag{8}$$

Satisfying the above constraints provides an upper bound on $\Lambda_{\rm Fl}$ as all the gauge couplings (except α_{10}) increase towards the UV above $\Lambda_{\rm HC}$. We should also remind the reader that $\Lambda_{\rm Fl}$ cannot be too low, otherwise we risk generating dangerous flavour-changing neutral currents when the partial compositeness four-fermion interactions are generated. Thus, we will generically require $\Lambda > 10^5$ TeV for generic flavour violating effects.

We show in figure 1 the one-loop running of the 4 gauge couplings above the electroweak scale, assuming $\Lambda_{\rm HC} = 10$ TeV for the model in table 2. In this case, the coupling that first crosses the upper limit in Eq. (8) is the QCD one α_3, shown in red. The thin horizontal line, representing the bound, is crossed at 10^9 GeV, which is therefore the upper limit on the value of $\Lambda_{\rm Fl}$. We thus add the χ's for the light generations at $\Lambda_{\rm Fl} = 10^{8.5}$ GeV and plot in dashed lines the running after the large-N_f resummation. We can see that all the couplings run to the UV fixed point (for $SU(2)$ it is not shown in the plot because of the larger value). The behaviour of α_3 after $U(1)$ saturates the asymptotic value is a numerical artefact, nevertheless it shows the impact of the $U(1)$ running on the other gauge couplings as a large $F_1(\alpha_1)$ will affect all the β-functions in Eq. (5). This example shows that a completely UV-safe composite Higgs and Dark Matter model with partial compositeness is indeed feasible.

3. Dark Matter phenomenology

The low energy description of this model contains electroweakly charged pNGBs generated by the condensation of the ψ underlying fermions. Out of the cosets shown in table 1, only the $SU(4)^2/SU(4)$ one enjoys a potentially stable pNGB Dark Matter candidate. The parity, which is a combination of charge conjugation and an $SU(4)$ rotation, has been studied in Ref. [25]. While the phenomenology of the model with bi-linear mass operators has been studied in Ref. [26], here we will focus on the case where the top mass is generated via partial compositeness. We will thus focus on two scenarios: one based on a holographic model and one based on the UV-complete model discussed in the previous section.

In both cases, the pNGBs that are odd under the dark parity transform as a second Higgs doublet, a real triplet, a charged and a neutral singlet. In general, the lightest state will be a mixture of the three neutral components. We will here discuss the general strategy we employed in the calculation, and show the explicit results in the following sections.

For the top partial compositeness, after appropriately choosing the representation of the fermionic operator, we write down the linear mixing of the right-handed top and of the left-handed doublet. This leaves two free parameters, i.e. the two pre-Yukawas, which are partly fixed by the value of the top mass. At one loop level, this mixing will also generate a potential for the pNGBs, including the Higgs. Here we use the standard technique of computing the loops and imposing maximal symmetry [44] to guarantee the finiteness of the contribution. Note that there is an alternative way, namely computing a basis of operators generated by the same pre-Yukawa, treated as spurions [45]. This second technique would be more appropriate if the top partners, i.e. the massive resonances associated with the fermionic operators, were heavy and/or strongly coupled. In fact, the loop calculation can only be trusted in a regime of light and weakly coupled composite fermions [46], which is not always the case in strongly interacting dynamics. It should also be noticed that the two techniques do not necessarily lead to the same results, as it was shown in a specific case in Ref. [47].

In this work, we also considered the contribution of the gauge loops and of bare masses for the technifermions, see Ref. [25]. In particular, the masses play an important role in determining the nature of the lightest stable pNGB. There are two possible masses that can be added: one for the left-handed doublet ψ_L and one for the right-handed doublet ψ_R (see table 2). These two masses explicitly break $SU(4)^2 \to SU(2)_L \times SU(2)_R$ (if $m_{\psi_L} = \pm m_{\psi_R}$, then a global $SU(4)$ is preserved). We can define, therefore, an asymmetry between the two as

$$\delta = \frac{m_{\psi_L} - m_{\psi_R}}{m_{\psi_L} + m_{\psi_R}}. \tag{9}$$

We observe that, in general, the lightest neutral state is dominantly the $SU(2)_L$ real triplet for $\delta < 0$, and the singlet for $\delta > 0$, while an equal mixture occurs for $\delta \sim 0$. Interestingly this pattern does not depend crucially on the top mass mechanism.

To partly fix the free parameters of the model, we first impose that the electroweak scale is correctly generated at the minimum of the potential and then impose the correct value of the Higgs and top masses. This fixes only 3 of the free parameters of the model. We then scan over the remaining ones to check for points that satisfy the correct relic abundance.

3.1. Results for a holographic model

The first case we analyse corresponds to a model inspired by holography. In fact, in order to preserve the dark parity, we need to choose an appropriate representation for the top partners under the global $SU(4) \times SU(4)$ symmetry. The simplest choice is a representation that is real under charge conjugation, i.e. $(4, \bar{4})$. With this choice, the dark parity maps this representation on itself as it acts as a transpose on the $SU(4)^2$ matrix. As we will see in the next section, this choice is not possible in the UV complete models studied in the previous section.

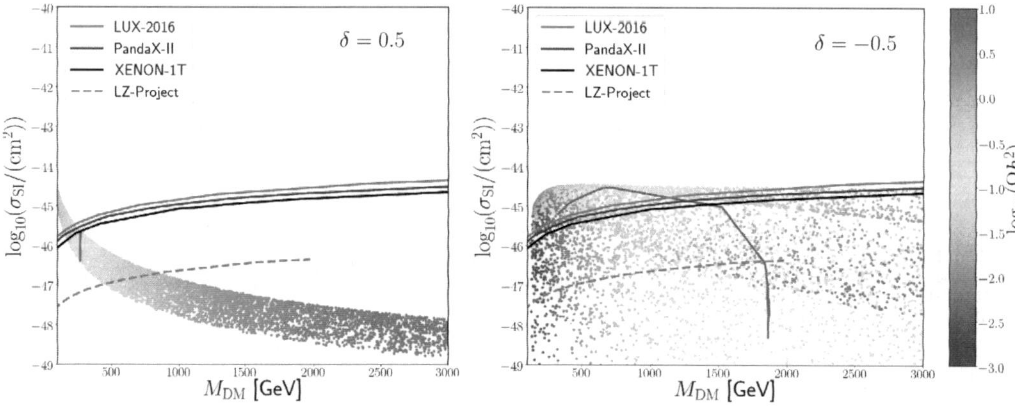

Figure 2. Scan of the re-weighted Spin-Independent cross section compared to current and future Direct Detection constraints [48, 49, 50] for $\delta = 0.5$ (left) and $\delta = -0.5$ (right). The colour encodes the value of the relic density, where the green line indicates the points saturating the Planck value. [Plots from the first version of Ref. [34]]

IOP Conf. Series: Journal of Physics: Conf. Series **1271** (2019) 012014 doi:10.1088/1742-6596/1271/1/012014

The results of the scan are shown in figure 2, where we show the Spin-Independent cross section re-weighted to the correct value of the relic density versus Direct Detection exclusions. The colour of the points encodes the value of the relic density, with the green lines marking points that saturate the Planck bound. Points that tend to the blue hues are under-dense, thus not excluded. We see that in the case of the dominantly singlet ($\delta > 0$), the relic density is saturated by a relatively low mass, with $M_{DM} \approx 250$ GeV. This is due to the typically low annihilation cross section, which is dominated by $\eta\eta \to \gamma\gamma, ZZ, W^+W^-$ channels. Direct Detection just touches the allowed region, which will however be completely excluded by the next generation experiments. Note that Direct Detection is only due to a trilinear coupling of the scalar Dark Matter candidate to the Higgs.

For $\delta < 0$, the Dark Matter candidate is dominantly a triplet. While a low mass region starting at the same value is present, with annihilation again dominantly in W^+W^-, for points where the $\eta\eta \to t\bar{t}$ channel is dominant larger masses up to 1.8 TeV are needed. However, Direct Detection (XENON1T) already excludes masses up to 1.5 TeV, thus only the high mass end is left available.

Note that the value of f is not fixed here, as it is correlated to the value of the Dark Matter mass. In the scan we have $f = v/\sin\theta$, with θ varying within the range $[0.003, 0.3]$. The preferred values of the masses fall in the range that is typically preferred by electroweak precision bounds too.

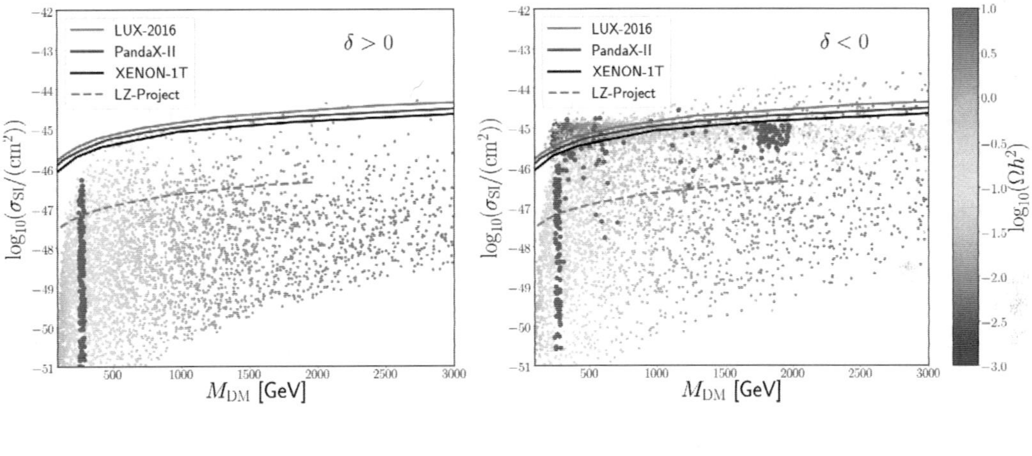

Figure 3. Same as figure 2, but for the UV complete model M10. [Plots from the revised version of Ref. [34]]

3.2. Results for the safe model

We now turn our attention to the UV-complete model M10 (in fact these results also apply to M11). In this case, the representation of the top partners cannot be chosen to be $(4, \bar{4})$ due to the quantum numbers of the underlying fermions. In fact, to make a baryon $\psi\psi\chi$, we need to combine $16 \times 16 \times 10$, while the vacuum is characterised by $16 \times \overline{16}$. It turns out, therefore, that the allowed representations are $(4, 4) \oplus (\bar{4}, \bar{4})$ or $(6, 1) \oplus (1, 6)$. These representations are clearly not mapped onto themselves by the dark parity if taken individually, thus one needs to make sure that a symmetric coupling to the charge-conjugate representations are also present. [2] While this seems contrived, it is actually a natural choice in terms of the underlying model.

[2] The dark parity would map $(4, 4) \leftrightarrow (\bar{4}, \bar{4})$ and $(6, 1) \leftrightarrow (1, 6)$.

The easiest way to see this is to write the four-fermion couplings in terms of Weyl spinors: we define by a super-script l and r the left- and right-handed components of the Dirac spinors from table 2. In the case of the $(6,1) \oplus (1,6)$ representation, for instance, the four-fermion interactions for the right-handed top t^r would thus read

$$y_1 \frac{(t^{r,C} \chi_u^{3,l})(\psi_R^{u,l} \psi_R^{d,l})}{\Lambda^2} + y_2 \frac{(t^{r,C} \chi_u^{3,l})(\psi_R^{u,r} \psi_R^{d,r})}{\Lambda^2} . \tag{10}$$

The dark parity would be preserved if $y_1 = y_2$. Note that the only difference between the two operators is the chirality of the ψ–spinors, thus a symmetric scenario can be easily achieved if the operator is generated by a scalar mediator that couples to both chiralities of the spinors.

The results of the scan for the choice $(4,4) \oplus (\bar{4},\bar{4})$ is shown in figure 3. Similar to the plot in the previous section, we show points with different hues indicating the relic abundance, with the one tending to blue being under-abundant. The points in dark green saturate the correct value. Once again, we observe that for dominantly singlet Dark Matter candidate, i.e. $\delta > 0$, the points have a light mass, close to $M_{DM} \approx 250$ GeV. However, contrary to the previous case, the SI cross section can be much smaller in this case, thus evading current and future experiments.

For the dominantly triplet case, $\delta < 0$, a similar mass range as before is accessed, between 250 GeV and 2 TeV. However, we see that now the large mass points lie close to the exclusion limit and will be probed by future Direct Detection experiments, while the low mass points allow for very small and undetectable Spin Independent cross sections.

4. Conclusions and Outlook

Models where the electroweak symmetry breaking is dynamically generated and the Higgs emerges as a composite pNGB are still a valid alternative to the elementary Higgs nature of the SM. In these proceedings we have addressed the issue of UV completing this scenario, which is currently dominantly studied at the effective model level. We presented the first serious attempt to define a theory all the way up to the Planck scale.

Our main construction is based on the observation that, in order to generate masses via partial compositeness to all the SM fermions, a microscopic fermion-gauge theory needs to contain a large number of fermions charged under the SM gauge symmetries as well as the confining HC interactions. This in turns allows us to use large-N_f techniques. This resummation hints at the presence of an interacting UV fixed point in the gauge running. As such, the theory can be extrapolated up to arbitrarily high energies. A key constraint on the model building is the fact that none of the gauge couplings should be larger than the fixed value once the large multiplicity of fermions enters in the game. Thus, there is an upper limit on the value of the mass of such fermions. We showed in an explicit example based on a confining $SO(10)_{\text{HC}}$ interaction that this is indeed allowed. We showed that the new fermions can be added at a scale close to 10^9 GeV, above which all gauge couplings run to a safe point. In this work we did not address the origin of the four-fermion interactions responsible for the partial compositeness. However, we showed that they can be easily generated via scalar interactions. As long as the scalar mediator are added close to the highest scale in the model, i.e. the threshold for the many fermions, the theory remains natural (i.e., no unjustified hierarchies are present). The next step in validating this model would be to check that the Yukawa couplings between the scalar mediators and the underlying fermions also run to a UV fixed point. This will be the scope of our future explorations.

We have also addressed the phenomenology of the Dark Matter candidate in scenarios where the dominant coupling is the top partial compositeness. We find two interesting regions: one where the mass of the Dark Matter is ≈ 250 GeV, and one where it can extend up to 2 TeV. Interestingly, next generation Direct Detection experiments can probe most of the parameter space of the model. However, in particular at low mass, there are regions of small Spin-Independent cross section on nuclei. However, the LHC may have a chance to cover this region of the parameter space. We leave this exploration for a future study.

Acknowledgments
Y.W. is supported by the Natural Sciences and Engineering Research Council of Canada (NSERC). T.M. is supported in part by project Y6Y2581B11 supported by 2016 National Postdoctoral Program for Innovative Talents. G.C. acknowledges partial support from the China-France LIA FCPPL and the Labex Lyon Institute of the Origins - LIO.

References

[1] Weinberg S 1976 *Phys. Rev.* **D13** 974–996
[2] Dimopoulos S and Susskind L 1979 *Nucl. Phys.* **B155** 237–252 [2,930(1979)]
[3] Kaplan D B and Georgi H 1984 *Phys. Lett.* **B136** 183–186
[4] Maldacena J M 1999 *Int. J. Theor. Phys.* **38** 1113–1133 [Adv. Theor. Math. Phys.2,231(1998)] (*Preprint* hep-th/9711200)
[5] Contino R, Nomura Y and Pomarol A 2003 *Nucl. Phys.* **B671** 148–174 (*Preprint* hep-ph/0306259)
[6] Hosotani Y 1983 *Phys. Lett.* **126B** 309–313
[7] Marzocca D, Serone M and Shu J 2012 *JHEP* **08** 013 (*Preprint* 1205.0770)
[8] Panico G and Wulzer A 2016 *Lect. Notes Phys.* **913** pp.1–316 (*Preprint* 1506.01961)
[9] Cacciapaglia G and Sannino F 2014 *JHEP* **04** 111 (*Preprint* 1402.0233)
[10] Kaplan D B 1991 *Nucl. Phys.* **B365** 259–278
[11] Contino R and Pomarol A 2004 *JHEP* **11** 058 (*Preprint* hep-th/0406257)
[12] Cacciapaglia G, Marandella G and Terning J 2009 *JHEP* **06** 027 (*Preprint* 0802.2946)
[13] Ferretti G and Karateev D 2014 *JHEP* **03** 077 (*Preprint* 1312.5330)
[14] Barnard J, Gherghetta T and Ray T S 2014 *JHEP* **02** 002 (*Preprint* 1311.6562)
[15] von Gersdorff G, Pontón E and Rosenfeld R 2015 *JHEP* **06** 119 (*Preprint* 1502.07340)
[16] Ayyar V, DeGrand T, Golterman M, Hackett D C, Jay W I, Neil E T, Shamir Y and Svetitsky B 2018 *Phys. Rev.* **D97** 074505 (*Preprint* 1710.00806)
[17] Bennett E, Hong D K, Lee J W, Lin C J D, Lucini B, Piai M and Vadacchino D 2018 *JHEP* **03** 185 (*Preprint* 1712.04220)
[18] Lee J W, Bennett, Hong D K, Lin C J D, Lucini B, Piai M and Vadacchino D 2018 *PoS* **LATTICE2018** 192 (*Preprint* 1811.00276)
[19] Witzel O, Hasenfratz A and Rebbi C 2018 *13th Conference on the Intersections of Particle and Nuclear Physics (CIPANP 2018) Palm Springs, California, USA, May 29-June 3, 2018* (*Preprint* 1810.01850)
[20] Vecchi L 2017 *JHEP* **02** 094 (*Preprint* 1506.00623)
[21] Espriu D, Palanques-Mestre A, Pascual P and Tarrach R 1982 *Z. Phys.* **C13** 153
[22] Palanques-Mestre A and Pascual P 1984 *Commun. Math. Phys.* **95** 277
[23] Gracey J A 1996 *Phys. Lett.* **B373** 178–184 (*Preprint* hep-ph/9602214)
[24] Frigerio M, Pomarol A, Riva F and Urbano A 2012 *JHEP* **07** 015 (*Preprint* 1204.2808)
[25] Ma T and Cacciapaglia G 2016 *JHEP* **03** 211 (*Preprint* 1508.07014)
[26] Wu Y, Ma T, Zhang B and Cacciapaglia G 2017 *JHEP* **11** 058 (*Preprint* 1703.06903)
[27] Cai C, Cacciapaglia G and Zhang H H 2018 (*Preprint* 1805.07619)
[28] Cacciapaglia G, Cai H, Deandrea A and Kushwaha A 2019 (*Preprint* 1904.09301)
[29] Chala M 2013 *JHEP* **01** 122 (*Preprint* 1210.6208)
[30] Ballesteros G, Carmona A and Chala M 2017 *Eur. Phys. J.* **C77** 468 (*Preprint* 1704.07388)
[31] Balkin R, Ruhdorfer M, Salvioni E and Weiler A 2017 *JHEP* **11** 094 (*Preprint* 1707.07685)
[32] Balkin R, Ruhdorfer M, Salvioni E and Weiler A 2018 *JCAP* **1811** 050 (*Preprint* 1809.09106)
[33] Belyaev A, Cacciapaglia G, Cai H, Ferretti G, Flacke T, Parolini A and Serodio H 2017 *JHEP* **01** 094 (*Preprint* 1610.06591)
[34] Cacciapaglia G, Vatani S, Ma T and Wu Y 2018 (*Preprint* 1812.04005)
[35] Antipin O and Sannino F 2018 *Phys. Rev.* **D97** 116007 (*Preprint* 1709.02354)
[36] Banks T and Zaks A 1982 *Nucl. Phys.* **B196** 189–204
[37] Dietrich D D and Sannino F 2007 *Phys. Rev.* **D75** 085018 (*Preprint* hep-ph/0611341)
[38] Sannino F 2009 *Phys. Rev.* **D79** 096007 (*Preprint* 0902.3494)
[39] Holdom B 1981 *Phys. Rev.* **D24** 1441
[40] Kowalska K and Sessolo E M 2018 *JHEP* **04** 027 (*Preprint* 1712.06859)

IOP Conf. Series: Journal of Physics: Conf. Series **1271** (2019) 012014 doi:10.1088/1742-6596/1271/1/012014

[41] Antipin O, Dondi N A, Sannino F, Thomsen A E and Wang Z W 2018 *Phys. Rev.* **D98** 016003 (*Preprint* 1803.09770)
[42] Alanne T and Blasi S 2018 *JHEP* **08** 081 [Erratum: JHEP09,165(2018)] (*Preprint* 1806.06954)
[43] Alanne T and Blasi S 2018 (*Preprint* 1808.03252)
[44] Csaki C, Ma T and Shu J 2017 *Phys. Rev. Lett.* **119** 131803 (*Preprint* 1702.00405)
[45] Alanne T, Bizot N, Cacciapaglia G and Sannino F 2018 *Phys. Rev.* **D97** 075028 (*Preprint* 1801.05444)
[46] Contino R, Marzocca D, Pappadopulo D and Rattazzi R 2011 *JHEP* **10** 081 (*Preprint* 1109.1570)
[47] Bizot N, Cacciapaglia G and Flacke T 2018 *JHEP* **06** 065 (*Preprint* 1803.00021)
[48] Cui X *et al.* (PandaX-II) 2017 *Phys. Rev. Lett.* **119** 181302 (*Preprint* 1708.06917)
[49] Akerib D S *et al.* (LUX) 2017 *Phys. Rev. Lett.* **118** 251302 (*Preprint* 1705.03380)
[50] Aprile E *et al.* (XENON) 2018 (*Preprint* 1805.12562)

IOP Conf. Series: Journal of Physics: Conf. Series **1271** (2019) 012015 doi:10.1088/1742-6596/1271/1/012015

Thermal Model Fits in p-p, p-Pb and Pb-Pb Collisions

Jean Cleymans[1], Boris Hippolyte[2], Masimba Paradza[1], Natasha Sharma[3]
[1]UCT-CERN Research Centre and Department of Physics, University of Cape Town, Rondebosch 7701, South Africa
[2]Institut Pluridisciplinaire Hubert Curien and Université de Strasbourg Institute for Advanced Study, CNRS-IN2P3, Strasbourg, France
[3]Department of Physics, Panjab University, Chandigarh 160014, India

Abstract. Recent work on the particle composition (hadrochemistry) of the final state in proton-proton (p-p), proton-lead (p-Pb) and lead-lead (Pb-Pb) collisions as a function of the charged particle multiplicity ($\mathrm{d}N_{\mathrm{ch}}/\mathrm{d}\eta$) is reviewed. It is argued that for high multiplicities (at least about 20 charged hadrons in the mid-rapidity interva) consistent results are obtained in the thermal model.

1. Use of Thermal Concepts in Heavy-Ion Collisions

The final state produced in heavy ion collisions at the Large Hadron Collider (LHC) is characterized by a large number of hadrons. In central collisions at $\sqrt{s_{NN}} = 5.02$ TeV [1] a total of about 30 000 particles are produced as can be seen from figure 1. To analyze the properties of such a large number of particles it is quite natural to use concepts from statistical mechanics e.g. use energy density, particle density, pressure, temperature, chemical composition, etc ... As it turns out these concepts are useful at all beam energies.

In high energy collisions applications of the thermal-statistical model in the form of the hadron resonance gas model have been successful (see e.g. [2, 3] for two recent publications) in describing the composition of the final state e.g. the yields of pions, kaons, protons and other hadrons.

In this presentation the focus will be on the results of an analysis performed recently [4, 5] about the validity of statistical concepts. In particular it has been concluded there that if the multiplicity exceeds 20 hadrons at mid-rapidity then the use of statistical concepts is justified, irrespective of whether it is a p-p, a p-Pb or a Pb-Pb collision. In high energy collisions applications of the thermal-statistical model in the form of the hadron resonance gas model have been successful (see e.g. [2, 3] for two recent publications) in describing the composition of the final state e.g. the yields of pions, kaons, protons and other hadrons.

2. The Theoretical Basis for the Thermal Model

Using the Cooper-Frye formula [6], the momentum distribution of particles of type i is determined by:

$$E\frac{dN_i}{d^3p} = \frac{g_i}{(2\pi)^3}\int d\sigma_\lambda p^\lambda \exp\left(-\frac{p^\mu u_\mu}{T} + \frac{\mu_i}{T}\right), \tag{1}$$

where the integration is done over the freeze-out surface σ_λ, u_μ is the four-velocity, T and μ_i are the freeze-out temperature and chemical potential respectively Integrating this over the momenta one obtains

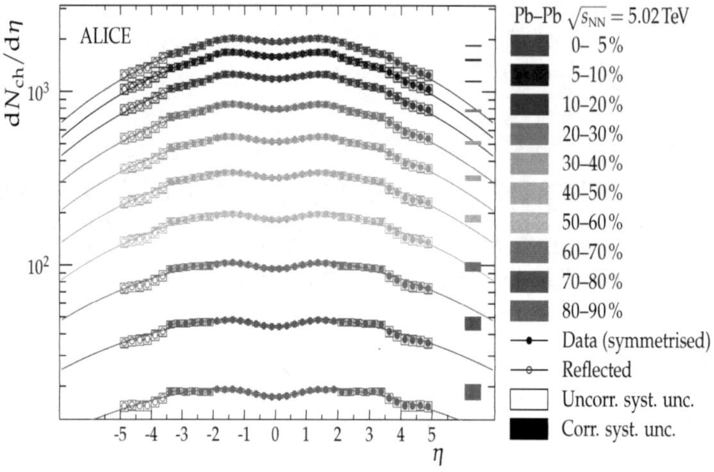

Figure 1. The number of charged particles in Pb-Pb collisions at $\sqrt{s_{NN}}$ = 5.02 TeV as a function of pseudo-rapidity [1].

$$N_i = \frac{g_i}{(2\pi)^3}\int d\sigma_\lambda \int \frac{d^3p}{E} p^\lambda \exp\left(-\frac{p^\mu u_\mu}{T} + \frac{\mu_i}{T}\right)$$

The second integral in the above expression, namely the one over the momenta must be a four-vector, since only u^λ is available as a four-vector, it leads to

$$N_i = \int d\sigma_\lambda u^\lambda n_i^0(T,\mu)$$

where n_i^0 is the density in a fireball at rest. If the temperature and chemical potential are the same along the freeze-out curve, then one obtains

$$N_i = n_i^0(T,\mu)\int d\sigma_\lambda u^\lambda$$

i.e. integrated (4π) multiplicities are the same as for a single fireball at rest (apart from the volume).

Therefore after integration over p_T and y one has:

$$\frac{N_i}{N_j} = \frac{n_i^0}{n_j^0} \tag{2}$$

where, as above, n_i^0 and n_j^0 are the particle yields for particle types i and j as calculated in a fireball at rest. For this result to hold, the freeze-out temperature has to be the same for all particles (which may not always be the case).

This does not mean that the freeze-out has to be instantaneous. The only requirement is that the freeze-out temperature has to be the same along the freeze-out curve.

A well known model for the expansion of matter in a heavy ion collision combimes Bjorken scaling [7] with a transverse expansion, in this case it can be shown that one obtains after integration over p_T [8, 9, 10].

$$\frac{dN_i/dy}{dN_j/dy} = \frac{n_i^0}{n_j^0} \tag{3}$$

where N_i^0 is the particle yield as calculated in a fireball at rest. Effects of hydrodynamic flow cancel out in ratios. The volume is given by $\pi R^2\tau$ corresponding to a cylindrical expansion lasting for a time τ in the longitudinal direction.

3. Uncertainties in the Thermal Model

Uncertainties are related to the incomplete knowledge of hadron species and resonance properties as reflected in the Particle Data Booklet [11]. Particle yields are determined from:

$$N_i = \sum_j N_j Br(j \to i). \tag{4}$$

Hence one must know how hadronic resonances decay.

As an example, the final yield of π^+'s is given by

$$N_{\pi^+} = N_{\pi^+}(\text{thermal}) + N_{\pi^+}(\text{resonance decays}) \tag{5}$$

depending on the temperature, over 80% of observed pions are due to resonance decays. In these descriptions use is made of the grand canonical ensemble and the canonical ensemble with exact strangeness conservation. In addition the use of the canonical ensemble with exact baryon, strangeness and charge conservation were also concidered.

The yields produced in heavy-ion collisions have been the subject of intense discussions over the past few years and several proposals have been made in view of the fact that the number of pions is underestimated while the number of protons is overestimated. Several proposals to improve on this have been made recently:

- Incomplete hadron spectrum [12],
- chemical non-equilibrium at freeze-out [13, 14, 15],
- modification of hadron abundances in the hadronic phase [16, 17, 18],
- separate freeze-out for strange and non-strange hadrons [19, 20, 21, 22],
- excluded volume interactions [23],
- energy dependent Breit-Wigner widths [24],
- use the phase shift analysis to take into account repulsive and attractive interactions [25, 26],
- use the K-matrix formalism to take interactions into account [27].

These proposals improve the agreement with the observed yields and furthermore, some of them change the chemical freeze-out temperature, T_{ch} in only a minimal way like those presented recently in [24, 26]. In the present analysis Therefore the basic structure of the thermal model was kept with a single freeze-out temperature and focus on the resulting thermal parameters T_{ch}, γ_s and the radius. All calculations were done using the latest version of THERMUS [28] [1].

The results [4, 5] show some interesting new features:

- the grand canonical ensemble, the ensemble with strict strangeness conservation and the one with strict baryon number, strangeness and charge conservation agree very well for the particle composition in Pb-Pb collisions, they also agree well for p-Pb collisions but marked differences for p-p collisions are present. These differences disappear as the multiplicity of charged particles increases in the final state. Thus, p-p collisions with high multiplicities agree with what is seen in large systems like p-Pb and Pb-Pb collisions. Quantitatively this agreement starts when there are at least 20 charged hadrons in the mid-rapidity interval being considered. It also throws doubt on the applicability of the thermal model as applied to p-p collisions with low multiplicity.
- The convergence of the results in the three ensembles lends support to the idea that one reaches a thermodynamic limit where the results are independent of the ensemble being used.

[1] B. Hippolyte and Y. Schutz, https://github.com/thermus-project/THERMUS

4. Ensembles considered in the thermal model

Three different ensembles based on the thermal model were compared. In the following equations use is made of the Boltzmann approximation, but the actual numerical calculations use quantum statistics as far as possible.

- Grand canonical ensemble (GCE), the conservation of quantum numbers is implemented using chemical potentials. The quantum numbers are conserved on the average. The partition function depends on thermodynamic quantities and the Hamiltonian describing the system of N hadrons:

$$Z_{GCE} = \mathrm{Tr}\left[e^{-(H-\mu N)/T}\right] \tag{6}$$

which, in the framework of the thermal model considered here, leads to

$$\ln Z_{GCE}(T,\mu,V) = \sum_i g_i V \int \frac{d^3p}{(2\pi)^3} \exp\left(-\frac{E_i-\mu_i}{T}\right) \tag{7}$$

in the Boltzmann approximation, g_i is the degeneracy factor of hadron i, V is the volume of the system, μ_i is the chemical potential associated with the hadron. The yield is given by:

$$N_i^{GCE} = g_i V \int \frac{d^3p}{(2\pi)^3} \exp\left(-\frac{E_i}{T}\right), \tag{8}$$

where we have put the chemical potentials equal to zero, as relevant for the beam energies at the Large Hadron Collider considered here. The decays of resonances have to be added to the final yield

$$N_i^{GCE}(\mathrm{total}) = N_i^{GCE} + \sum_j Br(j \to i) N_j^{GCE}. \tag{9}$$

- Canonical ensemble with exact implementation of strangeness conservation, we will refer to this as the strangeness canonical ensemble (SCE). There are chemical potentials for baryon number B and charge Q but not for strangeness:

$$Z_{SCE} = \mathrm{Tr}\left[e^{-(H-\mu N)/T}\delta_{(S,\sum_i S_i)}\right] \tag{10}$$

The delta function imposes exact strangeness conservation, requiring overall strangeness to be fixed to the value S, in this paper we will only consider the case where overall strangeness is zero, $S = 0$. This change leads to [29]:

$$Z_{SCE} = \frac{1}{(2\pi)} \int_0^{2\pi} d\phi e^{-iS\phi} Z_{GCE}(T,\mu_B,\lambda_S) \tag{11}$$

where the fugacity factor is replaced by

$$\lambda_S = e^{i\phi} \tag{12}$$

$$N_i^{SCE} = V \frac{Z_i^1}{Z_{S=0}^C} \sum_{k,p=-\infty}^{\infty} a_3^p a_2^k a_1^{-2k-3p-s} I_k(x_2) I_p(x_3) I_{-2k-3p-s}(x_1), \tag{13}$$

where $Z_{S=0}^C$ is the canonical partition function

$$Z_{S=0}^C = e^{S_0} \sum_{k,p=-\infty}^{\infty} a_3^p a_2^k a_1^{-2k-3p} I_k(x_2) I_p(x_3) I_{-2k-3p}(x_1),$$

where Z_i^1 is the one-particle partition function calculated for $\mu_S = 0$ in the Boltzmann approximation. The arguments of the Bessel functions $I_s(x)$ and the parameters a_i are introduced as,

$$a_s = \sqrt{S_s/S_{-s}} \ , \ x_s = 2V\sqrt{S_s S_{-s}}, \tag{14}$$

where S_s is the sum of all $Z_k^1(\mu_S = 0)$ for particle species k carrying strangeness s. As previously, the decays of resonances have to be added to the final yield

$$N_i^{SCE}(\text{total}) = N_i^{SCE} + \sum_j Br(j \to i) N_j^{SCE}. \tag{15}$$

- Canonical ensemble with exact implementation of B, S and Q conservation, we will refer to this as the full canonical ensemble (FCE). In this ensemble there are no chemical potentials. The partition function is given by:

$$Z_{FCE} = \text{Tr}\left[e^{-H/T}\delta_{(B,\sum_i B_i)}\delta_{(Q,\sum_i Q_i)}\delta_{(S,\sum_i S_i)}\right] \tag{16}$$

$$Z_{FCE} = \frac{1}{(2\pi)^3}\int_0^{2\pi} d\alpha e^{-iB\alpha}\int_0^{2\pi} d\psi e^{-iQ\psi}\int_0^{2\pi} d\phi e^{-iS\phi} Z_{GCE}(T, \lambda_B, \lambda_Q, \lambda_S) \tag{17}$$

where the fugacity factors have been replaced by

$$\lambda_B = e^{i\alpha}, \quad \lambda_Q = e^{i\psi}, \quad \lambda_S = e^{i\phi}. \tag{18}$$

As before, the decays of resonances have to be added to the final yield

$$N_i^{FCE}(\text{total}) = N_i^{FCE} + \sum_j Br(j \to i) N_j^{FCE}. \tag{19}$$

In this case the analytic expression becomes very lengthy and we refrain from writing it down here, it is implemented in the THERMUS program [28].

In all three case we have also taken into account the strangeness saturation factor γ_s [30] which enters as a multiplicative factor, raised to the power of the strangeness content, in the particle yields. Keeping this factor fixed at one does not change the fixed message, only the resulting value of χ^2 is increased indicating a worsening of the fits.

These three ensembles are applied to p-p collisions at 7 TeV in the central region of rapidity [31], to p-Pb collisions at 5.02 TeV [32, 33] and to Pb-Pb collisions at 2.76 TeV [34, 35, 36] with particular focus on the dependence on the charged particle multiplicity. It is well known that in this kinematic region, one has particle - antiparticle symmetry and therefore there is no net baryon density and also no net strangeness. The different ensembles nevertheless give different results because of the way they are implemented. A clear size dependence is present in the results of the ensembles. In the thermodynamic limit they should become equivalent. Clearly there are other ensembles that could be investigated and also other sources of finite volume corrections.

A similar analysis was done in [37, 38, 39] for p-p collisions at 200 GeV but without the dependence on charged multiplicity.

For p-Pb and Pb-Pb collisions the Ω measurements were included in the analysis, so that six particle species were considered for p-Pb and Pb-Pb. It was also checked checked explicitly that for the five bins in p-p collisions where the Ω has also been measured, there is no difference in the outcome for the values of T_{ch}, γ_s and the radius.

As shown in [40, 41] the ϕ meson is not described very well and has not been included.

IOP Conf. Series: Journal of Physics: Conf. Series **1271** (2019) 012015 doi:10.1088/1742-6596/1271/1/012015

5. Comparison of different statistical ensembles.

In figure 2 we show the chemical freeze-out temperature as a function of the multiplicity of hadrons in the final state [31]. As explained in the previous section the freeze-out temperature has been calculated using three different ensembles. The highest values are obtained using the canonical ensemble with exact conservation of three quantum numbers, baryon number B, strangeness S and charge Q, all of them being set to zero as is appropriate for the central rapidity region in p-p collisions at 7 TeV. In this ensemble the temperature drops strongly from the lowest to the highest multiplicity.

The lowest values for $T_{\rm ch}$ are obtained when using the grand canonical ensemble, in this case the conserved quantum numbers are again zero. The results are clearly different from those obtained in the previous ensemble, especially in the low multiplicity intervals. They gradually approach each other and they become equal at the highest multiplicities.

For comparison with the previous two cases we also calculated $T_{\rm ch}$ using the canonical ensemble with only strangeness S being exactly conserved using the method presented in [29]. In this case the results are close to those obtained in the grand canonical ensemble, with the values of $T_{\rm ch}$ always slightly higher than in the grand canonical ensemble. Again for the highest multiplicity interval the results become equivalent.

The canonical BSQ ensemble leads to results which are incompatible with those obtained from Lattice Gauge Theory [42] which indicate that hadrons cannot exist above the critical temperature which has been estimated to be about 156$\pm$1.5 MeV. As can be seen in figure 2, even though all the ensembles produce different results, for high multiplicities the results converge to a common value close to 160 MeV. In figure 3 we show results for the strangeness saturation factor γ_s [30]. In this case we obtain again quite substantial differences in each one of the three ensembles considered. The highest values being found in the canonical ensemble with exact strangeness conservation. Note that the values of γ_s become compatible with unity, i.e. with chemical equilibrium for all light flavors. It is to be noticed the for multiplicities the value of γ_s is slightly above unity (by almost 10%) but is compatible with full chemical equilibrium within one standard deviation. In figure 4 the radius at chemical freeze-out obtained in the three ensembles is presented. As in the previous figures, the results become independent of the ensemble chosen for the highest multiplicities while showing clear differences for low multiplicities.

The results show that there is a strong correlation between some of the parameters. The very high temperature obtained in the canonical BSQ ensemble (FCE) correlates with the small radius in the same ensemble. Particle yields increase with temperature but a small volume decreases them, hence the correlation between the parameters.

The fits to the hadronic yields obtained in p-p collisions at 7 TeV in five different centrality bins are shown in figure4 and 5. The upper panels show the yields while the lower panels show the ratios of the measured data divided by the fit values for the three different ensembles considered here. The three lines corresponding to the fits are often very close to each other and overlap, hence they are not always visible on the figures.

6. Discussion and Conclusions

In this paper we have investigated three different ensembles to analyze the variation of particle yields with the multiplicity of charged particles produced in proton-proton collisions at the center-of-mass energy of $\sqrt{s}$ = 7 TeV [31], p-Pb collisions at 5.02 TeV [32, 33] and Pb-Pb collisions at 2.76 TeV [34, 35, 36].

The basic structure of the thermal model as presented in [28] was kept and the focus was on the resulting thermal parameters T_{ch}, γ_s and the radius and their dependence on the final state multiplicity. It is to be noted in this regards that recent improvements on the treatment of the particle yields do not lead to substantial changes of the chemical freeze-out temperature, T_{ch} [24, 26]. The results show two new interesting features:

- a comparison of the grand canonical ensemble, the ensemble with strict strangeness conservation and the one with strict baryon number, strangeness and charge conservation agree very well for large systems like p-Pb and Pb-Pb, but show marked differences for p-p collisions. These differences tend

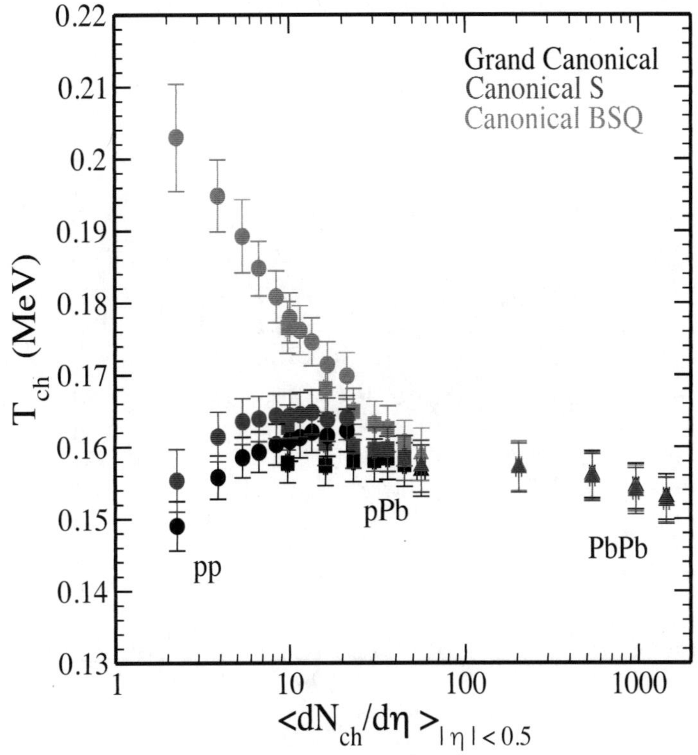

Figure 2. The chemical freeze-out temperature, T_{ch}, obtained for three different ensembles. The black points are obtained using the grand canonical ensemble, the blue points use exact strangeness conservation while the red points have built-in exact baryon number, strangeness and charge conservation. Circles are for p-p collisions at 7 TeV [31], squares are for p-Pb collisions at 5.02 TeV [32, 33] while triangles are for Pb-Pb collisions at 2.76 TeV [34, 35, 36].

to disappear as the multiplicity of charged particles increases in the final state of p-p collisions. This supports the fact that p-p collisions with high multiplicities agree with what is seen in large systems like Pb-Pb. Quantitatively this starts happening when there are more than 20 charged hadrons in the mid-rapidity interval being considered. It also throws doubt on the applicability of the thermal model in low multiplicity p-p collisions.

- The convergence of the results in the three ensembles lends support to the notion a thermodynamic limit is reached where results are independent of the ensemble being used.

It is of interest to note that all three ensembles lead to the same results when the multiplicity of charged particles $\mathrm{d}N_{\mathrm{ch}}/\mathrm{d}\eta$ exceeds 20 at mid-rapidity. This could be interpreted as reaching the thermodynamic limit since the three ensembles lead to the same results. It would be of interest to extend this analysis to higher beam energies and higher multiplicity intervals.

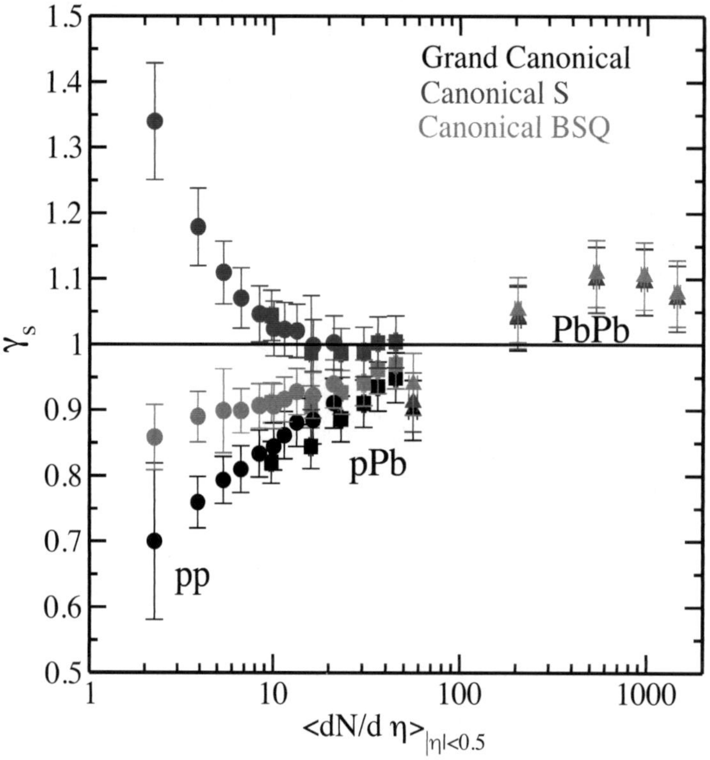

Figure 3. The strangeness saturation factor γ_s obtained for three different ensembles. The black points were obtained using the grand canonical ensemble, the blue points uses exact strangeness conservation while the red points have built-in exact baryon number, strangeness and charge conservation. Circles are for p-p collisions at 7 TeV [31], squares are for p-Pb collisions at 5.02 TeV [32, 33] while triangles are for Pb-Pb collisions at 2.76 TeV [34, 35, 36].

Acknowledgments
One of us (J.C.) gratefully thanks the National Research Foundation of South Africa for financial support. N.S. acknowledges the support of SERB Ramanujan Fellowship (D.O. No. SB/S2/RJN- 084/2015) of the Department of Science and Technology of India. B.H. acknowledges the support of the Université de Strasbourg Institute for Advanced Study.

References

[1] Adam J *et al.* (ALICE) 2017 *Phys. Lett.* **B772** 567–577 (*Preprint* 1612.08966)
[2] Andronic A, Braun-Munzinger P, Redlich K and Stachel J 2017 (*Preprint* nucl-th/1710.09425)
[3] Becattini F, Bleicher M, Steinheimer J and Stock R 2017 (*Preprint* hep-ph/1712.03748)
[4] Sharma N, Cleymans J, Hippolyte B and Paradza M 2019 *Phys. Rev.* **C99** 044914 (*Preprint* 1811.00399)
[5] Sharma N, Cleymans J and Hippolyte B 2018 (*Preprint* 1803.05409)
[6] Cooper F and Frye G 1974 *Phys. Rev.* **D10** 186
[7] Bjorken J D 1983 *Phys. Rev.* **D27** 140–151
[8] Cleymans J 1998 *Proceedings, 3rd International Conference on Physics and astrophysics of quark-gluon plasma (ICPA-QGP '97)* pp 55–64 (*Preprint* nucl-th/9704046)
[9] Broniowski W and Florkowski W 2001 *Phys. Rev. Lett.* **87** 272302 (*Preprint* nucl-th/0106050)

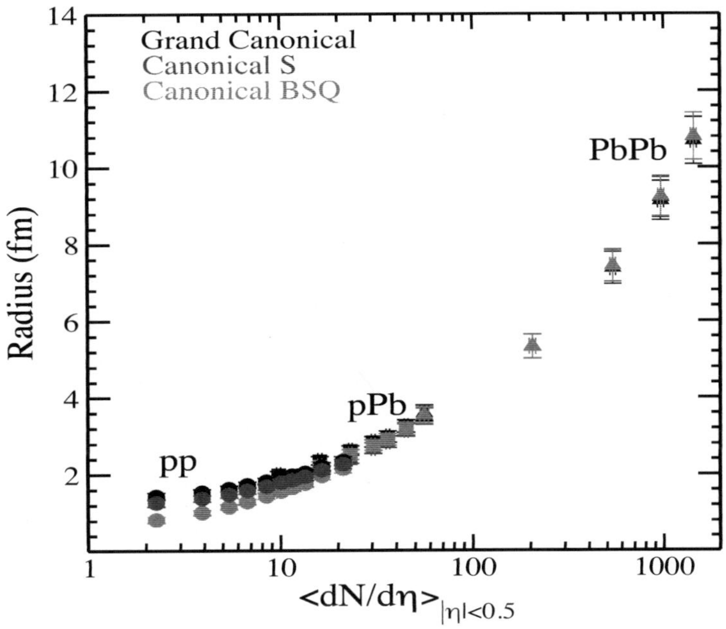

Figure 4. The chemical freeze-out radius obtained for three different ensembles. The black points were obtained using the grand canonical ensemble, the blue points uses exact strangeness conservation while the red points have built-in exact baryon number, strangeness and charge conservation. Circles are for p-p collisions at 7 TeV [31], squares are for p-Pb collisions at 5.02 TeV [32, 33] while triangles are for Pb-Pb collisions at 2.76 TeV [34, 35, 36].

[10] Akkelin S V, Braun-Munzinger P and Sinyukov Yu M 2002 *Nucl. Phys.* **A710** 439–465 (*Preprint* `nucl-th/0111050`)
[11] Patrignani C *et al.* (Particle Data Group) 2016 *Chin. Phys.* **C40** 100001
[12] Noronha-Hostler J and Greiner C 2014 *Nucl. Phys.* **A931** 1108–1113 (*Preprint* `1408.0761`)
[13] Petrán M, Letessier J, Petráček V and Rafelski J 2013 *Phys. Rev.* **C88** 034907 (*Preprint* `1303.2098`)
[14] Begun V, Florkowski W and Rybczynski M 2014 *Phys. Rev.* **C90** 014906 (*Preprint* `1312.1487`)
[15] Begun V, Florkowski W and Rybczynski M 2014 *Phys. Rev.* **C90** 054912 (*Preprint* `1405.7252`)
[16] Steinheimer J, Aichelin J and Bleicher M 2013 *Phys. Rev. Lett.* **110** 042501 (*Preprint* `1203.5302`)
[17] Becattini F, Bleicher M, Kollegger T, Schuster T, Steinheimer J and Stock R 2013 *Phys. Rev. Lett.* **111** 082302 (*Preprint* `1212.2431`)
[18] Becattini F, Steinheimer J, Stock R and Bleicher M 2017 *Phys. Lett.* **B764** 241–246 (*Preprint* `1605.09694`)
[19] Chatterjee S, Godbole R M and Gupta S 2013 *Phys. Lett.* **B727** 554–557 (*Preprint* `1306.2006`)
[20] Bellwied R, Borsanyi S, Fodor Z, Katz S D and Ratti C 2013 *Phys. Rev. Lett.* **111** 202302 (*Preprint* `1305.6297`)
[21] Chatterjee S, Dash A K and Mohanty B 2017 *J. Phys.* **G44** 105106 (*Preprint* `1608.00643`)
[22] Bellwied R 2016 *J. Phys. Conf. Ser.* **736** 012018
[23] Alba P, Vovchenko V, Gorenstein M I and Stoecker H 2018 *Nucl. Phys.* **A974** 22–34 (*Preprint* `1606.06542`)
[24] Vovchenko V, Gorenstein M I and Stoecker H 2018 *Phys. Rev.* **C98** 034906 (*Preprint* `1807.02079`)
[25] Dash A, Samanta S and Mohanty B 2018 (*Preprint* `1806.02117`)
[26] Andronic A, Braun-Munzinger P, Friman B, Lo P M, Redlich K and Stachel J 2018 (*Preprint* `1808.03102`)
[27] Dash A, Samanta S and Mohanty B 2018 *Phys. Rev.* **C97** 055208 (*Preprint* `1802.04998`)
[28] Wheaton S, Cleymans J and Hauer M 2009 *Comput. Phys. Commun.* **180** 84–106 (*Preprint* `hep-ph/0407174`)
[29] Braun-Munzinger P, Cleymans J, Oeschler H and Redlich K 2002 *Nucl. Phys.* **A697** 902–912 (*Preprint* `hep-ph/0106066`)
[30] Letessier J, Tounsi A, Heinz U W, Sollfrank J and Rafelski J 1995 *Phys. Rev.* **D51** 3408–3435 (*Preprint*

hep-ph/9212210)
[31] Adam J *et al.* (ALICE) 2017 *Nature Phys.* **13** 535–539 (*Preprint* 1606.07424)
[32] Abelev B B *et al.* (ALICE) 2014 *Phys. Lett.* **B728** 25–38 (*Preprint* 1307.6796)
[33] Adam J *et al.* (ALICE) 2016 *Phys. Lett.* **B758** 389–401 (*Preprint* 1512.07227)
[34] Abelev B *et al.* (ALICE) 2013 *Phys. Rev.* **C88** 044910 (*Preprint* 1303.0737)
[35] Abelev B B *et al.* (ALICE) 2013 *Phys. Rev. Lett.* **111** 222301 (*Preprint* 1307.5530)
[36] Abelev B B *et al.* (ALICE) 2014 *Phys. Lett.* **B728** 216–227 [Erratum: Phys. Lett.B734,409(2014)] (*Preprint* 1307.5543)
[37] Abelev B I *et al.* (STAR) 2007 *Phys. Rev.* **C75** 064901 (*Preprint* nucl-ex/0607033)
[38] Becattini F, Castorina P, Milov A and Satz H 2011 *J. Phys.* **G38** 025002 (*Preprint* 0912.2855)
[39] Becattini F, Castorina P, Milov A and Satz H 2010 *Eur. Phys. J.* **C66** 377–386 (*Preprint* 0911.3026)
[40] Vislavicius V and Kalweit A 2016 (*Preprint* nucl-ex/1610.03001)
[41] Sharma N, Cleymans J and Kumar L 2018 *Eur. Phys. J.* **C78** 288 (*Preprint* 1802.07972)
[42] Bazavov A *et al.* 2018 (*Preprint* 1812.08235)

IOP Conf. Series: Journal of Physics: Conf. Series **1271** (2019) 012016 doi:10.1088/1742-6596/1271/1/012016

Unification of gauge and Yukawa couplings

Aldo Deandrea

Université de Lyon, France; Université Lyon 1, CNRS/IN2P3, UMR5822 IPNL, F-69622 Villeurbanne Cedex, France.

E-mail: deandrea@ipnl.in2p3.fr

Abstract. In this talk we describe how unification of gauge and Yukawa couplings can be obtained in gauge-Higgs unification models in extra-dimensions using a toy model. It is usually considered that such unification is difficult to obtain using simple group theory arguments. We consider a toy model based on the $SU(3)$ symmetry and show that including the renormalisation group running at one loop for the couplings improve the result. The gauge couplings unify asymptotically at high energies, and this may result from the presence of an UV fixed point. The Yukawa coupling is enhanced at low energies, showing that a genuine unification of gauge and Yukawa couplings may be achieved. Some other arguments and directions of study are also mentioned to tackle the problem of producing a scalar potential allowing to accommodate the measured Higgs mass at 125 GeV.

1. Introduction

The Higgs boson discovery at the LHC and the measurements of its mass and couplings allows probing of the electroweak (EW) symmetry breaking sector of the Standard Model (SM). This vast experimental program is still ongoing, but has already allowed to test at least partially the origin of symmetry breaking in the SM and its extensions. Fundamental scalar field mass terms are not protected by quantum symmetry, therefore many extensions of the SM focus on finding a more satisfactory origin of the scalar sector. Another intriguing property implemented in Beyond the Standard Model (BSM) models is the unification of gauge couplings at high energies, due to the renormalisation group evolution of the couplings. This brought to the development of Grand Unified Theories (GUT). Moreover the fact that the mass of the top quark is close to the electroweak scale may suggest that the Yukawa coupling of the top quark can have a similar origin.

In the following we shall follow the attractive ideas developed in Gauge-Higgs Unification (GHU) models [1, 2, 3]. Extra-dimensional models contain a peculiar class of scalar fields, arising as additional polarisations of vector gauge fields from the extra compact space. If the Higgs is identified as such a scalar, its couplings are related to the gauge couplings. A mass terms for the Higgs boson is forbidden by gauge invariance in the extra-dimensions solving the problem of the presence of a massive and fundamental scalar field mentioned above. If the gauge symmetry is broken by boundary conditions in a suitable way, a massless scalar is present in the physical spectrum of the theory when reduced to four space-time dimensions, and its potential is radiatively generated and finite [4, 5]. This is an attractive property as it addresses at the same time gauge-Yukawa unification and naturalness of the scalar sector.

However attempts in this sense are challenging as one has to find a gauge group, $\mathcal{G}_{\rm GHU}$, that predicts the measured values of the SM couplings and the Higgs mass value. Indeed most

Table 1. Gauge and bulk fermion Yukawa couplings in the SU(3) GHU model compared to the SM values at M_Z.

	$\mathrm{SU}(2)_L$	$\mathrm{U}(1)_Y$	Yuk.	$\mathrm{SU}(3)_c$
	g	g'	y	g_s
SU(3) GHU	g_{GHU}	$\sqrt{3}\, g_{\mathrm{GHU}}$	$g_{\mathrm{GHU}}/\sqrt{2}$	-
SM	0.66	0.35	1.0	1.2

of the models seem to give incorrect predictions [6]. This conclusion can be at least partially modified if the energy evolution of the couplings is taken into account. The extra-dimensional model is actually an effective theory, and its unified predictions are only valid close to the cut-off scale of the theory while the experimental values are measured at the EW scale, and the couplings change due to the running between the two scales. This fact is studied in extra-dimensional GUTs [7, 8]. The running in extra-dimensions is not logarithmic but follows a power law [7, 9, 10]. It is therefore much faster than in four dimensions. Taking into account the running, the tree-level predictions can be strongly modified and the low energy values of the SM couplings may match the experimental values for some choices of the gauge group and matter representations. In the case of the top-quark Yukawa coupling, the running tends to reduce the tension between the value of the large Yukawa coupling and the small gauge coupling at low energy.

2. The minimal SU(3) toy model

The simplest group allowing to embed the EW symmetry and the Higgs boson is $\mathcal{G}_{\mathrm{GHU}} = \mathrm{SU}(3)_W$ [11]. This group is of rank two and contains an $SU(2) \times U(1)$ subgroup that can be identified with the gauged EW group of the SM. We insert in the model a quark as a bulk field, while the other SM fermions can be added as localised fields [6, 11, 12]. Their couplings to the bulk Higgs will be suppressed, thus implying for those fermion fields masses below the EW scale. Gauge and Yukawa couplings, in terms of the unified SU(3) coupling g_{GHU}, are given in table 1. For the Yukawa coupling we consider the top quark as our bulk field because it has the largest one. It is clear that the tree-level GHU predictions are different from the SM values, however, they only apply at the cut-off of the effective theory, which may be different from the EW scale. We consider the running effects with a single extra-dimension compactified on an interval $S^1/\mathbb{Z}_2$. The boundary conditions are such that the initial group is broken to the EW one. The spectrum of the theory contains massless gauge bosons and a massless scalar associated to the broken generators. The bulk fermion transforming as the fundamental of $\mathrm{SU}(3)_W$ is assigned boundary conditions such that only two massless fermions appear and we identify them with the third generation quark doublet and down-type singlet (the other fermions are assumed to be localised). At low energy the spectrum of the theory matches the one of the SM. The running of the couplings is modified by the presence of the Kaluza-Klein (KK) fields once the mass thresholds are met, starting at $m_{\mathrm{KK}} = 1/R$ (where R is the compactification radius). The results presented in this talk are those obtained in [12]. The running of the gauge and Yukawa couplings is shown in figure 1 as a function of the energy scale μ, normalised as in table 1:

$$\{g_1, g_2, g_3, g_y\} = \left\{ \frac{g'}{\sqrt{3}}, g, g_s, \sqrt{2}\, y \right\} . \tag{1}$$

The normalisations follow from the structure of the $\mathrm{SU}(3)_W$ matrices, while the QCD coupling is unrelated in this simple SU(3) model, even if it possible to include it using a larger symmetry

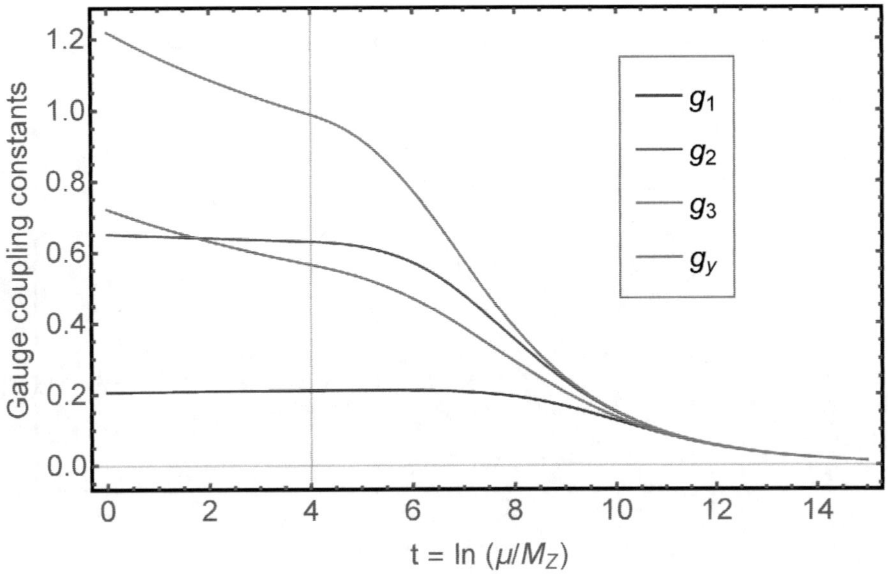

Figure 1. Running of the normalised gauge and Yukawa couplings for the SU(3) model, with $1/R = 5$ TeV. The first KK mode is at the scale $t_{\mathrm{KK}} \sim 4.0$.

group. The couplings follow the SM evolution up to the scale of the first KK resonances:

$$t_{\mathrm{KK}} = \ln \frac{1}{M_Z R}\,. \tag{2}$$

and from there the running is modified by the extra-dimensions and the gauge couplings asymptotically tend to the same value.

In figure 2 we show an estimate of the 5-dimensional loop factor based on naive dimensional analysis (NDA) [13, 14]:

$$\alpha_i^{\mathrm{NDA}}(\mu) \sim \frac{g_i^2(\mu)}{8\pi}\mu R\,. \tag{3}$$

The loop factor, which can be thought of as a 5D 't Hooft coupling, can be used as an indication of the energy where the extra-dimensional theory becomes unreliable. The value stays actually small suggesting that the theory may have a more extended validity than usually estimated.

The strong coupling also falls very close as actually the GHU model contains two SU(3) gauge structures, one associated to QCD and the other one to the EW gauge sector, and the bulk fermion is a bi-fundamental. This allows a $\mathbb{Z}_2$ exchange symmetry between the two sectors at high energy that implies equal couplings.

The initial value of the Yukawa coupling ($y(m_Z) = 0.51$), is selected to achieve unification in the UV. This value depends mildly on the scale of the extra-dimension $1/R$. The running of the Yukawa coupling does not follow the gauge ones at high energy, as the compactification of the extra-dimension singles out the scalar component of the gauge field in the bulk. The effect of the running shows that the value of the Yukawa coupling at low energy is larger than the values at unification, $y = g_2/\sqrt{2}$, even if the enhancement is not enough to explain the Yukawa coupling of the top, $y = 1$ in this toy model. Two loop corrections, the embedding of the top in a more realistic model, or even a different choice of the compactified geometry, may further improve the agreement.

We also performed the running in a different set-up, following the same technique. Using a larger representation that contains a singlet with the correct hypercharge to match the right-

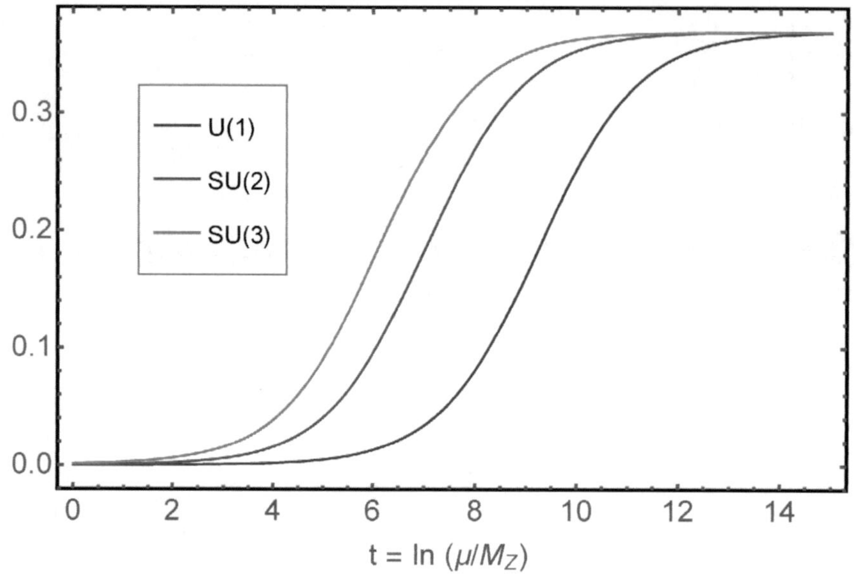

Figure 2. 5D NDA loop factor as a function of the energy, for $1/R = 5$ TeV.

handed top the minimal possibility is to use a 2-index symmetric representation. However, the NDA loop factor estimate in that case goes to non-perturbative values well before unification. This seems to indicate that only models with small representations of the bulk gauge symmetries can be realistic candidates to implement GHU.

3. Conclusions

Simple models of GHU where both the EW gauge couplings and the top Yukawa unify can be built and the running of couplings from the EW scale to the extra-dimension scale is an important ingredient in order to obtain reliable results. We have described a model in five-dimensions, compactified on $S^1/\mathbb{Z}_2$, with gauge groups $SU(3)_c\times$ $SU(3)_W$ and a bulk fermion transforming as a bi-fundamental which constitutes a toy example for including a fermion with a large Yukawa coupling at low energy. The fermions contained in the bulk fermion have the quantum numbers of a down-type quark, and the effective Yukawa coupling is enhanced at low energies due to the running. This structure is good enough to describe the EW gauge sector unified in $SU(3)_W$. The running allows to match the value of the Weinberg angle at the EW scale, as well as large Yukawa couplings. The values of the couplings appear as an attractor in the UV, providing asymptotic unification. The QCD gauge coupling also unifies, suggesting that the double-SU(3) structure may be embedded in a larger group. Another important test in more realistic models is the calculation of the Higgs mass, as indeed typical GHU models fail to obtain a large enough Higgs boson mass to reach the experimentally observed value, unless extra ingredients are added.

Acknowledgements

I wish to thank my collaborators on this subject: Ammar Abdalgabar, Mohammed Omer Khojali, Alan S. Cornell and Giacomo Cacciapaglia, for the many fruitful discussions and common work. AD is partially supported by the Labex-LIO (Lyon Institute of Origins) under grant ANR-10-LABX-66 and FRAMA (FR3127, Fédération de Recherche "André Marie Ampère").

References

[1] Y. Hosotani, Phys. Lett. **126B** (1983) 309.
[2] H. Hatanaka, T. Inami and C. S. Lim, Mod. Phys. Lett. A **13** (1998) 2601 [hep-th/9805067].
[3] G. R. Dvali, S. Randjbar-Daemi and R. Tabbash, Phys. Rev. D **65** (2002) 064021 [hep-ph/0102307].
[4] A. Masiero, C. A. Scrucca, M. Serone and L. Silvestrini, Phys. Rev. Lett. **87** (2001) 251601 [hep-ph/0107201].
[5] I. Antoniadis, K. Benakli and M. Quiros, New J. Phys. **3** (2001) 20 [hep-th/0108005].
[6] C. Csaki, C. Grojean and H. Murayama, Phys. Rev. D **67**, 085012 (2003) [hep-ph/0210133].
[7] K. R. Dienes, E. Dudas and T. Gherghetta, Phys. Lett. B **436** (1998) 55 [hep-ph/9803466].
[8] D. Ghilencea and G. G. Ross, Phys. Lett. B **442** (1998) 165 [hep-ph/9809217].
[9] D. M. Ghilencea, Phys. Rev. D **70** (2004) 045011 [hep-th/0311187].
[10] T. Varin, J. Welzel, A. Deandrea and D. Davesne, Phys. Rev. D **74** (2006) 121702 [hep-ph/0610130].
[11] C. A. Scrucca, M. Serone and L. Silvestrini, Nucl. Phys. B **669**, 128 (2003) [hep-ph/0304220].
[12] A. Abdalgabar, M. O. Khojali, A. S. Cornell, G. Cacciapaglia and A. Deandrea, Phys. Lett. B **776** (2018) 231 doi:10.1016/j.physletb.2017.11.033 [arXiv:1706.02313 [hep-ph]].
[13] S. Weinberg, Physica A **96** (1979) 327.
[14] A. Manohar and H. Georgi, Nucl. Phys. B **234** (1984) 189.
[15] G. Bhattacharyya, A. Datta, S. K. Majee and A. Raychaudhuri, Nucl. Phys. B **760** (2007) 117 [hep-ph/0608208].
[16] H. Gies, Phys. Rev. D **68** (2003) 085015 [hep-th/0305208].
[17] T. R. Morris, JHEP **0501** (2005) 002 [hep-ph/0410142].
[18] A. S. Cornell, A. Deandrea, L. X. Liu and A. Tarhini, Mod. Phys. Lett. A **28**, no. 11, 1330007 (2013) [arXiv:1209.6239 [hep-ph]].
[19] R. Contino, L. Pilo, R. Rattazzi and E. Trincherini, Nucl. Phys. B **622** (2002) 227 [hep-ph/0108102].
[20] M. Carena, T. M. P. Tait and C. E. M. Wagner, Acta Phys. Polon. B **33** (2002) 2355 [hep-ph/0207056].
[21] G. Cacciapaglia, C. Csaki and S. C. Park, JHEP **0603**, 099 (2006) [hep-ph/0510366].
[22] R. Contino, Y. Nomura and A. Pomarol, Nucl. Phys. B **671** (2003) 148 [hep-ph/0306259].
[23] Y. Hosotani and M. Mabe, Phys. Lett. B **615** (2005) 257 [hep-ph/0503020].
[24] Y. Grossman and M. Neubert, Phys. Lett. B **474** (2000) 361 [hep-ph/9912408].

IOP Conf. Series: Journal of Physics: Conf. Series **1271** (2019) 012017 doi:10.1088/1742-6596/1271/1/012017

Cornering top-philic dark matter with colliders and cosmology: the importance of QCD corrections

Benjamin Fuks[1,2]

[1] Laboratoire de Physique Théorique et Hautes Energies (LPTHE), UMR 7589, Sorbonne Université et CNRS, 4 place Jussieu, 75252 Paris Cedex 05, France
[2] Institut Universitaire de France, 103 boulevard Saint-Michel, 75005 Paris, France

E-mail: fuks@lpthe.jussieu.fr

Abstract. Constraining dark matter models necessitates accurate predictions for large set of observables originating from collider physics, cosmology and astrophysics. We consider two classes of top-philic dark matter models where the dark sector is coupled to the Standard Model via the top quark and study the complementarity of dark matter relic density, direct and indirect detection and collider searches in exploring the model parameter space. We moreover investigate how higher-order corrections affect the results.

1. Introduction

Dark matter is convincingly evidenced by numerous measurements and observations, like the cosmic microwave background, the flattening of the galaxy rotation curves or gravitational lensing. These many hints triggered a huge endeavour to detect dark matter and study its properties in experiments that range from direct and indirect probes to collider searches. However, despite this huge experimental effort, there is currently no indication about the true nature of dark matter. As a consequence, dark matter scenarios are more and more constrained by many complementary datasets. We illustrate the impact of this complementarity for models in which dark matter is top-philic, *i.e.* when it dominantly couples to top quarks [1, 2]. We moreover assess the impact of QCD higher-order corrections on the predictions, both for collider processes and dark matter annihilation in the universe. In Section 2, we focus on cosmological constraints, whilst Section 3 is dedicated to collider constraints. We conclude by combining these two sets of constraints in Section 4.

2. Probing top-philic dark matter with cosmology

For concreteness, we consider a simplified top-philic dark matter scenario in which the dark matter particle is a scalar S of mass m_S and interacts with the top quark t through a Yukawa interaction $\tilde{y}_t$ involving a vector-like quark T of mass m_T. We moreover impose a $\mathbb{Z}_2$ discrete symmetry under which all Standard Model fields are even and new physics states odd, which guarantees dark matter stability and forbids any mixing of the heavy quark with the top quark. This model is described by the Lagrangian

$$\mathcal{L} = \mathcal{L}_{\rm SM} + i\bar{T}\not{D}T - m_T\bar{T}T + \frac{1}{2}\partial_\mu S\partial^\mu S - \frac{1}{2}m_S S^2 + \Big[\tilde{y}_t\, S\; \bar{T}P_R t + {\rm h.c.}\Big]\ , \qquad (1)$$

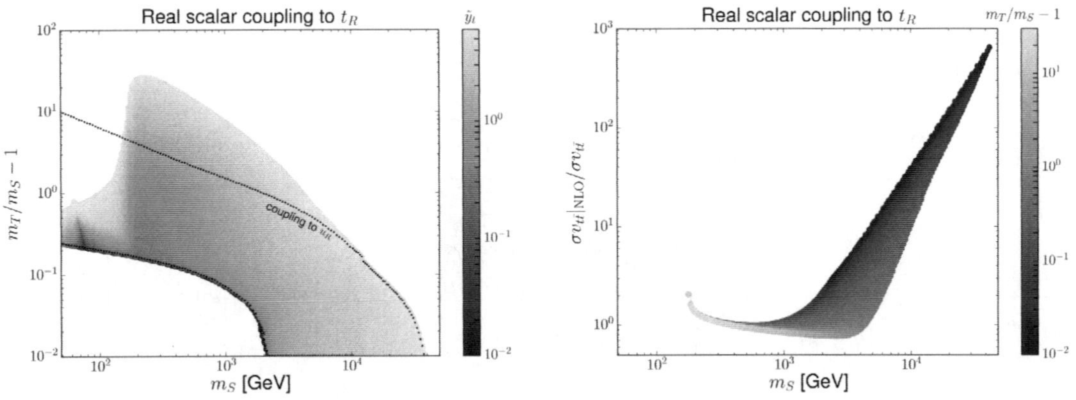

Figure 1. Left: Region of the parameter space in which the measured relic abundance can be accommodated by fixing $\tilde{y}_t$ (given through the colour code). The results are presented in the $(m_S, r-1)$ plane. For comparison, the dotted black contour represents the viable parameter space region obtained when the top mass is neglected. Right: Ratio of the NLO annihilation cross section into top-antitop pairs to the LO one. Figures from Ref. [2].

where P_R denotes the right-handed chirality projector, $\mathcal{L}_{\rm SM}$ the Standard Model Lagrangian, and where any coupling of dark matter to the Higgs sector is considered vanishing. The model parameter space is then defined by three parameters that we choose to be the dark matter mass m_S, the compression factor of the spectrum $r - 1 = m_T/m_S - 1$ and the Yukawa coupling $\tilde{y}_t$.

2.1. Relic abundance

In order to determine the dark matter relic abundance, we assume standard cosmology and a thermal freeze-out mechanism to describe dark matter dynamics in the early universe. The annihilation cross section is evaluated by including all tree-level $2 \to 2$ diagrams, to which we supplement contributions from dark matter annihilations into a three-body tWb system, Sommerfeld enhancement, annihilations into gluons and photons as well as next-to-leading-order (NLO) QCD corrections to the annihilation cross section into top-antitop pairs. Whilst threshold corrections should in principle be included as affecting the production of a top-antitop system at rest [3], they are expected to yield subleading corrections due to the absence of any toponium bound state. They are thus omitted. In the left panel of Figure 1, we present numerical predictions obtained with MicrOMEGAs [4], after having modified the program to accomodate the above-mentioned features. We show the region of the $(m_S, r-1)$ plane for which there exists a value of the Yukawa coupling $\tilde{y}_t$, indicated through the colour code, that yields a relic density $\Omega h^2 = 0.12$ compatible with Planck data [5]. Our results ignore scenarios featuring a $\tilde{y}_t$ coupling smaller than 10^{-4} and larger than 6, the former being associated with a too naive treatment of co-annihilations by MicrOMEGAs [6] and the latter corresponding to a potential breakdown of perturbation theory.

For very heavy dark matter with $m_S > 5$ TeV, dark matter annihilations into top quarks dominate. The top mass m_t however plays a subleading role, the viable region of the parameter space (coloured area) matching the contour obtained when m_t is neglected (dotted black). In this kinematical regime, the annihilation cross section is entirely driven by NLO effects, as illustrated on the right panel of Figure 1 where we present K-factors defined as the ratio of the NLO result to the leading-order (LO) one for all viable scenarios. The huge K-factors representative of the $m_S > 5$ TeV region are driven by virtual internal bremsstrahlung (VIB)

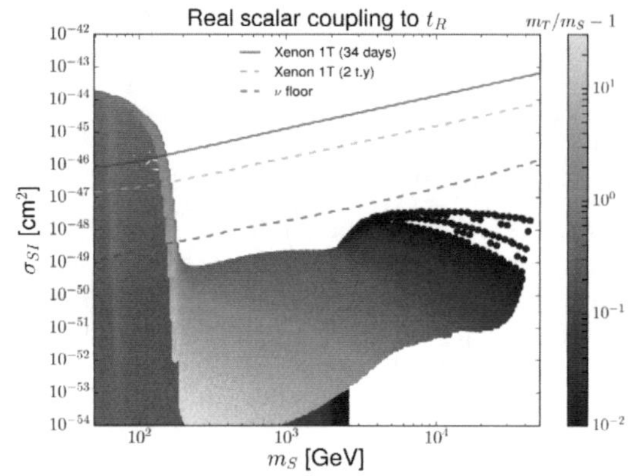

Figure 2. Spin-independent dark matter scattering cross section as a function of the dark matter mass m_S. For each scenario, the $\tilde{y}_t$ value is set to reproduce the observed relic density and the compressibility factor is depicted by the colour code. We indicate current (solid red) and future (dashed orange) 90% confidence level exclusions by the Xenon 1T experiment [10,11]. The dashed red line corresponds to the neutrino floor. Figure from Ref. [2].

diagrams [7–9]. Final-state radiation is indeed helicity suppressed (as proportional to m_t/m_S), in contrast to VIB emission that is proportional to m_T/m_S and thus enhanced as the spectrum is compressed in this parameter space region (see the colour code in the figure). For moderate dark matter masses $m_t < m_S < 5$ TeV, annihilations into top quarks still dominate but the behaviour is this time driven by tree-level S-wave contributions, the NLO corrections being small (K-factors of $\mathcal{O}(1)$). In the massless limit, annihilations into quark pairs are negligible and the relic density is driven by loop-induced annihilations into gluons (dotted black contour in the left panel of Figure 1). In contrast, in the massive case, significant annihilations into quarks allow for larger $r-1$ choices to accommodate the measured relic density (coloured contour). In the light dark matter case ($m_S < m_t$), the relic density is governed by annihilations into a tWb system or gluons, or through co-annihilations with the mediator for more compressed scenarios. Co-annihilations play an important role near $m_T + m_S \simeq m_t$ as they are resonantly enhanced, which corresponds to the dark blue region in the left panel of Figure 1 ($m_S \sim 70-80$ GeV). Annihilations into gluons are only relevant when all co-annihilation channels are closed (*i.e.* for very light dark matter), and annihilations into a tWb system only contribute close to threshold for $m_S \in [(m_t + m_W)/2, m_t]$.

2.2. Direct detection

In the scenario studied in this section, dark matter scattering off nuclei proceeds via loop-induced interactions with the gluon field strength tensor $G_{\mu\nu}$, that could be modelled effectively [12],

$$\mathcal{L}_g = \frac{\tilde{y}_t^2}{48\big[m_T^2 - m_S^2\big]} \frac{\alpha_s}{\pi} S^2 G_a^{\mu\nu} G^a_{\mu\nu} \ . \tag{2}$$

Accounting for a nucleus of mass m_A and comprised of Z protons p and $(A-Z)$ neutrons n, the spin-independent dark matter scattering cross section $\sigma_{\rm SI}$ is derived from the coherent sum of the proton and the neutron contributions,

$$\sigma_{\rm SI} = \frac{m_A^2}{\pi(m_S+m_A)^2}\Big[Zf_p + (A-Z)f_n\Big]^2 \quad \text{with} \quad \frac{f_N}{m_N} = -\frac{\tilde{y}_t^2}{54\big[m_T^2 - m_S^2\big]}\Big[1 - \sum_{q=u,d,s} f_{T_q}^{(N)}\Big] \ . \tag{3}$$

IOP Conf. Series: Journal of Physics: Conf. Series **1271** (2019) 012017 doi:10.1088/1742-6596/1271/1/012017

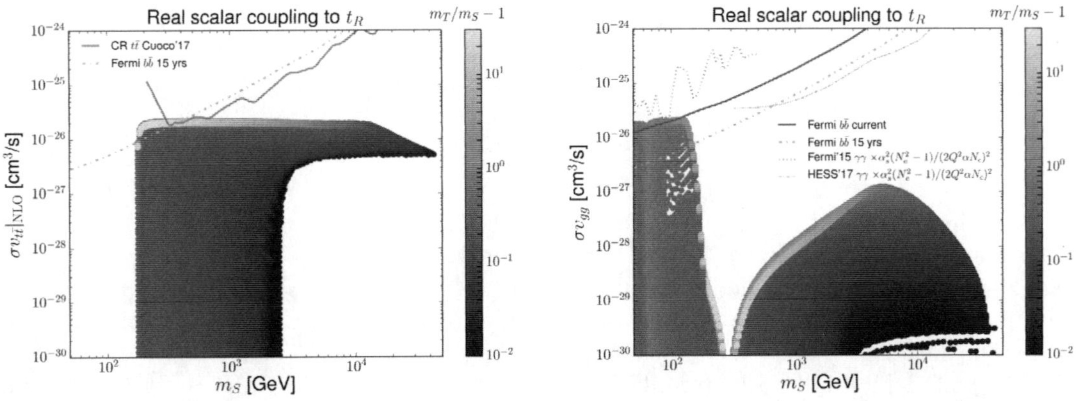

Figure 3. NLO $SS \to t\bar{t}$ (left) and loop-induced LO $SS \to gg$ (right) annihilation cross sections at zero velocity. We indicate the limits obtained from cosmic rays (solid light green), and from the analysis of current (solid dark green) and future (dot-dashed orange) Fermi-LAT dwarf spheroidal galaxy data in the $b\bar{b}$ channel. Upper limits obtained from the gamma-ray analysis of H.E.S.S. (double-dotted dashed grey) and Fermi-LAT (dotted grey) are also shown. For each scenario, the $\tilde{y}_t$ value is set to reproduce the observed relic density and the compressibility factor is depicted by the colour code. Figures from Ref. [2].

This expression depends on the coupling f_N of S to a nucleon N of mass m_N, that is related to the quark mass fractions $f_{T_q}^{(N)}$ deduced from lattice calculations [13].

In Figure 2, we present the m_S-dependence of $\sigma_{\rm SI}$ for all viable scenarios featuring a relic density in agreement with Planck data. Below the top threshold, the largest cross section values are obtained for setups in which the relic density is driven by annihilations into gluons and for which $\tilde{y}_t$ is large. Above the top threshold, the $\tilde{y}_t$ value necessary to get the right relic abundance drops, and so does $\sigma_{\rm SI}$. On different lines, $St \to Tg$ co-annihilations impact the results for $m_S \gtrsim 2.5$ TeV, yielding an enhancement of the scalar interactions with gluons in $\mathcal{L}_g$. For most scenarios, $\sigma_{\rm SI}$ lies below the neutrino floor and is thus not reachable by direct detection experiments. For spectrum featuring light dark matter annihilating into gluons, many scenarios are however within the present (34 days of exposure) and future (2.1 years of data acquisition) reach of Xenon 1T [10, 11].

2.3. Indirect detection

In Figure 3 we present, for all viable scenarios, predictions for the dark matter annihilation cross section at zero velocity for a top-antitop ($\sigma v_{t\bar{t}}$, left panel) and gluon-pair (σv_{gg}, right panel) final state. The respective cross sections are evaluated at the NLO and loop-induced LO accuracy in QCD. As already found in the previous subsections, annihilations into gluons are only relevant below the top threshold, and annihilations into top-antitop pairs dominate for $m_S > m_t$. We moreover recall that NLO effects largely affect the predictions for $m_S > 5$ TeV (see Section 2.1).

Dark matter annihilations into top-antitop pairs can be constrained from antiproton cosmic ray fluxes [14] (solid light green line), and annihilations into both top quarks and gluons can be constrained by the analysis of gamma-ray Fermi-LAT dwarf spheroidal data in the $b\bar{b}$ mode [15] (solid dark green line). Similarly, the future gamma-ray sensitivity to the model can be extrapolated after accounting for the prospects of 15 years of Fermi-LAT running [16] (dot-dashed orange lines). Although there is no direct Fermi-LAT results for annihilations into the considered final states, the subsequent secondary gamma-ray spectrum is expected to exhibit a similar behaviour as for annihilations into a $b\bar{b}$ system [2]. An estimate of the limits can therefore be extracted through a reweighting procedure [17],

$$\sigma v_{gg,t\bar{t}} = \sigma v_{b\bar{b}} \frac{N_\gamma^{b\bar{b}}}{N_\gamma^{gg,t\bar{t}}} , \tag{4}$$

where N_γ^X is the number of photons originating from an X final state that we estimate from the hadronisation model implemented in the Pythia 8 package [18].

H.E.S.S. [19] (double-dot-dashed grey line) and Fermi-LAT gamma-ray data [20] (dashed grey line) can potentially constrain the model when considering direct dark matter annihilations into photons. The latter is expected to be enhanced in two regimes, *i.e.* for light dark matter ($m_S < m_t$) where loop-induced annihilations into photons could be important (most other channels being closed) and in the multi-TeV regime where radiative emission is dominated by VIB diagrams. The corresponding cross sections, associated with dark matter annihilations into photons and into a $t\bar{t}\gamma$ system, can be deduced from

$$\sigma v_{\gamma\gamma} = \frac{4Q^4\alpha^2 N_c^2}{\alpha_S^2 (N_c^2 - 1)} \sigma v_{gg} \qquad \text{and} \qquad \sigma v_{t\bar{t}\gamma} = \frac{2N_c Q^2 \alpha}{(N_c^2 - 1)\alpha_s} \sigma v_{t\bar{t}g} , \tag{5}$$

where $N_c = 3$ denotes the number of colours, α and α_S the electromagnetic and strong coupling, and Q the electric charge of the heavy quark T. The predictions can be confronted to the results of Fermi-LAT at energies around and below m_t as well as to the results of H.E.S.S. for larger dark matter masses (following the procedure of Ref. [21]).

As shown in Figure 3, present and future gamma-ray and cosmic-ray data turns out to be sensitive to only a small fraction of the model parameter space.

3. Collider constraints on top-philic dark matter

In full generality, top-philic dark matter models can be probed at colliders via several processes [1]. First, dark matter can be produced from (virtual or resonant) mediator exchanges, potentially together with one or more Standard Model particles. Whilst dark matter is stable and leaves a missing energy signature in the detector, the extra Standard Model objects are visible and provide handles on new physics. Next, depending on the details of the model, mediator production can also resonantly leads to a purely Standard Model final state, which offers hence another set of probes of the model.

The dark matter model detailed in Section 2 does not give rise to the most general phenomenology as it cannot be probed through resonance searches in Standard Model final states. We therefore focus on a different model that can be probed through both dark matter searches and resonance searches. We illustrate this feature, as well as the impact of the QCD corrections, by studying, for the sake of the example, LHC Run 1 results. The final results on the model of the previous section will nevertheless include LHC Run 2 data and will be shown in Section 4. We consider a simplified top-philic dark matter model built by supplementing the Standard Model with a Dirac fermionic dark matter field χ of mass m_X and a scalar mediator Y_0 of mass m_Y. The interactions are described by the Lagrangian

$$\mathcal{L} = \mathcal{L}_{\rm SM} + \mathcal{L}_{\rm kin} - \Big(g_t \frac{y_t}{\sqrt{2}} \bar{t}t + g_X \bar{\chi}\chi \Big) Y_0 , \tag{6}$$

in which we denote by g_t and g_X the coupling strengths of the mediator to the Standard Model and dark sector respectively. We have assumed that the mediator couples to fermions with a strength proportional to the corresponding Yukawa coupling y_f, so that only the top quark can be considered, and $\mathcal{L}_{\rm kin}$ includes new physics kinetic and mass terms.

IOP Conf. Series: Journal of Physics: Conf. Series **1271** (2019) 012017 doi:10.1088/1742-6596/1271/1/012017

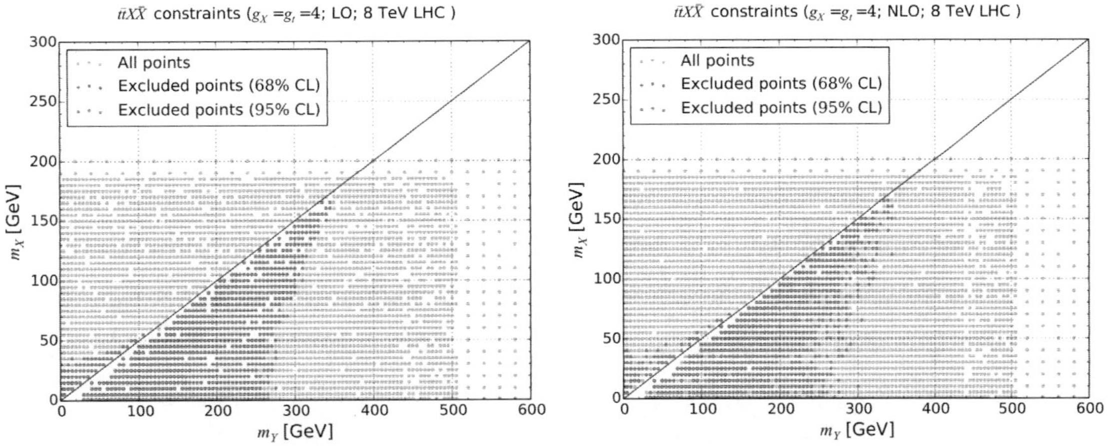

Figure 4. LHC Run 1 sensitivity to the considered model (with $g_t = g_X = 4$) based on the CMS dark matter analysis in the top-antitop plus missing energy channels. Signal simulations are achieved at LO (left) and NLO (right) in QCD. Figures from Ref. [1].

For our collider simulations, we have implemented the above Lagrangian into FEYNRULES [22], which we have jointly used with NLOCT [23] to generate an NLO UFO model [24]. We have made use of MG5_aMC@NLO [25] to generate hard-scattering events obtained by the convolution of LO and NLO matrix elements with the NNPDF 2.3 set of parton distribution functions [26]. The resulting events are matched with parton showers as modeled by PYTHIA 8 [18], that is also used to describe hadronisation. The response of the LHC detectors is handled by DELPHES 3 [27] that is tuned appropriately [28], and signal efficiencies and exclusion contours are estimated by reinterpreting the results of various LHC analyses with MADANALYSIS 5 [29–31].

3.1. Searches with missing transverse energy

Top-philic dark matter can be explored at colliders through a signature comprised of a top-quark pair and missing energy. We focus on the corresponding Run 1 CMS analysis [32]. In order to examine the impact of the QCD corrections on the sensitivity, we first fix the new physics couplings $g_t = g_X = 4$ and scan over the mediator and dark matter masses m_Y and m_X. The results are shown in Figure 4 when LO (left panel) and NLO (right panel) signal simulations are considered, and mass configurations excluded at the 68% (green) and 95% (magenta) confidence level (CL) are indicated. All excluded points (at the 95% CL) lie in the triangular low-mass region where the mediator is heavy enough to resonantly decays into a pair of dark matter particles, which corresponds to the region with the largest signal cross section. Scenarios featuring mediator masses ranging up to about 200–250 GeV are excluded, especially if the mediator decays close to threshold ($m_Y \sim 2m_X$). Further from threshold, the mediator width can become large by virtue of the large coupling choice, so that the excluded region is not exactly triangular and the signal fiducial cross section acquires a dark matter mass dependence.

As NLO corrections have a mild effect on the shape of the relevant signal distributions [33], their impact is entirely driven by changes in the signal rates. We observe that in the low-mass region (where the K-factor is found to be of about 1.10), the exclusion contours do not feature any significant effect. For larger m_Y values, the K-factors are of about 1 and the exclusion region is identical at LO and NLO. NLO corrections are however expected to significantly reduce theoretical uncertainties. In order to study this, we select the three benchmark scenarios

IOP Conf. Series: Journal of Physics: Conf. Series **1271** (2019) 012017 doi:10.1088/1742-6596/1271/1/012017

Table 1. Benchmark scenarios used to determine the NLO impact on the sensitivity of the Run 1 CMS search for dark matter in the top-antitop plus missing energy channel. LO and NLO cross sections are shown together with the CLs exclusions, all results including scale uncertainties.

	(m_Y, m_X)	$\sigma_{\rm LO}$ [pb]	$\rm CL_{LO}$ [%]	$\sigma_{\rm NLO}$ [pb]	$\rm CL_{NLO}$ [%]
I	(150, 25) GeV	$0.658^{+34.9\%}_{-24.0\%}$	$98.7^{+0.8\%}_{-13.0\%}$	$0.773^{+6.1\%}_{-10.1\%}$	$95.0^{+2.7\%}_{-0.4\%}$
II	(40, 30) GeV	$0.776^{+34.2\%}_{-24.1\%}$	$74.7^{+19.7\%}_{-17.7\%}$	$0.926^{+5.7\%}_{-10.4\%}$	$84.2^{+0.4\%}_{-14.4\%}$
III	(240, 100) GeV	$0.187^{+37.1\%}_{-24.4\%}$	$91.6^{+6.4\%}_{-18.1\%}$	$0.216^{+6.7\%}_{-11.4\%}$	$86.5^{+8.6\%}_{-5.5\%}$

presented in Table 1 together with the corresponding LO and NLO signal cross sections and the CLs exclusions. We additionally indicate the scale uncertainty bands that have been obtained by varying the central scale, set to the average transverse mass of all final state particles, by a factor of 2 up and down. Using NLO predictions leads to a significant reduction of the uncertainties in the total cross section, compared with the LO case, which propagates down to the CLs exclusions. NLO predictions therefore allow us to draw more reliable conclusions on the viability of a parameter space point.

Mono-X signals, in which a pair of dark matter particles is produced together with a Standard Model particle X, can also be relevant for obtaining bounds on dark matter models. Amongst all these searches, we consider the monojet one given the relative magnitude of the strong and electroweak couplings. As an example, we use the Run 1 CMS monojet analysis [34] and signal simulations at LO (as loop-induced). In our recasting procedure, we conservatively select the signal region driving the strongest bound. The results are shown in the left panel of Figure 5 for $g_t = g_X = 4$, in which we represent scenarios excluded at the 68% (green) and 95% (magenta) CL. Most excluded scenarios lie again in the triangular low-mass region of the parameter space, in which the mediator resonantly decays into a pair of dark matter particles. Except for the small subset of points excluded at the 68% CL for setups where $m_Y < 2m_X$, the boundaries of the excluded region are derived from the significant reduction of the monojet fiducial cross section that rapidly falls with m_Y once the top threshold is crossed. The LHC Run 1 is therefore not sensitive to mediator heavier than 500 GeV. In the offshell $m_Y < 2m_X$ regime, jets are harder [35], so that the signal selection efficiency is larger and such scenarios can be reached. The monojet search is overall found to be more constraining than the top-antitop plus missing energy one, especially for heavier mediator thanks to the larger monojet cross section.

3.2. Searches without missing transverse energy

Dark matter models can also be constrained by resonance searches when the new physics resonance decays into a pair of Standard Model particles, as mediators can sometimes decay back into the Standard Model sector. Dijet and diphoton resonance searches could in principle constrain the considered model. However, due to the double-loop suppression stemming from the mediator couplings to gluons and photons, new physics contributions to dijet and diphoton production are only relevant for $m_Y < 2\ \min(m_X, m_t)$, *i.e.* when the mediator cannot decay into top quarks and dark matter. Due to the large QCD background and trigger reasons, LHC dijet resonance searches target high invariant-mass dijet systems, so that the lowest mediator mass that could be reached by Run 1 data is of about 500 GeV, the corresponding cross section bound being 10 pb [36]. This consequently yields no constraint on the model. In contrast, despite the small mediator branching ratio into a diphoton final state, diphoton probes can potentially constrain the model due to the low level of associated backgrounds. In the right panel of Figure 5, we present, in the (m_Y, g_t) plane, the constraints derived from the results of

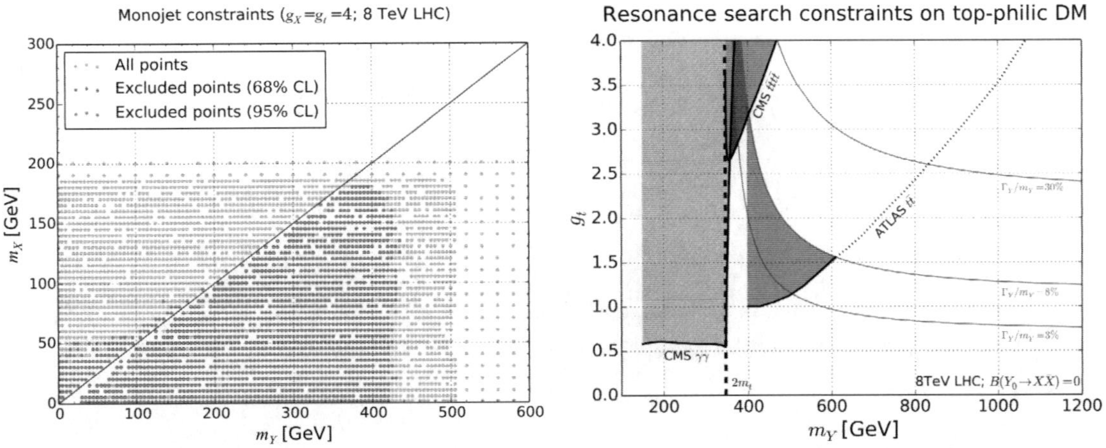

Figure 5. Left: LHC Run 1 monojet constraints on the considered top-philic dark matter model as obtained by reinterpreting the Run 1 CMS monojet search. Right: LHC Run 1 constraints issued from the reinterpretation of new physics searches in the diphoton (orange), $t\bar{t}$ (magenta) and $t\bar{t}t\bar{t}$ (blue) modes. Information on the mediator width to mass ratio is represented by green curves. Figures from Ref. [1].

the Run 1 CMS diphoton search [37] that investigates resonance masses ranging from 150 to 850 GeV. The bounds are obtained by comparing the experimental limits with predictions for single mediator production at the next-to-next-to-leading-order (NNLO) [38].

For scenarios with mediator masses above the top threshold, one relies instead on signatures featuring a $t\bar{t}$ resonance. In our setup, loop-induced single mediator production can enhance top-antitop production, in particular when there is a large coupling hierarchy ($g_t \gg g_X$) or mass hierarchy ($2m_t < m_Y < 2m_X$). We extract constraints from the results of the Run 1 ATLAS $t\bar{t}$ resonance search [39] that restricts the new physics contributions to the $t\bar{t}$ cross section as a function of the mediator mass. Comparing with NNLO signal cross section predictions, we show the obtained constraints in the right panel of Figure 5 (magenta) in the (m_Y, g_t) plane, dark matter being again decoupled. Mediators with masses lying in the [400,600] GeV range are found to be reachable for $g_t \in [1, 4]$. The exact form of the excluded region depends on m_Y and on the validity of the narrow-width approximation that must be enforced to recast the ATLAS limits (green curves). In addition, mediators can also be produced in association with a top-antitop pair, which yields a four-top signal after decay. Large new physics four-top contributions are forbidden [40], and recasting the limits extracted from the CMS analysis of Ref. [41] leads to the weak blue excluded region in Figure 5, as resulting from the steeply decreasing (NLO) signal cross section with m_Y.

4. Summary - the collider-cosmology complementarity

We now go back to the model of Section 2 and jointly apply all the constraints discussed in this paper. Our findings, reported in Figure 6, reflect the nice complementarity of cosmological and collider (LEP as well as LHC Run 1 and 2) constraints on the considered dark matter model. We compare the impact of experimental searches from varied origins by confronting their results to predictions including NLO QCD corrections, the latter being crucial for a good theoretical control. The red region is excluded by dark matter direct detection searches from the Xenon 1T [10] experiment, and the region inside the red dashed line is in principle testable by future direct detection searches as above the neutrino floor. The dark and light green regions are

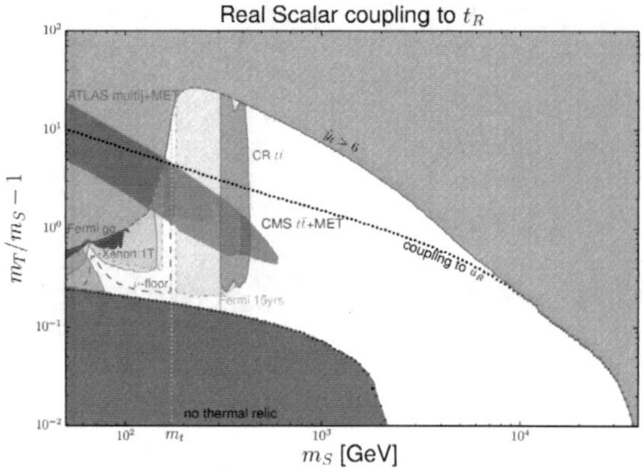

Figure 6. Phenomenologically-viable region of the top-philic dark matter model of Section 2, presented in the $(m_S, r-1)$ plane. The viable region of the parameter space (non-grey regions) is obtained by fixing the $\tilde{y}_t$ coupling to reproduce the observed relic density. We include constraints from direct and indirect dark matter detection as well as from colliders. Figure from Ref. [2].

excluded by Fermi-LAT gamma-ray and cosmic-ray constraints on low and large dark matter masses respectively [14, 15], while the orange area will be testable after 15 years of Fermi-LAT running [16]. Finally, collider searches in the top-antitop plus missing energy channel at LEP [42] and at the LHC [43, 44] are presented as the magenta and light blue contours respectively, and mono-X bounds [45, 46] are given in dark blue.

Our analysis reveals that despite the complementarity between the different sets of constraints, only a small fraction of the parameter space is currently tested. The prospects of all cosmological probes however demonstrate that the most fruitful strategy to further test the model on the longer run would be to increase the energy reach at colliders.

Acknowledgments

I am thankful to the organisers of the Kruger 2018 conference for the invitation and for putting such an inspiring meeting in place. I also warmly thank Alan Cornell for his hospitality at U. Witwatersrand. This study has been partly supported by French state funds managed by the Agence Nationale de la Recherche (ANR) in the context of the LABEX ILP (ANR-11-IDEX-0004-02, ANR-10-LABX-63).

References

[1] Arina C *et al.* 2016 *JHEP* **11** 111 (*Preprint* 1605.09242)
[2] Colucci S, Fuks B, Giacchino F, Lopez Honorez L, Tytgat M H G and Vandecasteele J 2018 *Phys. Rev.* **D98** 035002 (*Preprint* 1804.05068)
[3] Drees M and Hikasa K i 1990 *Phys. Lett.* **B240** 455 [Erratum: Phys. Lett.B262,497(1991)]
[4] Bélanger G, Boudjema F, Pukhov A and Semenov A 2015 *Comput. Phys. Commun.* **192** 322–329 (*Preprint* 1407.6129)
[5] Ade P A R *et al.* (Planck) 2014 *Astron. Astrophys.* **571** A16 (*Preprint* 1303.5076)
[6] Garny M, Heisig J, Lülf B and Vogl S 2017 *Phys. Rev.* **D96** 103521 (*Preprint* 1705.09292)
[7] Giacchino F, Lopez-Honorez L and Tytgat M H G 2013 *JCAP* **1310** 025 (*Preprint* 1307.6480)
[8] Toma T 2013 *Phys. Rev. Lett.* **111** 091301 (*Preprint* 1307.6181)
[9] Colucci S, Giacchino F, Tytgat M H G and Vandecasteele J 2018 *Phys. Rev.* **D98** 115029 (*Preprint* 1805.10173)
[10] Aprile E *et al.* (XENON) 2017 *Phys. Rev. Lett.* **119** 181301 (*Preprint* 1705.06655)
[11] Aprile E *et al.* (XENON) 2016 *JCAP* **1604** 027 (*Preprint* 1512.07501)
[12] Hisano J, Nagai R and Nagata N 2015 *JHEP* **05** 037 (*Preprint* 1502.02244)

[13] Ohki H, Takeda K, Aoki S, Hashimoto S, Kaneko T, Matsufuru H, Noaki J and Onogi T (JLQCD) 2013 *Phys. Rev.* **D87** 034509 (*Preprint* 1208.4185)
[14] Cuoco A, Heisig J, Korsmeier M and Krämer M 2017 (*Preprint* 1711.05274)
[15] Ackermann M *et al.* (Fermi-LAT) 2015 *Phys. Rev. Lett.* **115** 231301 (*Preprint* 1503.02641)
[16] Charles E *et al.* (Fermi-LAT) 2016 *Phys. Rept.* **636** 1–46 (*Preprint* 1605.02016)
[17] Bringmann T, Huang X, Ibarra A, Vogl S and Weniger C 2012 *JCAP* **1207** 054 (*Preprint* 1203.1312)
[18] Sjöstrand T, Ask S, Christiansen J R, Corke R, Desai N, Ilten P, Mrenna S, Prestel S, Rasmussen C O and Skands P Z 2015 *Comput. Phys. Commun.* **191** 159–177 (*Preprint* 1410.3012)
[19] Rinchiuso L, Moulin E, Viana A, Van Eldik C and Veh J (H.E.S.S.) 2017 *PoS* **ICRC2017** 893 (*Preprint* 1708.08358)
[20] Ackermann M *et al.* (Fermi-LAT) 2015 *Phys. Rev.* **D91** 122002 (*Preprint* 1506.00013)
[21] Ibarra A, Lee H M, Lopez Gehler S, Park W I and Pato M 2013 *JCAP* **1305** 016 [Erratum: JCAP1603,no.03,E01(2016)] (*Preprint* 1303.6632)
[22] Alloul A, Christensen N D, Degrande C, Duhr C and Fuks B 2014 *Comput. Phys. Commun.* **185** 2250–2300 (*Preprint* 1310.1921)
[23] Degrande C 2015 *Comput. Phys. Commun.* **197** 239–262 (*Preprint* 1406.3030)
[24] Degrande C, Duhr C, Fuks B, Grellscheid D, Mattelaer O and Reiter T 2012 *Comput. Phys. Commun.* **183** 1201–1214 (*Preprint* 1108.2040)
[25] Alwall J, Frederix R, Frixione S, Hirschi V, Maltoni F, Mattelaer O, Shao H S, Stelzer T, Torrielli P and Zaro M 2014 *JHEP* **07** 079 (*Preprint* 1405.0301)
[26] Ball R D *et al.* 2013 *Nucl. Phys.* **B867** 244–289 (*Preprint* 1207.1303)
[27] de Favereau J, Delaere C, Demin P, Giammanco A, Lemaître V, Mertens A and Selvaggi M (DELPHES 3) 2014 *JHEP* **02** 057 (*Preprint* 1307.6346)
[28] Dumont B, Fuks B, Kraml S, Bein S, Chalons G, Conte E, Kulkarni S, Sengupta D and Wymant C 2015 *Eur. Phys. J.* **C75** 56 (*Preprint* 1407.3278)
[29] Conte E, Fuks B and Serret G 2013 *Comput. Phys. Commun.* **184** 222–256 (*Preprint* 1206.1599)
[30] Conte E, Dumont B, Fuks B and Wymant C 2014 *Eur. Phys. J.* **C74** 3103 (*Preprint* 1405.3982)
[31] Conte E and Fuks B 2018 *Int. J. Mod. Phys.* **A33** 1830027 (*Preprint* 1808.00480)
[32] Khachatryan V *et al.* (CMS) 2015 *JHEP* **06** 121 (*Preprint* 1504.03198)
[33] Backovic M, Kramer M, Maltoni F, Martini A, Mawatari K and Pellen M 2015 *Eur. Phys. J.* **C75** 482 (*Preprint* 1508.05327)
[34] Khachatryan V *et al.* (CMS) 2015 *Eur. Phys. J.* **C75** 235 (*Preprint* 1408.3583)
[35] Mattelaer O and Vryonidou E 2015 *Eur. Phys. J.* **C75** 436 (*Preprint* 1508.00564)
[36] Khachatryan V *et al.* (CMS) 2016 *Phys. Rev. Lett.* **117** 031802 (*Preprint* 1604.08907)
[37] Khachatryan V *et al.* (CMS) 2015 *Phys. Lett.* **B750** 494–519 (*Preprint* 1506.02301)
[38] Andersen J R *et al.* (LHC Higgs Cross Section Working Group) 2013 (*Preprint* 1307.1347)
[39] Aad G *et al.* (ATLAS) 2015 *JHEP* **08** 148 (*Preprint* 1505.07018)
[40] Beck L, Blekman F, Dobur D, Fuks B, Keaveney J and Mawatari K 2015 *Phys. Lett.* **B746** 48–52 (*Preprint* 1501.07580)
[41] Khachatryan V *et al.* (CMS) 2014 *JHEP* **11** 154 (*Preprint* 1409.7339)
[42] Abbiendi G *et al.* (OPAL) 2002 *Phys. Lett.* **B545** 272–284 [Erratum: Phys. Lett.B548,258(2002)] (*Preprint* hep-ex/0209026)
[43] Sirunyan A M *et al.* (CMS) 2018 *Phys. Rev.* **D97** 032009 (*Preprint* 1711.00752)
[44] The CMS Collaboration (CMS) 2017 *CMS-PAS-SUS-16-052*
[45] Aaboud M *et al.* (ATLAS) 2016 *Phys. Rev.* **D94** 032005 (*Preprint* 1604.07773)
[46] Aaboud M *et al.* (ATLAS) 2016 *Eur. Phys. J.* **C76** 392 (*Preprint* 1605.03814)

IOP Conf. Series: Journal of Physics: Conf. Series **1271** (2019) 012018 doi:10.1088/1742-6596/1271/1/012018

Hydrodynamic flow in small systems

or: "How the heck is it possible that a system emitting only a dozen particles can be described by fluid dynamics?"

Ulrich Heinz[1a], **in collaboration with J. Scott Moreland**[b]

[a]Department of Physics, The Ohio State University, Columbus, OH 43210-1117, USA
[b]Department of Physics, Duke University, Durham, NC 27708-0305, USA

E-mail: heinz.9@osu.edu

Abstract. The "unreasonable effectiveness" of relativistic fluid dynamics in describing high energy heavy-ion and even proton-proton collisions will be demonstrated and discussed. Several recent ideas of optimizing relativistic fluid dynamics for the specific challenges posed by such collisions will be presented, and some thoughts will be offered why the framework works better than originally expected. I will also address the unresolved question where exactly hydrodynamics breaks down, and why.

1. Prologue

In recent years high-energy nuclear collisions at RHIC and the LHC have revealed strong indications for collective flow with hydrodynamic characteristics even in so-called "small"[2] collision systems (p-p, p-Au, d-Au, ^{3}He-Au, and p-Pb; see e.g. the reviews [1–3]). The question in the subtitle above is one that I get frequently asked in this context. Let me start by explaining that it is the wrong question to ask. To illustrate my point allow me to consider a world without quarks where the strong interaction is described by an SU(3) gauge theory ("QCD") which contains only gluons in its color-deconfined "gluon plasma" state and only glueballs (G) in its color-confined hadronic phase. In such a world a GG collision at LHC energies would create a gluon plasma with similar initial energy (e), entropy (s) and (if it allows for a quasiparticle description) total particle density (n) as the quark-gluon plasma created in a pp collision in our world at the real LHC. The equation of state (EoS) $p(e)$, speed of sound $c_s(e)$, and transport coefficients (such as the specific shear and bulk viscosities η/s, ζ/s) of this gluon plasma will be very similar to those of the quark-gluon plasma in our world where these quantities are all dominated by the interactions with and among gluons. So the dynamical evolution of the

[1] This work was supported in part by the U.S. Department of Energy (DOE), Office of Science, Office for Nuclear Physics, under Awards No. DE-SC0004286, DE-FG02-05ER41367, and through the Beam Energy Scan Theory (BEST) Collaboration, as well as by the National Science Foundation (NSF) within the framework of the JETSCAPE Collaboration under Award No. ACI-1550223. This work used the Extreme Science and Engineering Discovery Environment (XSEDE), which is supported by National Science Foundation grant number ACI-1548562. Travel support through a bilateral grant from FAPESP and The Ohio State University, as well as the hospitality of the Centre for Theoretical and Mathematical Physics (CTMP) at the University of Cape Town, are gratefully acknowledged.

[2] More about those quotation marks later.

gluon plasma created in GG collisions in this imaginary world will look qualitatively similar to that of the quark-gluon plasma created in pp collisions at the real LHC. If the total entropy of the collision fireball is enough to create, say, a dozen charged hadrons per unit pseudorapidity, corresponding to $dN/d\eta \simeq 20$ if you include neutrals, in our real world where most of these hadrons are pions with a rest mass of 140 MeV, it would not even suffice to create two glueballs G per unit rapidity, $dN_G/d\eta \simeq 2$, in the imaginary world if the lightest glueballs had a mass of 1.5 GeV or more.

Does this imply that the gluon plasma created in a GG collision with $dN_G/d\eta = 2$ in the imaginary world evolves less hydrodynamically than the quark-gluon plasma in a pp collision with $dN_{\rm ch}/d\eta = 12$ in the real world? Obviously not. That the former collision has much fewer particles in the final state than the latter is a cruel joke of Nature who forces the partition of the system's energy into a small number of very heavy final state hadrons in the glueball world while creating, under the same initial conditions, an order of magnitude more final-state hadrons in our real world. If pions where lighter (say 10 MeV instead of 140 MeV), that same pp collision would create about 300 hadrons per unit rapidity in its final state, the same order of magnitude as measured in off-center PbPb collisions at the LHC where few physicists doubt the validity of the hydrodynamic flow paradigm [4–7]. The quantization of emitted energy in heavy chunks implies that the underlying fluid dynamical behavior cannot be sampled continuously and suffers from finite number statistical fluctuations — even more so in the glueball world than in ours — such that its exploration requires averaging over many similar collision events (same collision system, centrality and collision energy) in order to sample the underlying physics with sufficient statistical precision. So, while the gluon plasma created in the GG collision of our imaginary glueball world may exhibit almost identical hydrodynamic flow patterns to the quark-gluon plasma in a pp collision at the LHC with the same initial entropy per unit rapidity, these patterns would be much harder to discern in the GG collision, due to much larger finite number statistical fluctuations in the final state. This doesn't mean, however, that no such patterns exists – it is just difficult to distill them from the strongly fluctuating observables.

I hope that this *Gedankenexperiment* convinces you that the absolute value of the number of final state hadrons per unit rapidity is a poor criterium for (pre-)judging the applicability of the hydrodynamic flow paradigm. Final state hadrons are only created at the end of the collision when the quark-gluon plasma hadronizes. Afterwards the hydrodynamic model quickly breaks down, due to the short-range nature of the residual "strong" interactions between color-neutral hadrons. The final-state hadrons are not responsible for the interactions that control the system's evolution towards local thermal equilibrium in its color-deconfined liquid stage – its EoS, speed of sound, and its transport properties. In its liquid state, the strong open-color interactions in the quark-gluon plasma may even largely invalidate its description in terms of well-defined quasiparticles, again making the number of particle degrees of freedom per unit rapidity a poor criterium for (pre-)judging its ability to develop hydrodynamic flow. The absolute value of the (initial) entropy per unit space-time rapidity, $dS/d\eta_s = \tau_0 \int d^2r_\perp s(\boldsymbol{r}_\perp, \eta, \tau_0)$, on the other hand, which is (on average) monotonically related to the final state charged hadron pseudorapidity density $dN_{\rm ch}/d\eta$ [8], remains well-defined even in strongly-coupled quantum field theories without good quasi-particles, and thus may be a better starting point for a breakdown criterium of the hydrodynamic paradigm (see, e.g., [7, 9, 10]).

2. The "unreasonable effectiveness" of hydrodynamics for nuclear collisions

In the last decade, relativistic dissipative ("viscous") fluid dynamics has become the workhorse of dynamical modeling of ultra-relativistic heavy-ion collisions [4–7]. In spite of the extraordinarily rapid expansion of the collision fireball, with dramatically different expansion rates along the beam direction (due to the inability of the two colliding nuclei to stop each other [11]) and in the transverse directions where the expansion is driven by pressure gradients and starts from zero, which generates large shear stresses, the hydrodynamic model has proven to possess high predictive power. (An early example is shown in Fig. 1.) It works even in "small"

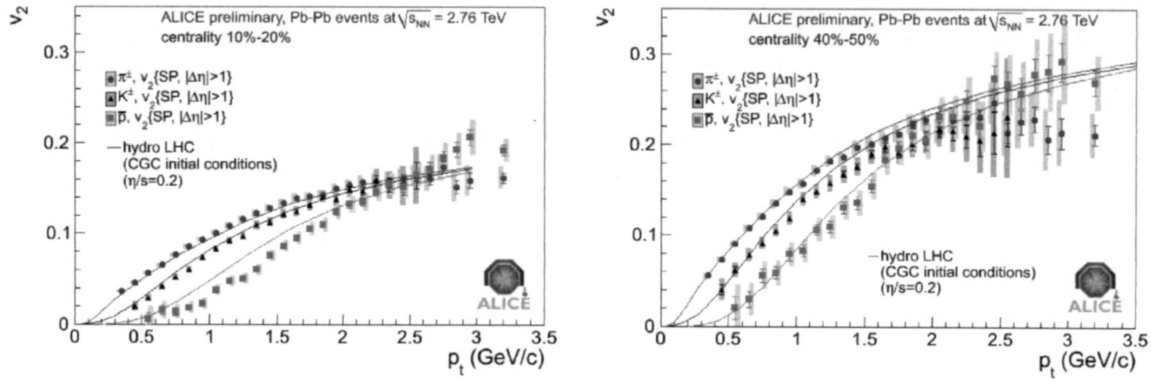

Figure 1. Differential elliptic flow for pions, kaons and protons in semi-central (left) and semi-peripheral (right) Pb-Pb collisions at the LHC, as reported by the ALICE Collaboration at the Quark Matter 2011 conference [12]. Solid lines show hydrodynamic ***pre***dictions from [13].

collision systems, such as p-Pb and p-p at the LHC [14] (see Fig. 2) or p-Au, d-Au, and ^{3}He-Au at RHIC [15], as long as subnucleonic fluctuations in the initial energy deposition are appropriately accounted for [16]. The largest uncertainties in comparing data from such small

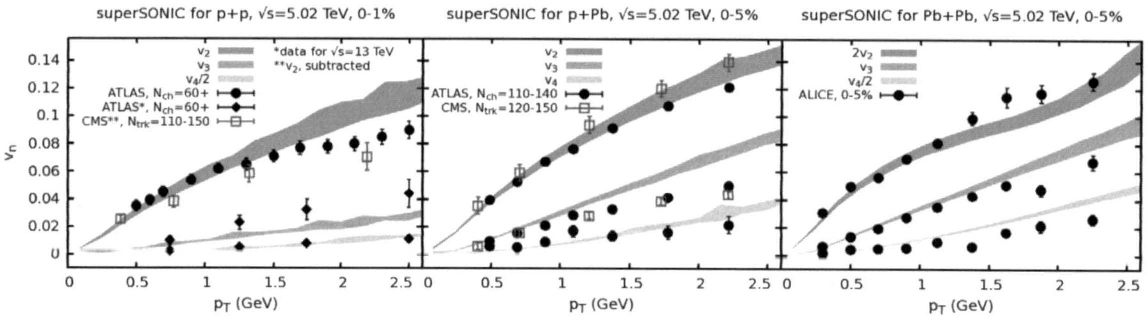

Figure 2. Differential elliptic (v_2), triangular (v_3), and quadrangular flow (v_4) for charged hadrons from p-p (left), p-Pb (middle) and PbPb (right) collisions at the LHC, compared with hydrodynamic model simulations using the superSONIC code package [17]. Figure taken from [14].

systems with fluid dynamical code packages (e.g. iEBE-VISHNU [18] (`https://u.osu.edu/vishnu`), superSONIC [17] (`https://sites.google.com/site/revihy/download`), or MUSIC [19] (`http://www.physics.mcgill.ca/music/`)) does not appear to arise from the applicability of the hydrodynamic model, but from our lack of precise knowledge of the internal structure of the nucleon, i.e. the distribution and event-by-event fluctuations of the gluon density inside protons and neutrons [16].

Let me explain now why I use quotation marks when writing about "small" collision systems. As argued above, a good starting point for (pre-)judging the applicability of hydrodynamics is the total entropy per unit space-time rapidity $dS/d\eta_s$ deposited in the collision zone. Although, for a given collision configuration, $dS/d\eta_s$ is monotonically related to the final charged hadron pseudorapidity density $dN_{\rm ch}/d\eta$, the proportionality constant depends on the additional entropy produced by viscous heating during the expansion, and the latter increases with the fireball expansion rate. Fig. 3 compares isotherms along the short and long directions of elliptically deformed fireballs created in peripheral Pb-Pb, central p-Pb, and high-multiplicity p-p collisions at $\sqrt{s_{\rm NN}} = 5.02\,$TeV with the same final charged hadron pseudorapidity density $dN_{\rm ch}/d\eta = 100$. (For comparison the right column shows the corresponding isotherms for less extreme p-p collisions with a five times smaller final multiplicity.) This comparison illustrates a number

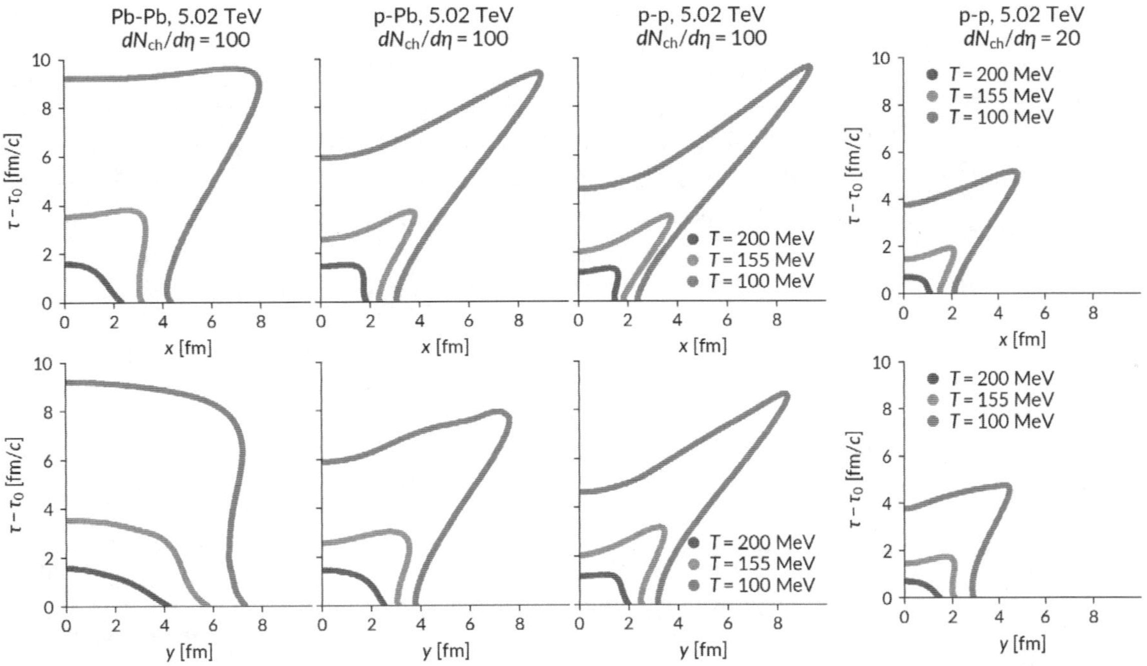

Figure 3. Isotherms of temperatures $T = 200$ (blue), 155 (orange) and 100 MeV (green) along the short (x, top row) and long (y, bottom row) directions of an ensemble-averaged fireball constructed from elliptically deformed and aligned fluctuating initial entropy density profiles from the T$_{\rm R}$ENTo model with exponent $p = 0$ [20] (similar to IP-GLASMA initial conditions) for Pb-Pb (left), p-Pb (center-left), and p-p collisions (center-right and right columns) at $\sqrt{s_{\rm NN}} = 5.02\,$TeV. The three left columns compare events from the three different collision systems with the same charged hadron pseudorapidity density $dN_{\rm ch}/d\eta = 100$ in the final state. The right column shows for comparison the corresponding isotherms for p-p collisions at the same collision energy but with five times smaller final pseudorapidity density $dN_{\rm ch}/d\eta = 20$. All events were evolved with iEBE-VISHNU [18] using transport coefficients and other model parameters determined by Bayesian model calibration [21].

of important points: First, the fireballs created in the so-called "small" collision systems p-Pb and p-p are only initially small, due to the small cross section of the proton. As long as roughly the same total entropy $dS/d\eta_s$ is deposited initially, they all have roughly the same (much larger) size at hadronization and at freeze-out. This is easy to understand: since constant temperature implies constant density, identical multiplicities must correspond to identical volumes. Therefore, events with the same final multiplicity $dN_{\rm ch}/d\eta$ have the same freeze-out volume, irrespective of how dilute or compact the fireball's initial configuration was.

Second, however, if the initial entropy $dS/d\eta_s$ is initially deposited within a smaller transverse area, the larger transverse pressure gradients in this more compact initial configuration drives stronger radial transverse flow [22,23], reflected in the larger rate of growth of the outer radius of the isotherms in the center-left and center-right columns of Fig. 3 compared to the left column.

The resulting higher expansion rate reduces the "Hubble-volume" of the expanding fireball whose dimensions ("HBT radii") can be measured with two-particle intensity interferometry [24]. At the same final multiplicity and volume, therefore, p-p collisions feature smaller HBT radii than p-Pb collisions, and p-Pb collisions have smaller HBT radii than Pb-Pb collisions. This effect has been observed and noted by the ALICE Collaboration (see Fig. 9 in [25]).

A popular criterium for the validity of fluid dynamics is the Knudsen number, defined as the ratio between the microscopic interaction length ("mean free path") and the macroscopic hydrodynamic length scale. In expanding systems, this macroscopic length scale is not given by the total radius of the fireball (related to the total freeze-out volume) but by its Hubble radius (which can be expressed through the expansion rate or through appropriately normalized space-time gradients of the energy or entropy density). This suggests that any phenomenological criterium for applicability of hydrodynamics should involve, in addition to the observed charged hadron multiplicity $dN_{\mathrm{ch}}/d\eta$ (as a proxy for the initial entropy density $dS/d\eta_s$), the HBT radii or, better, the cube root of the HBT volume $\left|\det(R^2_{ij})\right|^{1/6}$ (as a proxy for the Hubble radius).

As an aside I note that the higher expansion rate causes stronger viscous heating in the "small" collision systems. So, if the three systems shown in Fig. 3 had been initialized with the same initial entropy per unit rapidity $dS/d\eta_s$, the final entropy (and thus $dN_{\mathrm{ch}}/d\eta$) and final total volumes would be somewhat larger for p-Pb than for Pb-Pb, and still larger for p-p collisions. This would have further exacerbated the above-mentioned effects on the radial flow and HBT radii.

That initially denser collision systems develop stronger radial flow, as predicted by hydrodynamics, can be directly seen in Fig. 4. It shows pion, kaon and proton spectra from

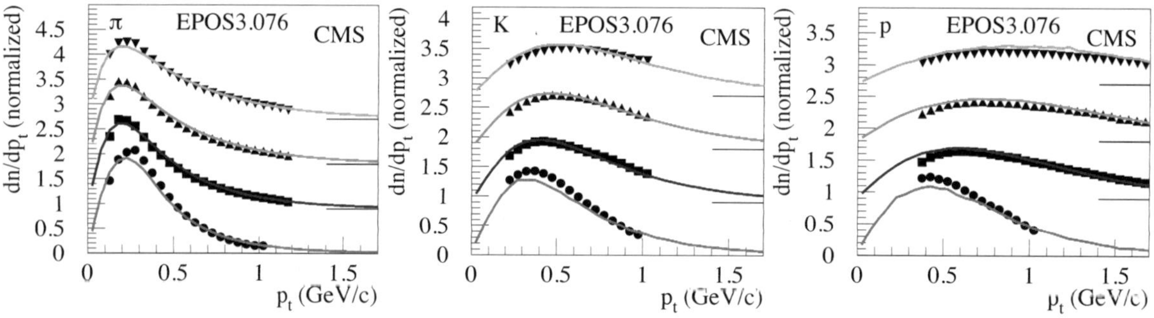

Figure 4. Transverse momentum spectra of pions, kaons and protons from p-Pb collisions at 5.02 TeV with different numbers of charged tracks in $|\eta| < 2.4$ (8 (dots), 84 (squares), 160 (triangles), 235 (inverted triangles)) as measured by CMS [26]. The lines show numerical simulations with the EPOS3.076 code which features a viscous fluid dynamic core [27]. Similar behavior was found in by CMS in p-p collisions at 0.9, 2.76 and 7 TeV [28]. Figure taken from [27].

5.02 TeV p-Pb collisions measured by CMS for four multiplicity bins. As the multiplicity increases, the spectra become flatter ("harder"), and the effect increases with particle rest mass. This is a hallmark of radial flow [29,30] which (i) increases with the initial entropy density and (ii) pushes hadrons out to larger p_T values by an amount that increases with their rest mass. For the same final multiplicity and collision energy, the effect is stronger in p-p collisions [28] than in p-Pb collisions [26]. The measured behavior can be quantitatively described by EPOS3.076 which has a viscous hydrodynamic core. That same model also describes the elliptic flow ("double ridge") discovered by CMS in high-multiplicity p-p collisions at 7 TeV [31] and later confirmed by both ATLAS [32] and CMS [33] in p-p collisions at 13 TeV, whose collective nature was demonstrated by CMS by measuring it with 4- and 6-particle cumulants [34].

IOP Conf. Series: Journal of Physics: Conf. Series **1271** (2019) 012018 doi:10.1088/1742-6596/1271/1/012018

3. Far-from-equilibrium hydrodynamics

The discussion above has established that (i) viscous fluid dynamics is phenomenologically very successful and yields quantitatively precise descriptions of and predictions for soft-hadron spectra and flow correlations in Au-Au at RHIC and Pb-Pb at the LHC while providing at least a semiquantitative description of the same observables in p-Au, d-Au, ^{3}He-Au, p-Pb and even high-multiplicity p-p collisions, while at the same time (ii) being characterized by large dissipative effects caused by the extremely rapid and anisotropic expansion of the heavy-ion collision fireballs. In fact, the approach is now being successfully used for extracting, with quantified uncertainties, key parameters characterizing the thermodynamic and transport properties of quark-gluon plasma from a global model-to-data comparison with advanced Bayesian statistical analysis tools [21, 35]. Why do these large dissipative corrections not destroy the precision and predictive power of the hydrodynamic approach?

In the last part of my presentation I will cover some work that my collaborators and I have performed over the last few years to address the particular challenges faced by hydrodynamic approaches when applied to relativistic heavy-ion collisions. These studies uncovered several surprises that showed that the hydrodynamic approach is much more robust and resilient than originally expected. We now understand that its applicability requires neither local thermalization (*i.e.* thermalized exponential momentum distributions in the local rest frame) nor even local momentum isotropy. This is a dramatic change in our understanding compared to 20 years ago when it was believed (certainly by me!) that the good agreement between ideal fluid dynamics and RHIC data [36] implied very short thermalization times of order $< 1\,\mathrm{fm}/c$ [37]. We now understand that this time characterizes the time scale of "hydrodynamization" at which the system enters the region of validity of a second-order viscous hydrodynamic approach, rather than real local thermalization at which the fluid would obey the laws of ideal fluid dynamics. In other words, ***dissipative hydrodynamics works even far from local thermal equilibrium, with quantitative precision.***

Ultra-relativistic heavy-ion collision dynamics pose two specific challenges to the applicability of dissipative fluid dynamics [38–40]: (i) a large shear-viscous stress, in the form of a large difference $P_\perp - P_L$ between the transverse and longitudinal pressures, caused by large initial anisotropies between the longitudinal and transverse expansion rates, and (ii) a possibly large bulk viscous pressure Π caused by critical dynamics near the quark-hadron phase transition. Optimized hydrodynamic approaches, such as anisotropic hydrodynamics [38, 40–45] can handle these challenges more efficiently than standard dissipative fluid dynamics.

Hydrodynamics is an effective theory whose form is independent of the strength of the microscopic interactions. Hydrodynamic equations can thus be derived from kinetic theory in a window of weak coupling and small pressure gradients where both approaches are simultaneously valid. Only the values of the transport coefficients and the equation of state depend on the microscopic coupling strength; for the strongly coupled quark-gluon plasma created in heavy-ion collisions, they must be obtained with non-perturbative methods.

In kinetic theory, the conserved macroscopic currents $j^\mu(x) = \langle p^\mu \rangle(x)$ (particle current) and $T^{\mu\nu}(x) = \langle p^\mu p^\nu \rangle(x)$ (energy-momentum tensor) are obtained by taking momentum moments $\langle O(p)\rangle(x) \equiv \frac{g}{(2\pi)^3}\int \frac{d^3p}{E_p} O(p) f(x,p)$ of the distribution function $f(x,p)$. Hydrodynamic equations are obtained by splitting the distribution function into a leading-order contribution f_0, parametrized through macroscopic observables as

$$f_0(x,p) = f_0\left(\frac{\sqrt{p_\mu \Omega^{\mu\nu}(x) p_\nu} - \tilde{\mu}(x)}{\tilde{T}(x)}\right), \tag{1}$$

and a smaller first-order correction δf ($|\delta f/f_0| \ll 1$):

$$f(x,p) = f_0(x,p) + \delta f(x,p). \tag{2}$$

In Eq. (1), $p_\nu \Omega^{\mu\nu}(x) p_\nu = m^2 + (1+\xi_\perp(x)) p^2_{\perp,\mathrm{LRF}} + (1+\xi_L(x)) p^2_{z,\mathrm{LRF}}$, where the hydrodynamic flow field $u^\mu(x)$ defines the local fluid rest frame (LRF). $\tilde{T}(x)$ and $\tilde{\mu}(x)$ are the effective temperature and chemical potential in the LRF, Landau matched to the energy and particle densities, e and n [39]. $\xi_{\perp,L}$ parametrize the momentum anisotropy in the LRF and are Landau matched to the transverse and longitudinal pressures, P_T and P_L [40, 44, 45]. The latter encode the bulk viscous pressure $\Pi = (2P_\perp + P_L)/3 - P_{\mathrm{eq}}$ and the largest shear stress component $P_\perp - P_L$. In anisotropic hydrodynamics, $P_\perp$ and P_L evolve macroscopically according to equations that reflect the competition between macroscopic anisotropic expansion (driving the system away from local equilibrium and momentum isotropy) and microscopic scattering (trying to restore them) [40].

Using the decomposition (2) we write $T^{\mu\nu} = T_0^{\mu\nu} + \delta T^{\mu\nu} \equiv T_0^{\mu\nu} + \Pi^{\mu\nu}$, $j^\mu = j_0^\mu + \delta j^\mu \equiv j_0^\mu + V^\mu$. Different hydrodynamic approaches can be characterized by the assumptions they make about the dissipative corrections and/or the approximations they use to derive their dynamics from the underlying Boltzmann equation:

1. Ideal hydrodynamics assumes local momentum isotropy, setting f_0 to be isotropic ($\xi_{\perp,L} = 0$) and all dissipative currents to zero: $\Pi^{\mu\nu} = V^\mu = 0$.

2. Navier-Stokes (NS) theory maintains local momentum isotropy at leading order and postulates instantaneous constituent relations for $\Pi^{\mu\nu}$ and V^μ by introducing viscosity and heat conduction as transport coefficients that relate these flows to their driving forces. It ignores the microscopic relaxation time that is needed for these flows to adjust to their Navier-Stokes values, leading to acausal signal propagation.

3. Israel-Stewart (IS) theory [46] improves on NS theory by evolving $\Pi^{\mu\nu}$ and V^μ dynamically, with evolution equations derived from moments of the Boltzmann equation, keeping only terms linear in the Knudsen number $\mathrm{Kn} = \lambda_{\mathrm{mfp}}/\lambda_{\mathrm{macro}}$.

4. Denicol-Niemi-Molnar-Rischke (DNMR) theory [47] improves IS theory by keeping nonlinear terms up to order Kn^2 and $\mathrm{Kn}\cdot\mathrm{Re}^{-1}$ when evolving $\Pi^{\mu\nu}$ and V^μ.

5. Third-order Chapman-Enskog expansion [48] keeps terms of up to third order when evolving $\Pi^{\mu\nu}$ and V^μ.

6. Anisotropic hydrodynamics (aHydro) [41,42] allows for a leading-order local momentum anisotropy ($\xi_{\perp,L} \neq 0$), evolved according to equations obtained from low-order moments of the Boltzmann equation, but ignores residual dissipative flows: $\Pi^{\mu\nu} = V^\mu = 0$.

7. Viscous anisotropic hydrodynamics (vaHydro) [43, 49] improves on aHydro by additionally evolving (using IS or DNMR theory) the residual dissipative flows $\Pi^{\mu\nu}$, V^μ generated by the deviation δf around the locally anisotropic leading distribution function f_0.

There exist a few highly symmetric situations for which the Boltzmann equation, in Relaxation Time Approximation (RTA), can be solved exactly. These include the Bjorken [11] and Gubser [50] flows which are (although highly idealized) relevant for heavy-ion collisions [51,52]. While for Bjorken expansion the expansion rate decreases with longitudinal proper time τ like $1/\tau$, allowing the system to asymptotically reach local momentum isotropy and thermal equilibrium, Gubser expansion includes an additional strong transverse flow which leads to an asymptotically constant expansion rate and exponentially growing Knuden number Kn [52], and thus to asymptotic free-streaming. The exact evolution of the macroscopic currents $T^{\mu\nu}$ and j^μ associated with these solutions can be compared with that predicted by any of the 7 different hydrodynamic approximations listed above and thus be used to assess the accuracy of the latter in these two opposite extremes of asymptotic evolution.

To illustrate the differences between the different hydrodynamic approximations, we briefly summarize the corresponding evolution equations for the shear stress. (For both Bjorken and Gubser flow with a conformal equation of state $\Pi^{\mu\nu}$ has only one independent component, the shear stress $\pi^{\eta\eta}$.) For Gubser flow (where all macroscopic quantities depend on only one space-

IOP Conf. Series: Journal of Physics: Conf. Series **1271** (2019) 012018 doi:10.1088/1742-6596/1271/1/012018

time variable, the de Sitter time ρ [50]) one finds the following [51–56] (a similar discussion for Bjorken flow can be found in [43, 57, 58]):[3]

1. Ideal hydrodynamics gives $\hat{T}_{\text{ideal}}(\rho) = \frac{\hat{T}_0}{\cosh^{2/3}(\rho)}$, combined with zero shear stress, $\hat{\pi}^{\eta\eta} = 0$.

2.-7. For all dissipative hydrodynamic frameworks the temperature evolves instead according to the differential equation $\frac{1}{\hat{T}}\frac{d\hat{T}}{d\rho} + \frac{2}{3}\tanh\rho = \frac{1}{3}\bar{\pi}^{\eta}_{\eta}(\rho)\tanh\rho$ [50,53] where $\bar{\pi} \equiv \hat{\pi}^{\eta}_{\eta}/(\hat{T}\hat{s})$. Differences between the approaches arise from their different evolution of the shear stress. In

2. NS theory the shear stress is given by the (instantaneous) constituent relation $\hat{\pi}^{\eta\eta}_{NS} = \frac{4}{15}\hat{\tau}_{\text{rel}}\tanh\rho$ [52] where $\hat{\tau}_{\text{rel}} = \text{const.}/\hat{T}$. In all second and higher-order hydrodynamic approaches the shear stress evolves instead according to a differential equation of the type

$$d\bar{\pi}^{\eta}_{\eta}/d\rho + \bar{\pi}^{\eta}_{\eta}/\hat{\tau}_{\text{rel}} = (a_1 + a_2\bar{\pi} - a_3\bar{\pi}^2)\tanh\rho - \frac{4}{3}\mathcal{F}(\bar{\pi}). \quad (3)$$

For the approaches **3., 4., 5.** (*i.e.* as long as the equations are derived by expanding around a locally isotropic distribution function $\xi_{\perp,L} = 0$) the function $\mathcal{F}$ vanishes: $\mathcal{F}(\bar{\pi}) = 0$. Only for **7.** vaHydro $\mathcal{F}(\bar{\pi})$ is nonzero; its specific form is found in [55].[4] For the constants (a_1, a_2, a_3) one finds:

3. IS theory: $(a_1, a_2, a_3) = \left(\frac{4}{15}, 0, \frac{4}{3}\right)$.

4. DNMR theory: $(a_1, a_2, a_3) = \left(\frac{4}{15}, \frac{10}{21}, \frac{4}{3}\right)$.

5. Third-order Chapman-Enskog expansion: $(a_1, a_2, a_3) = \left(\frac{4}{15}, \frac{10}{21}, \frac{412}{147}\right)$ [56].

6. Anisotropic hydrodynamics (aHydro): See footnote 4.

7. Viscous anisotropic hydrodynamics vaHydro: $(a_1, a_2, a_3) = \left(\frac{5}{12}, \frac{4}{3}, \frac{4}{3}\right)$ [55].

[3] All quantities with hats have been made unitless by multiplying with appropriate powers of the proper time τ.
[4] The aHydro study in [54] expresses the evolution of the shear stress in terms of the microscopic parameters $\xi_{\perp,L}$ and thus cannot be directly compared to the macroscopic evolution equation (3).

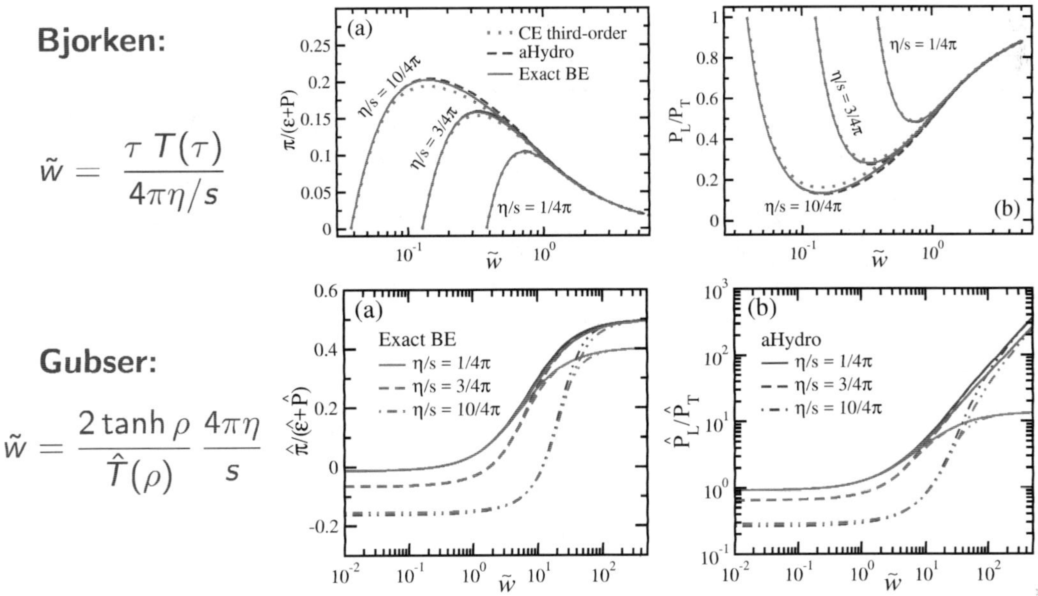

Figure 5. Normalized shear stress $\pi/(e+P_{\text{eq}}) = \frac{1}{4}(\pi/P_{\text{eq}})$ (left column) and pressure anisotropy $P_L/P_\perp$ (right column) for Bjorken (upper row) and Gubser flow (lower row), plotted as functions of a rescaled time variable $\tilde{w}$ which for Bjorken flow corresponds to the inverse Knudsen number, $\tilde{w} = \text{Kn}^{-1}$, and for Gubser flow directly to the Knudsen number, $\tilde{w} = \text{Kn}$ [56].

Figure 5 shows, for thermal equilibrium initial conditions, the time evolution of the normalized shear stress $\bar{\pi}$ and the pressure anisotropy $P_L/P_\perp$ for Bjorken and Gubser flows, for three systems with specific shear viscosities $4\pi\eta/s = 4\pi T\tau_{\rm rel}/5 = 1$, 3, and 10 [56]. For clarity, the exact solution of the Boltzmann equation (green solid lines) is compared only with the two best-performing hydrodynamic approximations, the third-order Chapman-Enskog expansion (red dotted lines) and anisotropic hydrodynamics (blue dashed lines) [where in this case, due to the high degree of symmetry of the flow, aHydro and vaHydro correspond to the same approximation [55]]. Similar comparisons for the other hydrodynamic approximations discussed in this contribution can be found in the literature [38,43,51–63]. Following [60] the time variables τ (for Bjorken flow) and ρ (for Gubser flow) are replaced by a scaling variable $\tilde{w}$, defined as the product of the macroscopic expansion rate ($=1/\tau$ in the case of Bjorken flow and $=2\tanh\rho$ for Gubser flow) with the microscopic relaxation time $\tau_{\rm rel}=4\pi\eta/(Ts)$ (Gubser flow), or as its inverse (Bjorken flow). The idea behind this rescaling is that for Bjorken flow the system approaches thermalization (*i.e.* a regime of small Knudsen numbers) at late times while for Gubser flow it becomes asymptotically free-streaming (*i.e.* approaches a regime of large Knudsen numbers at late times). The rate of this approach scales with the microscopic relaxation time.

Figure 5 shows that for Bjorken flow (top row) the solutions of the Boltzmann equation and of the two hydrodynamic approximations shown in the plot approach a common attractor [64] at late times where the system approaches local momentum isotropy and thermal equilibrium. Different initial conditions relax exponentially towards this attractor. However, the hydrodynamic models describe the exact Boltzmann dynamics well even at early times where, for $\eta/s=10/(4\pi)$, the systems moves very far away from equilibrium, as witnessed by the shear pressure becoming almost as large as the thermal pressure (corresponding to a large inverse Reynolds number $\mathrm{Re}^{-1}=\mathcal{O}(1)$).

In the bottom row of the figure one sees that the approach to a common late-time attractor persists in the case of Gubser flow, at least for anisotropic hydrodynamics while the third-order Chapman-Enskog approach approaches an incorrect asymptotic value for the inverse Reynolds number ($\pi/P_{\rm eq}\to 1.6$ instead of 2). Still, the third-order Chapman-Enskog approach performs much better than all other hydrodynamic approximation schemes that are based on expansions around local momentum isotropy. That anisotropic hydrodynamics correctly reproduces even the asymptotic free-streaming limit of Gubser flow is a striking counterexample to the folklore that hydrodynamics can only be applied to systems that are close to local momentum isotropy and thermal equilibrium. Still, the high quality of the anisotropic hydrodynamic description of Gubser flow by aHydro (blue-dashed lines) may be somewhat accidental in that Gubser symmetry may produce phase-space distributions that are particularly well adjusted to being decomposed as in Eqs. (1) and (2). Upcoming (3+1)-dimensional studies [40] will shed further light on this issue.

References

[1] Dusling K, Li W and Schenke B 2016 *Int. J. Mod. Phys.* **E25** 1630002 (*Preprint* 1509.07939)
[2] Li W 2017 *Nucl. Phys.* **A967** 59–66 (*Preprint* 1704.03576)
[3] Floris M and Li W 2018 *Adv. Ser. Direct. High Energy Phys.* **29** 313–341
[4] Heinz U 2010 *Landolt-Bornstein* **23** 240 (*Preprint* 0901.4355)
[5] Heinz U and Snellings R 2013 *Ann. Rev. Nucl. Part. Sci.* **63** 123–151 (*Preprint* 1301.2826)
[6] Gale C, Jeon S and Schenke B 2013 *Int. J. Mod. Phys.* **A28** 1340011 (*Preprint* 1301.5893)
[7] Romatschke P and Romatschke U 2017 (*Preprint* 1712.05815)
[8] Shen C, Qiu Z and Heinz U 2015 *Phys. Rev.* **C92** 014901 (*Preprint* 1502.04636)
[9] Basar G and Teaney D 2014 *Phys. Rev.* **C90** 054903 (*Preprint* 1312.6770)
[10] Kurkela A, Mazeliauskas A, Paquet J F, Schlichting S and Teaney D 2018 (*Preprint* 1805.01604)
[11] Bjorken J D 1983 *Phys. Rev.* **D27** 140–151
[12] Krzewicki M (ALICE) 2011 *J. Phys.* **G38** 124047 (*Preprint* 1107.0080)
[13] Shen C, Heinz U, Huovinen P and Song H 2011 *Phys. Rev.* **C84** 044903 (*Preprint* 1105.3226)

[14] Weller R D and Romatschke P 2017 *Phys. Lett.* **B774** 351–356 (*Preprint* 1701.07145)
[15] Aidala C *et al.* (PHENIX) 2019 *Nature Phys.* **15** 214–220 (*Preprint* 1805.02973)
[16] Welsh K, Singer J and Heinz U W 2016 *Phys. Rev.* **C94** 024919 (*Preprint* 1605.09418)
[17] Romatschke P 2015 *Eur. Phys. J.* **C75** 305 (*Preprint* 1502.04745)
[18] Shen C, Qiu Z, Song H, Bernhard J, Bass S and Heinz U 2016 *Comput. Phys. Commun.* **199** 61–85 (*Preprint* 1409.8164)
[19] Schenke B, Jeon S and Gale C 2011 *Phys. Rev. Lett.* **106** 042301 (*Preprint* 1009.3244)
[20] Moreland J S, Bernhard J E and Bass S A 2015 *Phys. Rev.* **C92** 011901 (*Preprint* 1412.4708)
[21] Bernhard J E, Moreland J S, Bass S A, Liu J and Heinz U 2016 *Phys. Rev.* **C94** 024907 (*Preprint* 1605.03954)
[22] Hirono Y and Shuryak E 2015 *Phys. Rev.* **C91** 054915 (*Preprint* 1412.0063)
[23] Kalaydzhyan T and Shuryak E 2015 *Phys. Rev.* **C91** 054913 (*Preprint* 1503.05213)
[24] Heinz U 1996 *Nucl. Phys.* **A610** 264C–277C (*Preprint* nucl-th/9608002)
[25] Adam J *et al.* (ALICE) 2015 *Phys. Rev.* **C91** 034906 (*Preprint* 1502.00559)
[26] Chatrchyan S *et al.* (CMS) 2014 *Eur. Phys. J.* **C74** 2847 (*Preprint* 1307.3442)
[27] Werner K, Guiot B, Karpenko I and Pierog T 2014 *Phys. Rev.* **C89** 064903 (*Preprint* 1312.1233)
[28] Chatrchyan S *et al.* (CMS) 2012 *Eur. Phys. J.* **C72** 2164 (*Preprint* 1207.4724)
[29] Lee K S, Heinz U and Schnedermann E 1990 *Z. Phys.* **C48** 525–541
[30] Schnedermann E, Sollfrank J and Heinz U 1993 *Phys. Rev.* **C48** 2462–2475 (*Preprint* nucl-th/9307020)
[31] Khachatryan V *et al.* (CMS) 2010 *JHEP* **09** 091 (*Preprint* 1009.4122)
[32] Aad G *et al.* (ATLAS) 2016 *Phys. Rev. Lett.* **116** 172301 (*Preprint* 1509.04776)
[33] Khachatryan V *et al.* (CMS) 2016 *Phys. Rev. Lett.* **116** 172302 (*Preprint* 1510.03068)
[34] Khachatryan V *et al.* (CMS) 2017 *Phys. Lett.* **B765** 193–220 (*Preprint* 1606.06198)
[35] Bernhard J E 2018-04-19 *Bayesian parameter estimation for relativistic heavy-ion collisions* Ph.D. thesis Duke U. (*Preprint* 1804.06469)
[36] Ackermann K H *et al.* (STAR) 2001 *Phys. Rev. Lett.* **86** 402–407 (*Preprint* nucl-ex/0009011)
[37] Heinz U and Kolb P F 2002 *Nucl. Phys.* **A702** 269–280 (*Preprint* hep-ph/0111075)
[38] Strickland M 2014 *Acta Phys. Polon.* **B45** 2355–2394 (*Preprint* 1410.5786)
[39] Heinz U, Bazow D, Denicol G S, Martinez M, Nopoush M, Noronha J, Ryblewski R and Strickland M 2015 [Nucl. Part. Phys. Proc.276-278,193(2016)] (*Preprint* 1509.05818)
[40] McNelis M, Bazow D and Heinz U 2018 *Phys. Rev.* **C97** 054912 (*Preprint* 1803.01810)
[41] Martinez M and Strickland M 2010 *Nucl. Phys.* **A848** 183–197 (*Preprint* 1007.0889)
[42] Florkowski W and Ryblewski R 2011 *Phys. Rev.* **C83** 034907 (*Preprint* 1007.0130)
[43] Bazow D, Heinz U W and Strickland M 2014 *Phys. Rev.* **C90** 054910 (*Preprint* 1311.6720)
[44] Tinti L 2016 *Phys. Rev.* **C94** 044902 (*Preprint* 1506.07164)
[45] Molnar E, Niemi H and Rischke D H 2016 *Phys. Rev.* **D94** 125003 (*Preprint* 1606.09019)
[46] Israel W and Stewart J M 1979 *Annals Phys.* **118** 341–372
[47] Denicol G S, Niemi H, Molnar E and Rischke D H 2012 *Phys. Rev.* **D85** 114047 [Erratum: Phys. Rev. D91, 039902 (2015)] (*Preprint* 1202.4551)
[48] Jaiswal A 2013 *Phys. Rev.* **C88** 021903 (*Preprint* 1305.3480)
[49] Bazow D, Heinz U W and Martinez M 2015 *Phys. Rev.* **C91** 064903 (*Preprint* 1503.07443)
[50] Gubser S S 2010 *Phys. Rev.* **D82** 085027 (*Preprint* 1006.0006)
[51] Denicol G S, Heinz U, Martinez M, Noronha J and Strickland M 2014 *Phys. Rev. Lett.* **113** 202301 (*Preprint* 1408.5646)
[52] Denicol G S, Heinz U, Martinez M, Noronha J and Strickland M 2014 *Phys. Rev.* **D90** 125026 (*Preprint* 1408.7048)
[53] Marrochio H, Noronha J, Denicol G S, Luzum M, Jeon S and Gale C 2015 *Phys. Rev.* **C91** 014903 (*Preprint* 1307.6130)
[54] Nopoush M, Ryblewski R and Strickland M 2015 *Phys. Rev.* **D91** 045007 (*Preprint* 1410.6790)
[55] Martinez M, McNelis M and Heinz U 2017 *Phys. Rev.* **C95** 054907 (*Preprint* 1703.10955)
[56] Chattopadhyay C, Heinz U, Pal S and Vujanovic G 2018 *Phys. Rev.* **C97** 064909 (*Preprint* 1801.07755)
[57] Florkowski W, Ryblewski R and Strickland M 2013 *Nucl. Phys.* **A916** 249–259 (*Preprint* 1304.0665)
[58] Florkowski W, Maksymiuk E, Ryblewski R and Strickland M 2014 *Phys. Rev.* **C89** 054908 (*Preprint* 1402.7348)
[59] Florkowski W, Ryblewski R and Strickland M 2013 *Phys. Rev.* **C88** 024903 (*Preprint* 1305.7234)
[60] Heller M P, Kurkela A, Spali?ski M and Svensson V 2018 *Phys. Rev.* **D97** 091503 (*Preprint* 1609.04803)
[61] Romatschke P 2018 *Phys. Rev. Lett.* **120** 012301 (*Preprint* 1704.08699)
[62] Alqahtani M, Nopoush M and Strickland M 2018 *Prog. Part. Nucl. Phys.* **101** 204–248 (*Preprint* 1712.03282)
[63] Strickland M, Noronha J and Denicol G 2018 *Phys. Rev.* **D97** 036020 (*Preprint* 1709.06644)
[64] Heller M P and Spalinski M 2015 *Phys. Rev. Lett.* **115** 072501 (*Preprint* 1503.07514)

IOP Conf. Series: Journal of Physics: Conf. Series **1271** (2019) 012019 doi:10.1088/1742-6596/1271/1/012019

Small System Corrections to Thermal Field Theory and pQCD Energy Loss

I Kolbé[1], W A Horowitz[1], and S Mogliacci[1]

[1]Department of Physics, R. W. James Building, University Road, University of Cape Town, Rondebosch, 7700, South Africa

E-mail: isobel.kolbe@gmail.com

Abstract. The ominous absence of partonic energy loss in small colliding systems is forcing the heavy-ion community to sharpen its theoretical tools. In particular, all the main forms of pQCD energy loss depend on the assumption that the brick of QGP is large, but it is clear that any droplet of QGP that may be created in high multiplicity $p/d + A$ or peripheral AA collisions cannot be much larger than the mean free path, $\lambda_{mfp}^{\text{QCD}} \sim 1$ fm. We present an illuminating and exhaustive numerical analysis of the recent extension of DGLV energy loss to all separation distances, and show a need for stronger theoretical control over the nature of the medium at early times and short separation distances. Prompted by this demand for an understanding of the small system corrections to medium properties that are traditionally computed in an infinite medium, we then present a thermal field theoretical calculation of the thermodynamic properties of a toy model. We consider a single, massless, scalar field and impose Dirichlet boundary conditions in order to investigate the qualitative effect of a boundary on the properties of the medium, finding significant deviations from the standard Stefan-Boltzmann results, even at relatively large L.

1. Introduction

A long program of high energy heavy-ion experiments at the Large Hadron Collider (LHC), complimented by similar experiments at the Relativistic Heavy Ion Collider (RHIC), have produced a wealth of data concerning the nature of what is believed to be a droplet of deconfined quark matter - the Quark Gluon Plasma (QGP). It is widely accepted that the QGP is produced in experiments involving collision systems that are considered to be large, such as nearly head-on lead-lead (PbPb) or gold-gold (AuAu) collisions (collectively, AA) [1–5]. The heavy-ion community's faith in the presence of the QGP in such large systems is due to the presence of a number of signatures of the QGP, most of which are well described within the paradigm of the production of a strongly coupled, almost perfect fluid. However, the heavy-ion community has now spent half a decade unable to resolve a major inconsistency in its understanding of the QGP; The inconsistency arises from the fact that many of the traditional signatures of the QGP are present also in smaller colliding systems such as proton-proton (pp), proton- or deuteron-gold and proton-lead (p/dA), and peripheral AA collisions, with the notable exception of one key observable - the partonic energy loss of color-charged objects traversing the QGP[6].

Resolving the question of the absence of energy loss in the apparent presence of the QGP requires a range of cross-checks and refinements. It is clear that one must attempt to understand the problem from both and experimental and a theoretical perspective. While major advances have been made on the experimental side, both in measuring the observed signatures more carefully (see, for instance, [7]) and understanding the biases related to measuring energy loss in the traditional manner [8], theoretical advances have been slow out of the blocks. Although some models have been put forward that might

IOP Conf. Series: Journal of Physics: Conf. Series **1271** (2019) 012019 doi:10.1088/1742-6596/1271/1/012019

explain the presence of a QGP signature in the absence of a QGP, see for instance [9], they are unable to satisfactory compliment the self-consistent picture of the presence of the QGP in large colliding systems.

A logical first step to understanding the absence of energy loss in small systems from a theoretical standpoint would be to simply compute the expected energy loss, since energy loss studies in large systems have lead to an understanding of the path-length dependence of energy loss [10, 11]. Such studies have been performed [12], but over-predict the energy loss in small systems. Their inability to reproduce the data might be explained by the simple fact that all major pQCD energy loss calculations assume that the system (L) is large compared to the mean free path (λ_{mfp}) of the parton.

In these proceedings we present two very different attempts at understanding the nature of a small droplet of QGP. First, we briefly review the recent computation of a correction to DGLV (Djordjevic, Gyulassy, Levai, Vitev) pQCD energy loss which relaxes an assumption whereby the first scattering of a hard parton traversing a QGP is assumed to be far from its production within the QGP. Armed with such a correction, we present a numerical investigation of the energy loss formula, finding a crucial dependence of the energy loss on the physics of short separation distances between the production of the hard parton and its first scattering.

In the second part of the proceedings we compute the thermodynamic properties of a toy model, consisting of a single, free, massless, scalar field, which has been subjected to a "geometrical confinement" in which Dirichlet boundary conditions are place on the field. We find that the fundamental properties of a quantum field are altered when it is confined to a small space.

2. pQCD Energy Loss in a Small System

In these proceedings the aim is simply to present the most recent findings of a calculation which involved computing the short separation distance correction to standard DGLV energy loss, and to offer them as motivation for the subsequent calculation in section 3. The reader is referred to the original DGLV calculation [13], the Masters thesis [14] in which details of the calculation of the correction are given and the upcoming publication already available online [15], for further details. For clarity, we treat the high-p_T eikonal parton produced at an initial point $(t_0, z_0, \mathbf{x}_0)$ inside a finite QGP brick, where we have used $\mathbf{p}$ to mean transverse 2D vectors, $\vec{\mathbf{p}} = (p_z, \mathbf{p})$ for 3D vectors and $p = (p^0, \vec{\mathbf{p}}) = [p^0 + p^z, p^0 - p^z, \mathbf{p}]$ for four vectors in Minkowski and light cone coordinates, respectively. As in the DGLV calculation, we consider the nth target to be a Gyulassy-Wang Debye screened potential [16] with Fourier and color structure given by

$$V_n = 2\pi\delta(q^0)\frac{4\pi\alpha_s}{\vec{q}_n^2 + \mu^2}e^{-i\vec{q}_n\cdot\vec{x}_n}T_{a_n}(R)\otimes T_{a_n}(n), \tag{1}$$

where the color exchanges are handled using the applicable $SU(N_c)$ generator, $T_a(n)$ in the d_n dimensional representation of the target, or $T_a(R)$ in the d_R dimensional representation of the high-p_T parent parton.

In light cone coordinates the four-momenta of the emitted gluon, the final high-p_T parton, and that exchanged with the medium Debye quasiparticle are, respectively,

$$k = \left[xP^+, \frac{m_g^2 + \mathbf{k}^2}{xP^+}, \mathbf{k}\right], \qquad p = \left[(1-x)P^+, \frac{M^2 + \mathbf{k}^2}{(1-x)P^+}, \mathbf{q} - \mathbf{k}\right], \qquad q = [q^+, q^-, \mathbf{q}], \tag{2}$$

where the initially produced high-p_T particle of mass M has large momentum $E^+ = P^+ = 2E$ and negligible other momentum components. Notice that we include the QCD analogue of the Ter-Mikayelian plasmon effect [17], a color dielectric modification of the gluon dispersion relation, with an effective emitted gluon mass m_g [13, 18]. See figure 1 for a visualization of the relevant momenta.

Following [13] we define $\omega \approx xE^+/2 = xP^+/2$, from which a shorthand for energy ratios will prove useful notationally: $\omega_0 \equiv \mathbf{k}^2/2\omega$, $\omega_i \equiv (\mathbf{k}-\mathbf{q}_i)^2/2\omega$, $\omega_{(ij)} \equiv (\mathbf{k}-\mathbf{q}_i-\mathbf{q}_j)^2/2\omega$, and $\tilde{\omega}_m \equiv (m_g^2+M^2x^2)/2\omega$.

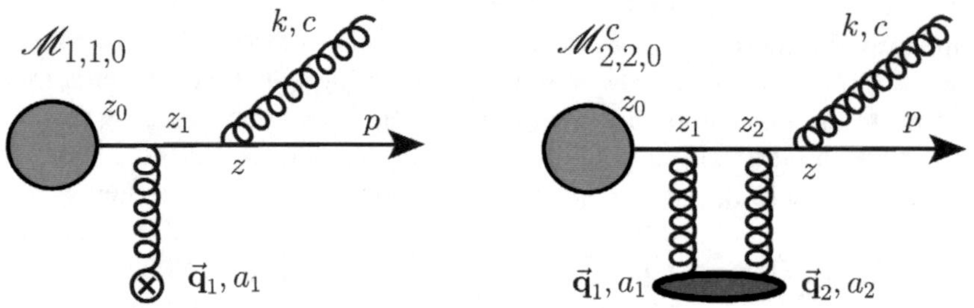

Figure 1: Following the diagrammatic numbering in [13], (a) $\mathcal{M}_{1,1,0}$ and (b) $\mathcal{M}^c_{2,2,0}$ are the only two diagrams that have non-zero small separation distance corrections in the large formation time limit. $\mathcal{M}^c_{2,2,0}$ is the double Born contact diagram, corresponding to the second term in the Dyson series in which two gluons are exchanged with the single scattering center.

We reevaluated the 12 diagrams contributing to the $N = 1$ in opacity energy loss amplitude [13] without the additional simplification of the large separation distance $\Delta z \gg 1/\mu$ assumption, and find

$$\Delta E^{(1)}_{ind} = \frac{C_R\alpha_s LE}{\pi\lambda_g}\int dx \int \frac{d^2\mathbf{q}_1}{\pi}\frac{\mu^2}{(\mu^2+\mathbf{q}_1^2)^2}\int\frac{d^2\mathbf{k}}{\pi}\int d\Delta z\, \rho(\Delta z)$$
$$\times\Bigg[-\frac{2(1-\cos\{(\omega_1+\tilde{\omega}_m)\Delta z\})}{(\mathbf{k}-\mathbf{q}_1)^2+m_g^2+x^2M^2}\left(\frac{(\mathbf{k}-\mathbf{q}_1)\cdot\mathbf{k}}{\mathbf{k}^2+m_g^2+x^2M^2}-\frac{(\mathbf{k}-\mathbf{q}_1)^2}{(\mathbf{k}-\mathbf{q}_1)^2+m_g^2+x^2M^2}\right)$$
$$+\frac{1}{2}e^{-\mu_1\Delta z}\Bigg\{\left(\frac{\mathbf{k}}{\mathbf{k}^2+m_g^2+x^2M^2}\right)^2\left(1-\frac{2C_R}{C_A}\right)\left(1-\cos\{(\omega_0+\tilde{\omega}_m)\Delta z\}\right)$$
$$+\frac{\mathbf{k}\cdot(\mathbf{k}-\mathbf{q}_1)}{(\mathbf{k}^2+m_g^2+x^2M^2)((\mathbf{k}-\mathbf{q}_1)^2+m_g^2+x^2M^2)}\left(\cos\{(\omega_0+\tilde{\omega}_m)\Delta z\}-\cos\{(\omega_0-\omega_1)\Delta z\}\right)\Bigg\}\Bigg]. \quad (3)$$

The reader will notice that equation (3) contains an integral on the first line over the distribution of scattering ceners, $\rho(\Delta z)$. One is free to choose this distribution. The original DGLV calculation used an exponential distribution which simplified the analytical investigation of the energy loss formula. We have studied different distribution functions that highlight the sensitivity of the energy loss formula to the physics of small separation distances. Our first choice attempts to avoid the bias toward small separation distances that the exponential distribution offers, by considering a step function (hereinafter the "Full Step Function", or "F"). In our second choice, we truncate this step function distribution and renormalize appropriately (so that all scatterings and subsequent energy loss occur for $1/\mu \lesssim \Delta z \leq L$), which we will call the "Truncated Step Function" or "T". Lastly, we endeavor to include the possibility of scattering but not losing energy by considering a truncated step function that has not been renormalized, the "Truncated Un-renormalized Function" or "TU".

In summary, the four scattering center distribution functions we consider in this article are given by

$$\rho_{\exp}(\Delta z) = \frac{2}{L}\exp\left(-\frac{2\Delta z}{L}\right), \qquad \rho_{\mathrm{F}}(\Delta z) = \frac{1}{L}\Theta(L-\Delta z),$$
$$\rho_{\mathrm{T}}(\Delta z) = \frac{1}{L-1/\mu}\Theta(\Delta z - 1/\mu)\Theta(L-\Delta z), \qquad \rho_{\mathrm{TU}}(\Delta z) = \frac{1}{L}\Theta(\Delta z-1/\mu)\Theta(L-\Delta z), \quad (4)$$

and are shown in figure 2a, for a brick of $L = 4$ fm.

As an illustrative example, we present in figure 2b, the relative energy loss of an $E = 100$ GeV bottom quark, as a function of the distance traveled through the medium, as computed using the Full

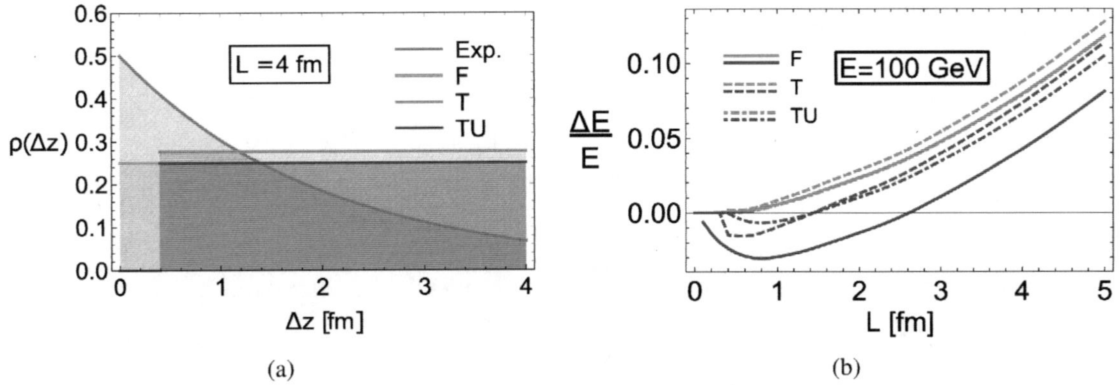

Figure 2: (a) The four scattering center distributions discussed in the text, see equation (4), for a brick of QGP of length $L = 4$ fm. (b): The relative energy loss of a bottom quark with energy $E = 100$ GeV, as a function of length of the path traversed through the brick of QGP, showing the original DGLV result (light) and the all separation distance result (dark), for three different choices of scattering center distribution.

(solid), Truncated (dashed), and Truncated Un-renormalized (dot-dashed) step functions, for both the original DGLV (light) and our all separation distance result (dark). Notice that, although the DGLV result is not sensitive to the choice of scattering center distribution, the all separation distance result is extremely sensitive to the physics of short separation distances. This sensitivity persists out to large (~ 5 fm) path lengths. The reader is referred to [15] for extensive discussion surrounding the mass and energy dependence of the all separation correction as well as an investigation into the nature of the sensitivity to short separation times. It is clear from extreme sensitivity of the energy loss to the physics of small separation distances that a better understanding of the nature of a small droplet of QGP is required.

3. Thermodynamics of a Small System

In order to investigate the nature of a quantum field that is subjected to spatial constraints, we present results from an investigation that involved considering a single, free, massless, scalar field upon which are placed Dirichlet boundary conditions. We then compute the thermodynamic properties of such a field. The reader is referred to our upcoming publication which is already online for details of the calculations [19]. For clarity we outline the calculation here.

In the standard thermal field theoretic formalism [20], the partition function may be written as a Feynman path integral

$$\mathcal{Z}(T,\,V) \propto \int_{\phi(0)}^{\phi(\beta)} [\mathcal{D}\phi] \exp\left\{-\int_0^\beta d\tau \int_V d^{D-1}x\,\mathcal{L}\right\}\Bigg|_{\phi(\beta)=\phi(0)} \tag{5}$$

In the usual Matsubara approach, one would now compute a Fourier decomposition of the field by compactifying both the temporal and the spatial dimensions onto circles. We propose compactifying instead those spatial dimensions that one would want to consider as having finite extent onto lenghts L_i. For paedegogical reasons, we will compactify one spatial dimension at a time, but the second and third compactifications follow simply and are presented in detail in [19]. We therefore consider the following Fourier decompisition of a single, free, massless scalar field which has finite extent in the x_1-direction.

$$\phi(\tau, z_1, \boldsymbol{x}) = \sum_{n\in\mathbb{Z}} \sum_{\ell_1\in\mathbb{N}} \sum_{\boldsymbol{k}\in\mathbb{Z}^{D-2}} \sqrt{\frac{2\beta}{L_1 R^{D-2}}}\, \exp\left\{i\omega_n\tau - i\omega_k\cdot\boldsymbol{x}\right\}\, \sin\left\{\omega_{\ell_1} z_1\right\}\, \tilde{\phi}_{n,\ell_1}(\omega_k) \tag{6}$$

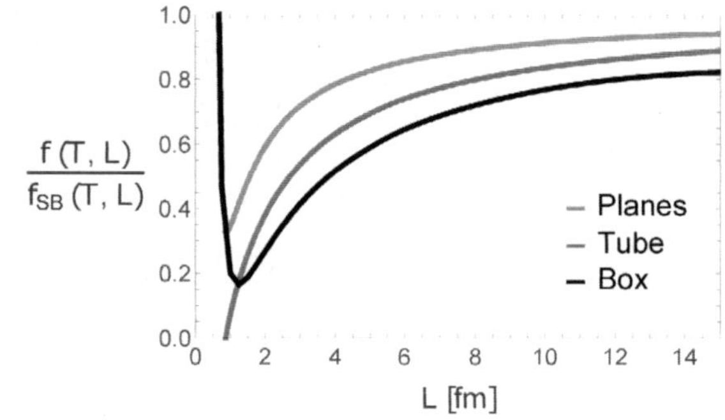

(a)

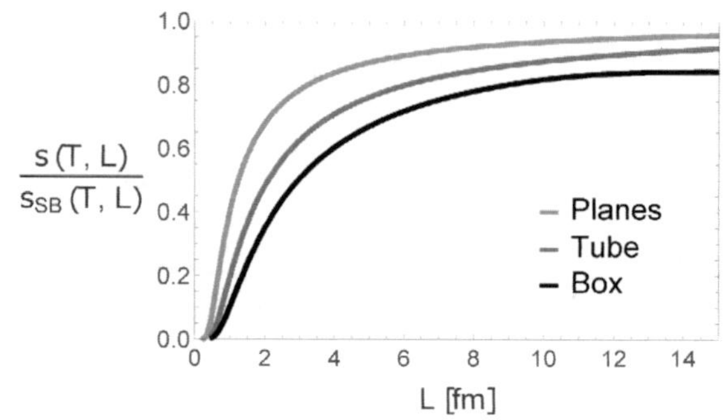

(b)

Figure 3: The free energy density (a) and entropy density (b) of a single, free, massless scalar field with one (orange), two (red) and three (black) spatial dimensions subjected to Dirichlet boundary conditions and rescaled to the Stefan-Boltzmann result.

One may now follow the usual procedure in which equation (6) is substituted into equation (5) and the partition function computed by taking all summations in infinite dimensions to integrals. We deviate from the usual procedure in that we retain those summations that are considered to occur in finite dimensions (here, in the case of only one compactified dimension, the summation over ℓ_1). These remaining summations must then be computed using a combination of Dimensional Regularisation and Epstein-Zeta Regularization. We have computed two numerically equivalent forms of the partition function that have different analytical forms and therefore different convergence properties. We present here only one of the results and only for the case of one compactified dimension, the reader is referred to [19] for more details.

The partition function may be used to compute any thermodynamic quantity. For instance, the free energy density and the entropy density

$$f = -\frac{T}{V}\,\mathrm{Ln}\mathcal{Z}, \qquad s = \frac{1}{V}\,\frac{\partial F}{\partial T}\bigg|_{\{L_k\}}. \tag{7}$$

The free energy density of a single, free, massless scalar field which is subjected to Dirichlet boundary conditions in one dimension is then given by

$$f^{(1)} = -\frac{\pi^2}{1440L_1^4} - \frac{T^2}{2L_1^2}\sum_{\ell=1}^{\infty}\left[\ell\,\mathrm{Li}_2\left(\mathrm{e}^{-\frac{\pi\ell}{TL_1}}\right)\right] - \frac{T^3}{2\pi L_1}\sum_{\ell=1}^{\infty}\left[\mathrm{Li}_3\left(\mathrm{e}^{-\frac{\pi\ell}{TL_1}}\right)\right]. \tag{8}$$

By compactifying one dimension, we are considering a situation in which the scalar field is confined between two infinite parallel plates. However, one may easily compute the cases for an infinite tube (compactifying two dimensions) or a fully compactified box. The reader is referred to [19] for the full expressions for the tube and box cases. We present the free energy density and the entropy density for these three cases in figure 3.

Further results from this study are presented in the present proceedings by S. Mogliacci. However, it is already clear from these first results that the nature of a quantum field is changed fundamentally by the presence of a boundary, since a correction of almost 20% is found in the box case, even for asymptotically large L.

4. Conclusions

The fact that partonic energy loss has not been observed in small systems while many other signatures of the QGP appear to be present, has not yet been satisfactorily understood. One major stumbling block to understanding the absence of energy loss in small colliding systems is the fact that there is very little theoretical control over finite systems.

We have presented the final results from a numerical study of a partonic energy loss formula that was derived in the DGLV formalism to include the effects of short separation distances between the production of a hard parton and its first scattering within the medium. We showed a strong dependence on the physics of separation distances on the order of the Debye screening length. We then presented preliminary results involving the study of a single, free, massless scalar field subjected to Dirichlet boundary conditions, revealing substantial deviations from the Stefan-Boltzmann limit, even for large systems.

These findings suggest that, not only are the properties of a droplet of QGP fundamentally different when the fields are subjected to finite boundary conditions, but also that the dynamics of energy loss are altered at short time scales. These effect need to be investigated quantitatively if the heavy-ion community is to understand the absence of partonic energy loss in small systems.

5. Acknowledgements

SM would like to acknowledge the financial support from the Claude Leon Foundation. IK and WAH wish to thank the SA-CERN Collaboration, the South African National Research Foundation, and the National Institute for Theoretical Physics for their support. IK wishes to further acknowledge Deutscher Akademischer Austauschdienst for supporting her work.

References

[1] Adcox K *et al.* (PHENIX) 2005 *Nucl. Phys.* **A757** 184–283 (*Preprint* nucl-ex/0410003)

[2] Adams J *et al.* (STAR) 2005 *Nucl. Phys.* **A757** 102–183 (*Preprint* nucl-ex/0501009)

[3] Aamodt K *et al.* (ALICE) 2011 *Phys. Lett.* **B696** 30–39 (*Preprint* 1012.1004)

[4] Chatrchyan S *et al.* (CMS) 2011 *Phys. Rev.* **C84** 024906 (*Preprint* 1102.1957)

[5] Aad G *et al.* (ATLAS) 2010 *Phys. Rev. Lett.* **105** 252303 (*Preprint* 1011.6182)

[6] Loizides C 2016 *Nucl. Phys.* **A956** 200–207 (*Preprint* 1602.09138)

[7] Adare A *et al.* (PHENIX) 2015 *Phys. Rev. Lett.* **115** 142301 (*Preprint* 1507.06273)

[8] Adam J *et al.* (ALICE) 2015 *Phys. Rev.* **C91** 064905 (*Preprint* 1412.6828)

[9] Mace M, Skokov V V, Tribedy P and Venugopalan R 2018 *Phys. Rev. Lett.* **121** 052301 (*Preprint* 1805.09342)

[10] Wiedemann U A 2010 521–562 [Landolt-Bornstein23,521(2010)] (*Preprint* 0908.2306)

[11] Majumder A and Van Leeuwen M 2011 *Prog. Part. Nucl. Phys.* **66** 41–92 (*Preprint* 1002.2206)

[12] Park C, Shen C, Jeon S and Gale C 2017 *Nucl. Part. Phys. Proc.* **289-290** 289–292 (*Preprint* 1612.06754)

[13] Djordjevic M and Gyulassy M 2004 *Nucl. Phys.* **A733** 265–298 (*Preprint* nucl-th/0310076)

[14] Kolbé I (*Preprint* 1509.06122)

[15] Kolbé I and Horowitz W A 2015 (*Preprint* 1511.09313)

[16] Gyulassy M and Wang X n 1994 *Nucl. Phys.* **B420** 583–614 (*Preprint* nucl-th/9306003)

[17] Djordjevic M and Gyulassy M 2003 *Phys. Rev.* **C68** 034914 (*Preprint* nucl-th/0305062)

[18] Wicks S, Horowitz W, Djordjevic M and Gyulassy M 2007 *Nucl. Phys.* **A784** 426–442 (*Preprint* nucl-th/0512076)

[19] Mogliacci S, Kolbé I and Horowitz W A 2018 (*Preprint* 1807.07871)

[20] Kapusta J I and Gale C 2011 *Finite-temperature field theory: Principles and applications* Cambridge Monographs on Mathematical Physics (Cambridge University Press) ISBN 9780521173223, 9780521820820, 9780511222801

IOP Conf. Series: Journal of Physics: Conf. Series **1271** (2019) 012020 doi:10.1088/1742-6596/1271/1/012020

Dark matter, collider searches and the early Universe

Farvah Mahmoudi

Univ. Lyon, Univ. Lyon 1, CNRS/IN2P3, Institut de Physique Nucléaire de Lyon, UMR5822, F-69622 Villeurbanne, France

E-mail: nazila@cern.ch

Abstract. Big-Bang nucleosynthesis (BBN) represents one of the earliest phenomena which can lead to observational constraints on the early Universe properties. Yet, it is well-known that many important mechanisms and phase transitions occurred before BBN. We will discuss the possibility to gain insight about the primordial Universe through studies of dark matter in cosmology, astroparticle physics and colliders. For this purpose we consider that dark matter is a thermal relic, and show that combining collider searches with dark matter observables can lead to strong constraints on the cosmological freeze-out period.

1. Introduction

Dark matter (DM) constitutes one of the strongest constraints on New Physics scenarios. However, DM detection searches suffer from large astrophysical uncertainties, namely on the DM distribution in the Galaxy and on cosmic ray propagation through the galactic medium. The calculation of the dark matter relic density also suffers from our lack of knowledge of early Universe phenomena. In the following we examine, in some detail, the impact of such uncertainties on the constraints on the minimal supersymmetric extension of the Standard Model (MSSM).

On the other hand, a reversed approach can be considered, in particular in the case of New Physics discoveries, to obtain constraints on the early Universe's properties. As an example of non-standard evolution, we consider models with a scalar field that may decay into beyond the Standard Model (BSM) particles. In any given BSM scenario, a deviation of the measured cold dark matter density from a calculation based on measurements of the model parameters and standard radiation dominated expansion would be a signature of novel phenomena in the very early Universe. One might argue that, if the calculated relic density is different from the measured dark matter density, the corresponding BSM scenario is disfavoured. Here, however, we propose to reverse this argument: if the calculated relic density is different from the measured dark matter density, it could be because of novel phenomena in the early Universe. This orthogonal point of view will become particularly important if new particles are discovered at colliders or in dark matter detection experiments: using dark matter observables, it is not possible to constrain BSM scenarios in isolation, but the constraints have to be applied simultaneously to a combination of BSM and cosmological scenarios.

2. Dark matter

2.1. Relic density

The dark matter abundance has been measured in the framework of the standard cosmological model, and the Planck Collaboration has provided a precise evaluation of the cold dark matter density [1]:

$$\Omega_c h^2 = 0.120 \pm 0.001 \ . \tag{1}$$

Constraints on new physics scenarios which propose dark matter candidates can therefore be obtained by comparing the computed dark matter density to the Planck value. The standard assumption to compute the dark matter density is to consider that dark matter particles are thermal relics, *i.e.* were in thermal equilibrium in the early Universe and we observe today only the surviving part. A second assumption is that there is a single thermal relic candidate contributing to the dark matter density, which is generally the case in BSM scenarios where dark matter particles have to be stable, electrically neutral and very weakly interacting.

With these assumptions, the relic density can be obtained by considering that all the new physics particles were originally in thermal equilibrium. Then the expansion of the Universe, which lowers the temperature, eventually breaks this equilibrium, and the evolution of the number densities of all the new particles can be obtained using Boltzmann equations, in which the expansion of the Universe introduces a friction-like term, and the collision terms include annihilations and co-annihilations of these new particles into SM particles. When the dark matter density is diluted enough so that the interactions become negligible, the relic density is frozen and becomes only diluted by the expansion of the Universe.

Comparing the obtained relic density to the very precise dark matter measurement, can lead to very strong constraints on new physics parameters. Several assumptions can nevertheless limit the constraining power of the relic density constraint as will be discussed in the following.

2.1.1. Higher order corrections The first uncertainties arise from the numerical calculations of the annihilation and co-annihilation cross sections. In the simplest cases the calculation of the relic density relies on a few decay channels, in the most compressed scenarios of the MSSM, more than 3000 channels can get involved, severely limiting the calculation speed of relic density. For this reason, the cross sections are generally considered at tree-level. Yet, in individual channels, higher-order corrections can lead to a 30% modification or more [2]. However, in most cases, the relic density calculated at tree-level differs by less than 10% from the one calculated at one-loop [3]. Therefore, in the general case, about a 10% uncertainty can be assigned to tree-level calculations of the relic density.

2.1.2. QCD equations of state A second limitation comes from the QCD equations of state. Indeed, computing the relic density requires the knowledge of the number of effective degrees of freedom of radiation, which lead to energy and entropy content of the Universe. While it was originally thought that the primordial plasma could be treated as an ideal gas above the QCD phase transition temperature, non-perturbative studies showed that at high temperature, the ideal gas approximation does not work, and different models for this plasma have been studied [4,5], leading to different sets of QCD equations of state. The consequences on the relic density are, however, rather mild and can modify it by a few percent.

2.1.3. Early Universe properties In the usual calculation of relic density, the expansion of the Universe is considered to be dominated purely by the radiation density. This hypothesis can, however, be falsified in many extensions of the standard model of cosmology [6, 7]. Similarly, entropy injection or non-thermal production of dark matter particles can modify the relic density [8,9]. These modifications of the standard model of cosmology can result in a change of the relic density by orders of magnitude, but are more likely to increase it. As a consequence, the uncertainties due to these effects are completely dominating the relic density calculation over the previous uncertainties.

2.2. Indirect detection

Dark matter particles hosted in galaxies are supposed to annihilate into SM particles to yield, after hadronisation and decay, nuclei, electrons, photons and neutrinos. Indirect detection experiments try to find an excess of those messengers on top of their astrophysical background. Even in the absence of signals, these experiments provide useful information about the nature of dark matter. The astrophysical background for antiparticle cosmic rays is composed of secondary particles, *i.e.* particles produced by the interaction of primary cosmic rays (mostly proton and helium nuclei) on the interstellar medium (mostly hydrogen and helium atoms). Hence, their background is feeble and relatively under control compared to other species. Antiprotons ($\bar{p}$) are the most abundant antinuclei in cosmic rays that could be produced by dark matter, and their spectral shape is distinguishable from the astrophysical background. For a dark matter mass larger than a few GeV, the flux of antiprotons features a cut off at the dark matter mass. The most accurate measurements of the $\bar{p}$ flux at the Earth was reported by the space-borne detector AMS-02 [10]. The discovery of an excess around 100 GeV was claimed a couple of years ago [11]. However, the astrophysical background suffers from theoretical uncertainties which make the significance of such an excess uncertain [12]. In any case, antiproton data provide strong constraints on the annihilation cross section of dark matter particles.

Compared to charged cosmic rays, gamma rays have the advantage of propagating straight ahead. This allows us to characterise the morphology of their sources and to observe regions where the dark matter particle density is expected to be large and to produce a sizeable flux. We consider here the results from the Fermi-LAT space-borne telescope, which covers the GeV energy range. Since the density of dark matter particles is peaked in the centre of the galaxy, the galactic centre is one of the best targets to look for a dark matter signal. Nevertheless, this region hosts important astrophysical activities, and it is difficult to estimate both the astrophysical background and foreground. On the other hand, dwarf spheroidal galaxies are considered as very interesting targets to look for a dark matter signal. Indeed, these systems are expected to: i) be dominated in mass by a DM component, ii) exhibit feeble stellar activities and have a low astrophysical background. Despite the fact that the dark matter distribution and concentration inside these objects is still under debate, they provide one of the best bounds on the average annihilating cross section $\langle \sigma v \rangle$.

2.2.1. Dark matter halo profiles Dark matter particles are assumed to be isotropically distributed in a spherical halo around the galactic centre, and several halo profiles are in particular considered: the NFW radial density profile of dark matter arising from cosmological simulations were parametrised by Navarro, Frenk and White [13]; the Einasto profile provides a better agreement with the latest simulations [14] and does not suffer from the central divergence of the NFW profile; the stellar activity occurring in the inner galaxy could sweep dark matter particles from the inner region, resulting in a core profile as observed in many galaxies, resulting in the Burkert profile [15].

2.2.2. Cosmic ray propagation As cosmic rays travel across the galaxy, they are affected by many processes as a result of their interactions with the galactic magnetic field. Moreover, cosmic rays can interact with the interstellar medium, leading to energy losses (including ionisation and Coulomb interactions) and their destruction. Finally, cosmic rays undergo the effect of the galactic wind produced by supernova remnant explosions in the galactic disc.

A semi-analytical method was used in [16] to derive the benchmark MIN, MED, and MAX propagation models. The MED model corresponds to the best fit to the boron over carbon (B/C) ratio, whereas the MIN and MAX sets of parameters define the lower and upper bounds for the primary $\bar{p}$ flux, consistent with the B/C ratio.

2.3. Direct detection

Direct dark matter searches aim at directly detecting WIMPs via tiny energy deposits when they scatter off target atomic nuclei in ultra-sensitive, low background detectors. No convincing dark matter signal has been detected so far, however, limits on the WIMP-nucleon cross section are set by comparing the measured differential recoil rate per unit detector mass to the theoretical rate.

Usually, the WIMP-nucleus cross section is decomposed into spin-independent (SI) and spin-dependent (SD) contributions in the zero momentum transfer limit. The SI form factors are experimentally well known from the study of elastic electronic scattering on nuclei and are reasonably well approximated by Helm form factors [17], while the SD form factors are obtained from nuclear shell model calculations [18].

The strongest limits on SI cross section, for $m_{DM} \gtrsim 10$ GeV, are given by xenon target experiments, and we consider in the following the results of XENON1T [19]. To constrain the SD WIMP-neutron cross section we consider the results from LUX [20].

2.3.1. Global and local dark matter densities All the experimental limits are calculated using the benchmark value $\rho_0 = 0.3$ GeV/cm^3 for the local DM density, but recent studies give a best fit value closer to 0.4 GeV/cm^3. The uncertainties on the local density value are still quite large, one of the main sources residing in the knowledge of the baryon density in the galaxy. There may also be a discrepancy between the value calculated from the study of the motion of nearby stars and the one calculated from a global fit of stellar dynamics over the galaxy, assuming a spherical dark matter halo. In our study, we will consider that the local DM density lies between 0.2 and 0.6 GeV/cm^3 (see [21] for a complete review) and will choose three different values to test the impact of those uncertainties on the exclusions in our sample of points: $\rho_0 = 0.2$, 0.4 and 0.6 GeV/cm^3.

2.3.2. Velocities Customarily, an isotropic Maxwellian distribution is assumed for the WIMP velocity distribution $f(\mathbf{v})$, with the Galactic disk rotation velocity v_{rot} being the most probable speed. It corresponds to the Standard Halo Model describing the dark matter halo as a non-rotating isothermal sphere [22]. The canonical value for v_{rot} is 220 km/s but it is believed that it can range from 200 to 250 km/s [23].

This velocity distribution is truncated at the escape velocity v_{esc} at which a WIMP can escape the galaxy potential well. Its value is subject to large uncertainties, $v_{esc} = 500 - 600$ km/s, with a benchmark value $v_{esc} = 544$ km/s [24]. However, for WIMP masses $m_{DM} > 10$ GeV, v_{min} is relatively low. The velocity distribution is then integrated over a large range of velocities and dR/dE_R is not sensitive to the tail of the distribution. Thus, the uncertainties on v_{esc} should not impact our analysis.

Other halo models have been proposed, such as the King Model, which describes the finite size of the halo and the gravitational interaction with ordinary matter in a more realistic way [25], or triaxial halo models [26]. In this study we will focus only on the uncertainties related to the Standard Halo Model, which is the most widely used in the literature.

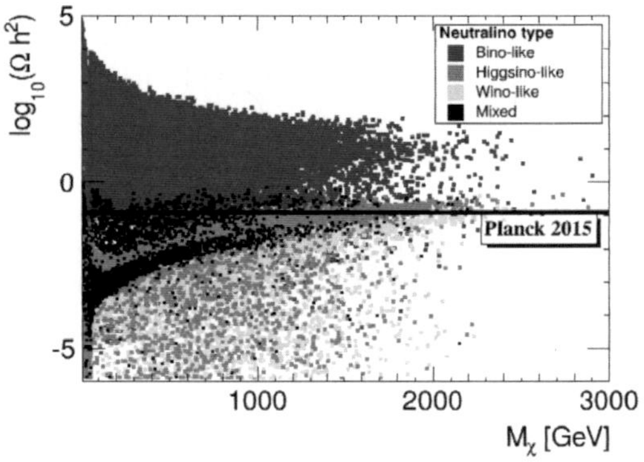

Figure 1: Neutralino relic density as a function of the neutralino 1 mass, for the different neutralino types. The central value of the Planck dark matter density is shown for comparison.

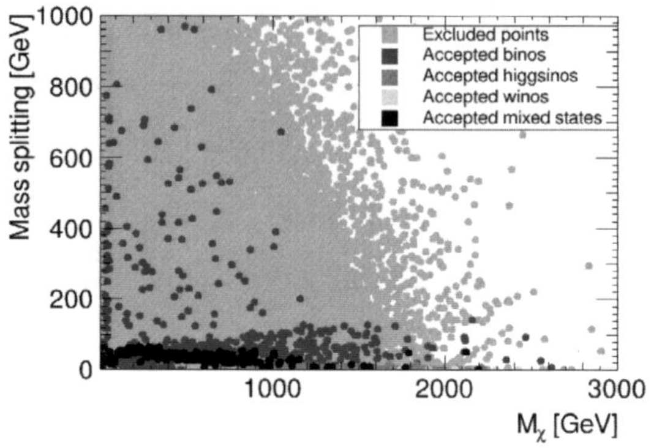

Figure 2: Points respecting the Planck relic dark matter density measurement in the mass splitting between the neutralino and the next lightest supersymmetric particle and the neutralino mass parameter plane.

3. Dark matter in the MSSM

3.1. MSSM Scans

We consider in this analysis the pMSSM, which is the most general R-parity and CP-conserving MSSM scenario with minimal flavour violation. The pMSSM points are generated with SOFTSUSY [27], with a flat random sampling over the 19 parameters with statistics of more than 20 million pMSSM points [28, 29]. All the masses are varied between 0 and 3 TeV, the trilinear couplings between -10 and 10 TeV, and $\tan\beta$ between 1 and 60. After checking the theoretical validity of each point, we impose that the neutralino be the lightest supersymmetric particle (LSP), as well as a light Higgs of mass between 122 and 128 GeV. We then apply different constraints from the dark matter and collider experiments, which are described below.

3.2. Dark matter constraints in the MSSM

We consider only model-points which have the lightest neutralino as the LSP and dark matter particle. In the following, the neutralino 1 (denoted χ) will be said to be bino-/wino-/Higgsino-like if it is composed of 90% of the bino-/wino-/Higgsino component, respectively, or mixed state otherwise. Bino-like χ are the most represented points in our sample, followed by the winos and Higgsinos, with an almost equal share of each component. The fraction of mixed states is negligible.

3.3. Relic density constraints

We first consider the relic density constraint. The value of the neutralino relic density is computed with SuperIso Relic [30–32]. In figure 1, the relic density is shown as a function of the neutralino 1 mass, for the different types [28]. Bino-like neutralinos 1 have in general large relic densities, above the Planck measurement. This can be explained by the smaller couplings of the binos with SM particles, which leads to smaller annihilation cross sections and therefore larger relic densities. On the other hand, the Higgsino-like χ give smaller relic densities which are close to the Planck measurements for χ masses around 1.3 TeV. The wino-like χ tend to have even smaller relic densities, and the Planck line is naturally reached for a mass of 2.7 TeV. The line at about 90 GeV in the figure corresponds to cross section enhancements through a Z-boson resonance, which lowers the relic density.

IOP Conf. Series: Journal of Physics: Conf. Series **1271** (2019) 012020 doi:10.1088/1742-6596/1271/1/012020

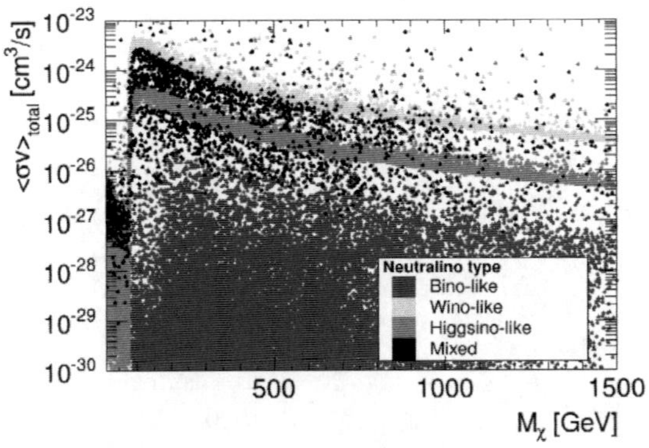

Figure 3: Total annihilation cross section as a function of the neutralino 1 mass for the different neutralino types.

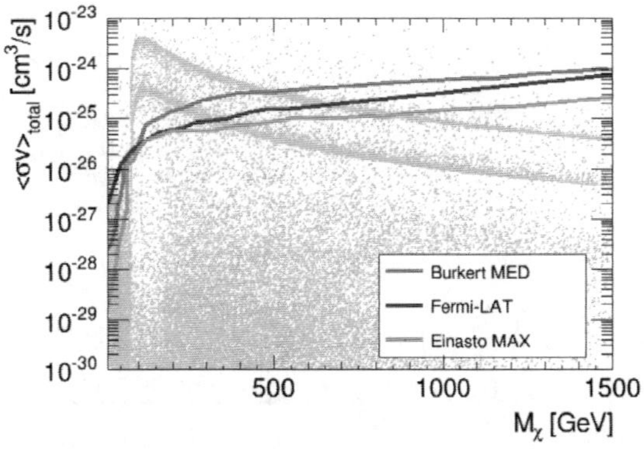

Figure 4: Points excluded by Fermi-LAT gamma ray and AMS-02 antiproton data in the total annihilation cross section vs. neutralino 1 mass parameter plane.

Imposing both the upper and lower relic density bounds generally leads to a selection of scenarios with co-annihilations, for which the mass splitting of the neutralino 1 with the next-to-lightest supersymmetric particle is small, or of scenarios where χ annihilations are enhanced through a resonance of the Z-boson or one of the neutral Higgs bosons. This is demonstrated in figure 2. The valid points require in general small mass splitting, apart from some spread binos with larger mass splittings, which have a heavy Higgs boson or Z-boson resonance. For the case of winos, the small mass splitting is due to a chargino with a mass very close to the χ mass. For the Higgsino case, both the chargino 1 and the neutralino 2 have masses close to the neutralino 1 mass.

As discussed in section 2.1, we consider only the upper bound of the Planck dark matter density interval, which favours light wino- and Higgsino-like χ, and bino-like χ with strong co-annihilations.

3.3.1. Indirect detection We consider the constraints from AMS-02 antiproton and Fermi-LAT gamma ray data, which probe specific dark matter annihilation channels. For both sets of constraints the most important parameters are the χ annihilation cross sections into specific channels. Annihilations to WW and $b\bar{b}$ are particularly interesting in the context of the pMSSM.

In figure 3, the total annihilation cross section times velocity $\langle\sigma v\rangle_{\rm tot}$ is shown as a function of the χ mass, for the different types. $\langle\sigma v\rangle_{\rm tot}$ is the sum of all the σv of the different channels. The wino- and Higgsino-like neutralino 1 regions form two separate strips. The different types of neutralinos 1 have specific main decay channels: binos annihilate mainly into $t\bar{t}$, $b\bar{b}$, and in a lesser extent into Wh, Zh and $\tau\tau$, Higgsinos into WW and ZZ, and winos into WW, when the decay channels are open. When the above-mentioned channels are closed because of a small χ mass, the χ mostly decays to $b\bar{b}$ and $\tau\tau$, and less frequently into $c\bar{c}$ and $s\bar{s}$, independently from their type [28]. As seen earlier, winos more strongly annihilate than the other χ types, followed by the Higgsinos. The binos, apart from the case of a resonant annihilation, are more weakly annihilating and are far below the experimental limits.

In figure 4, the exclusion by Fermi-LAT and AMS-02 is shown in the $\langle\sigma v\rangle_{\rm tot}$ vs. neutralino 1 mass parameter plane. In order to quantify the uncertainties related to indirect detection, we consider separately the most conservative limits, *i.e.* obtained using Burkert profile and MED propagation model, and the most stringent ones, *i.e.* using Einasto profile and

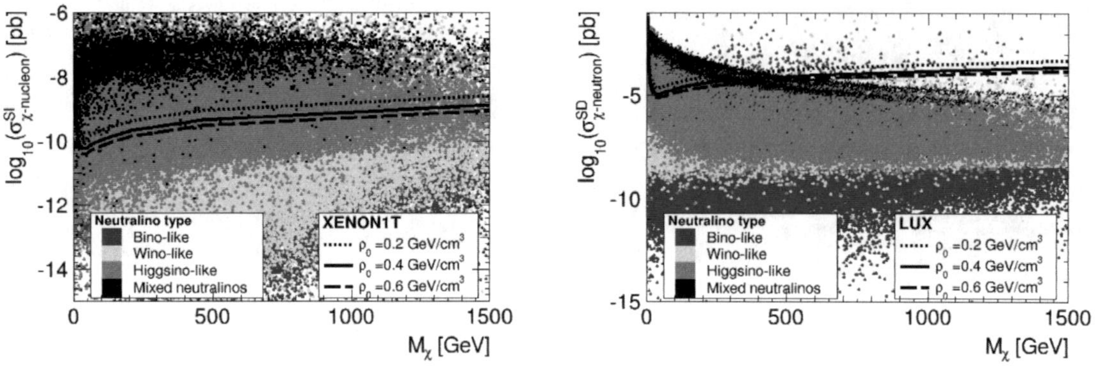

(a) (b)

Figure 5: Generalised (a) spin-independent and (b) spin-dependent neutralino scattering cross section as a function of the neutralino mass. The lines show the XENON1T (left) and LUX (right) 90% C.L. upper limit for three different values of the local dark matter density ρ_0.

Max propagation model. The conservative limits lead to the exclusion of neutralinos 1 with masses between 90 and 550 GeV, which are mainly wino-like. The stringent limits exclude points with χ masses between 0 and 850 GeV. In the small mass region, as well as for masses above 90 GeV, the stringent exclusion limit is strengthened by one order of magnitude in comparison to the conservative case. The stringent case excludes large zones of the wino strip, and of the Higgsino one in a lesser extent. AMS-02 alone brings very strong constraints in the stringent case, beyond the Fermi-LAT limits.

3.3.2. Direct detection Contrary to relic density and indirect detection, which mainly depend on the annihilation and co-annihilation cross sections, direct detection relies on the scattering cross section of neutralino 1 with nucleons. Direct detection is therefore complementary to indirect detection and relic density.

In figure 5, the generalised SI and SD WIMP-nucleon cross sections are shown as functions of the neutralino mass, for the different neutralino 1 types. Higgsinos are in general more strongly interacting than the winos, leading to larger cross sections. In order to assess the consequences of the uncertainties on the obtained constraints, the limits of the XENON1T and LUX experiments are superimposed, for three values of the local dark matter density, namely $\rho_0 = 0.2$, 0.4 and 0.6 GeV/cm^3. Between the conservative line corresponding to $\rho_0 = 0.2$ GeV/cm^3 and the most stringent limit obtained for $\rho_0 = 0.6$ GeV/cm^3, there is at most a factor 3 difference. While this is a large factor, in the context of pMSSM, it does not change much the excluded region, which contains mainly Higgsino-like neutralinos 1.

4. Modified cosmological scenarios

Big-Bang nucleosynthesis (BBN) provides constraints on the properties of the early Universe. Measurements of the nuclear abundances are in agreement with the assumption that the standard cosmological model is correct up to high energies (i.e., that the early Universe evolution is driven by radiation), which limits possible deviations from the standard cosmological model. However, probing cosmology at temperatures higher than tens of MeV is currently impossible. On the other hand, experiments at colliders, and particularly at the LHC, can probe the standard model

IOP Conf. Series: Journal of Physics: Conf. Series **1271** (2019) 012020 doi:10.1088/1742-6596/1271/1/012020

of particle physics at the TeV scale. In the following we combine results from colliders and DM observations to set constraints on the cosmological properties of the Universe before BBN.

As an example we consider the cosmological scenario with a scalar field decaying into radiation and SUSY particles. We perform a scan over the reheating temperature T_{RH} and the initial scalar field density parametrised as the ratio between the scalar field density and the photon density at $T = T_{\rm init}$, $\kappa_\phi = \frac{\rho_\phi}{\rho_\gamma}(T = T_{\rm init})$, and calculate the relic density of different pMSSM points [33, 34]. We consider different values of the parameter $\eta = b\left(\frac{1\ \rm GeV}{m_\phi}\right)$, where b is the branching ratio of the scalar field into WIMPs and m_ϕ the scalar field mass, in order to study the effect of non-thermal production of SUSY particles on the relic density. In each case we derive constraints on the scalar field parameters for our sample of pMSSM19 points so as to investigate the influence of the neutralino properties on the limits derived from the relic DM density.

We start integrating the Boltzmann equations at a temperature $T_{\rm init} = 40$ GeV. For our sample of pMSSM19 points, we use $T_{\rm init} = 1.5 \times T_{\rm fo}$, where $T_{\rm fo}$ is the freeze-out temperature in the standard cosmological model. These choices were made in order to reduce the computation time while starting the calculation sufficiently long before freeze-out and the decay of the scalar field.

4.1. Point with a large relic density

We first investigate the case where the neutralino has a relic density that is too large in the standard cosmological model.

As illustrated in figure 6(a), when T_{RH} is small, $\widetilde{\Sigma}^*$ can remain at its maximum during a large range of temperatures before its decrease due to the decay of the scalar field [33]. The neutralino and scalar field densities decrease during this period with a slope -5, as expected when $\widetilde{\Sigma}^*$ is at its maximum. Figure 6(b) shows that for a large value of T_{RH}, however, the fields are diluted over a smaller range of temperatures and the total decrease is reduced.

Figure 7 shows the effect of varying κ_ϕ and T_{RH} on the relic density. Points respecting the Planck constraints, which we will refer to as *accepted points*, lie along a thin line in the $\log_{10}(\kappa_\phi)/\log_{10}(T_{RH})$ plane. They follow a line of slope ~ 1 at small T_{RH} that changes slightly at $T_{RH} \sim 150$ MeV to a slope 1.5. This transition is the result of the quark/hadronic phase transition, which lowers the number of radiation degrees of freedom. In particular, below $T \sim 150$ MeV, pions become non-relativistic and no longer contribute to the radiation density. This feature is independent of the WIMP and scalar field properties, and is present in all the following results.

The line of accepted points becomes vertical at $T_{RH} \sim T_{\rm fo}$, which is to be expected when the scalar field decays completely during neutralino thermal equilibrium, as there is no possible modification of the relic density. Thus, we can derive a maximum value of the reheating temperature $T_{RH} \lesssim T_{\rm fo}$. One can also note that if $T_{RH} < T_{RH}^{\rm BBN\,lim} \sim 6$ MeV, the scalar field density is too large during BBN, and using AlterBBN [35, 36], we showed that the model is excluded by BBN constraints. This constraint is very general, as it is also independent of the WIMP properties, and thus applicable to any WIMP model. This limit gives us a lower bound for the reheating temperature, as well as a minimum value for the initial scalar field density κ_ϕ using $T_{RH} = T_{RH}^{\rm BBN\,lim}$. For our benchmark point we can deduce $\kappa_\phi \gtrsim 0.1$, but this minimum value will depend on the nature of the WIMP.

No enhancement of the relic density is possible when $\eta = 0$. At small T_{RH} and large κ_ϕ, where the scalar field density could have increased the freeze-out temperature via its relation with the Hubble parameter, and thereby increased the relic density, the densities are in fact already significantly reduced by dilution. Therefore, in order to increase the relic density, it is necessary to consider non-thermal production of the WIMP, i.e., $\eta > 0$. In our case, the region

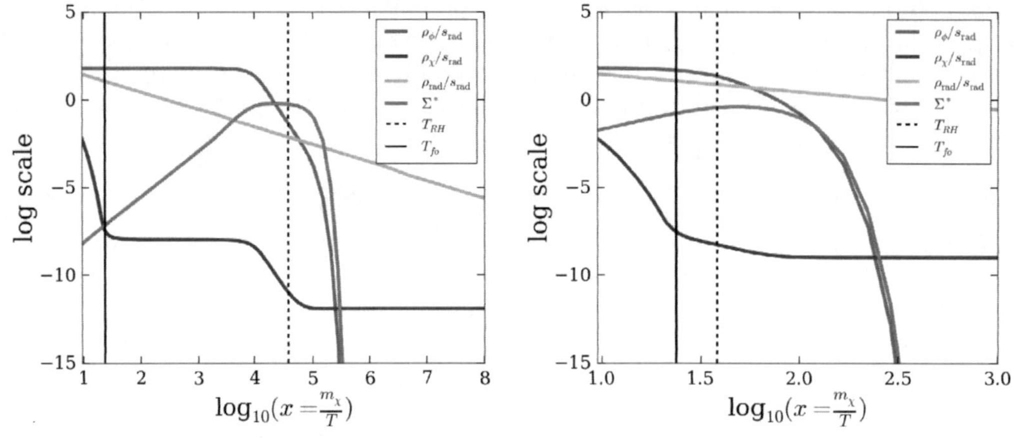

(a) (b)

Figure 6: The evolution of the scalar field, neutralino and radiation densities normalised to the radiation entropy density, and of the injection of entropy $\widetilde{\Sigma}^*$, as a function of $x = m_\chi/T$. (a): $T_{RH} = 0.01$ GeV, $\kappa_\phi = 100$, $T_{\rm init} = 40$ GeV. (b): $T_{RH} = 10$ GeV, $\kappa_\phi = 100$, $T_{\rm init} = 40$ GeV.

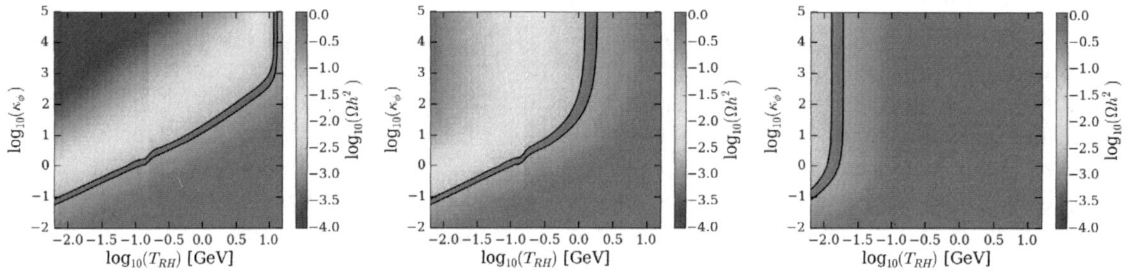

(a) $\eta = 0$ (b) $\eta = 10^{-12}$ (c) $\eta = 10^{-10}$

Figure 7: The effect of varying η on $\log_{10}(\Omega h^2)$ for the benchmark point with a large relic density, indicated by the colour code in the legend.

of interest will be at small T_{RH} and large κ_ϕ, where the relic density is strongly reduced by dilution. The scalar field decay into SUSY particles provides an additional contribution to the relic density, and the DM density measured by Planck may be reached with the appropriate value of η. We test four different values of η in figure 7, and notice that the larger η is the more the line of accepted points is shifted towards small T_{RH}.

In the limit of large κ_ϕ and small T_{RH}, we find that the evolution of the relic density with respect to η and T_{RH} can be approximated by:

$$\Omega h^2 \approx \eta \left(\alpha \, T_{RH} + \beta\right) , \tag{2}$$

where α and β are numerical factors that depend, *a priori*, on the WIMP properties. When η goes to zero, the relic density vanishes, which is expected since in this region of the parameter space the dilution due to the entropy injection is dominant, in the absence of non-thermal production. One can also note that the effects of the dilution and of the non-thermal production equilibrate in such a way that the above expression does not depend on κ_ϕ. For our benchmark

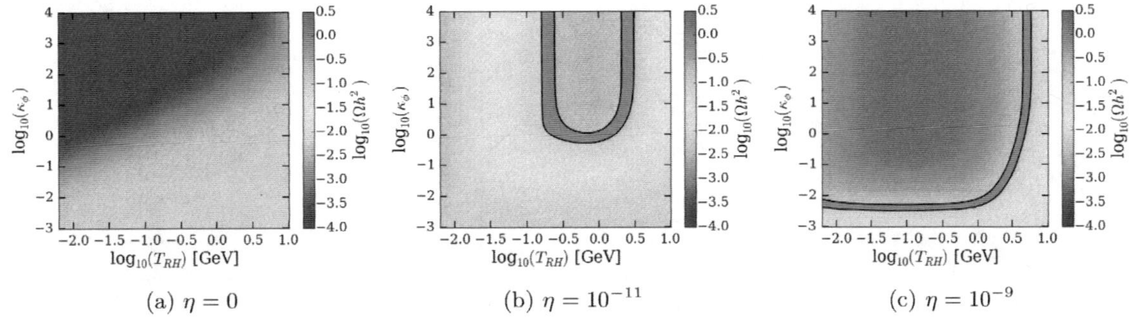

Figure 8: The effect of varying η on $\log_{10}(\Omega h^2)$ for a benchmark point with a small relic density, indicated by the colour code in the legend.

point we find that $\alpha \approx 7.68 \times 10^{10}$ GeV^{-1} and $\beta \approx 2.62 \times 10^7$. This parametrisation enables us to find the value of η required to get the correct relic density for a given reheating temperature. On the other hand, a maximum value of η can be calculated by considering the reheating temperature where the BBN constraints start excluding the model ($T_{RH}^{\text{lim}} \approx 6 \times 10^{-3}$ GeV):

$$\eta_{\text{Max}} = \frac{\Omega h^2{}_{\text{DM}}^{\text{upper lim}}}{\alpha T_{RH}^{\text{lim}} + \beta} \,. \tag{3}$$

For our benchmark point we calculate $\eta_{\text{Max}} \approx 2.93 \times 10^{-10}$. Thus, in this scenario, the branching ratio into SUSY particles must be very small, which can be traced back to our choice of a scalar field with a substantial initial density. We note also that the variation in η does not modify the constraints on κ_ϕ and T_{RH} that we derived in the case $\eta = 0$. Strong constraints on the scalar field parameters can therefore be derived, namely 6 MeV $\lesssim T_{RH} \lesssim T_{\text{fo}}$, $\kappa_\phi \gtrsim 0.1$ and $\eta \lesssim 2.93 \times 10^{-10}$.

4.2. Point with a small relic density

As discussed previously, no enhancement of the relic density is possible when only entropy injection is considered. Therefore, one needs to allow the scalar field to decay into BSM particles. We show in figure 8 the result of scans over T_{RH} and κ_ϕ for a benchmark point with three different values of η. In each scenario the region of accepted points forms a U shape in the κ_ϕ $/T_{RH}$ plane. The vertical right limit corresponds to $T_{RH} \sim T_{\text{fo}}$, and does not move significantly as η increases. The vertical left limit, however, is shifted to the left along the T_{RH} axis and the horizontal limit is shifted downwards towards lower values of κ_ϕ. The constraints on T_{RH} that we deduced for the previous benchmark point hold also in this case: $T_{RH}^{\text{BBN lim}} \lesssim T_{RH} \lesssim T_{\text{fo}}$. However, it is difficult to find limits on κ_ϕ and η as stringent as the ones we found previously.

The largest effect is in the case where the scalar field decays entirely into BSM particles and not into radiation. Thus, if a decay produces two SUSY particles, for example, $b = 2$ and $m_\phi > 2m_\chi$, so $\eta < 1/m_\chi$. In such a case, all the SUSY particles produced by the scalar field decay, starting from the neutralino freeze-out, constitute an overall contribution to the relic density that has to be added to the value of the relic density in the standard model, i.e., $Y = Y_{\text{stand}} + Y_\phi^{\text{T}=\text{T}_{\text{fo}}}/m_\chi$. Therefore, one has a constraint on the scalar field density at freeze-out.

4.3. pMSSM19 sample

In the following we study how the constraints on the scalar field depend on the WIMP properties disregarding the case of a relic density that is too small, as the constraints deduced in this case

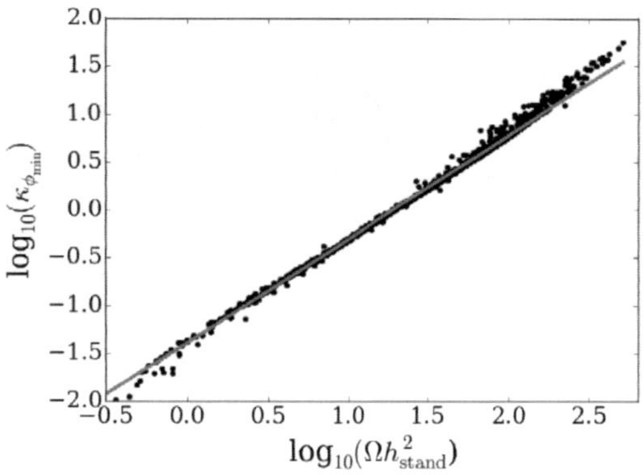

Figure 9: The values of κ_ϕ required to reduce the relic density to the measured DM density with $T_{RH} = T_{RH}^{\rm BBN\,lim}$ and $T_{\rm init} = 40$ GeV as a function of the relic density calculated in the standard model of cosmology.

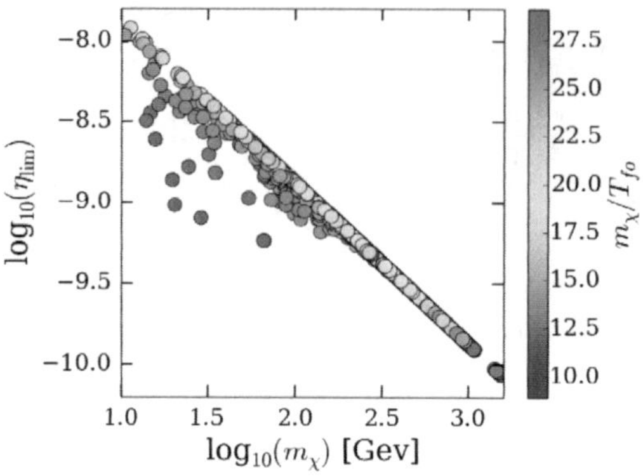

Figure 10: The maximum value of the parameter η for the pMSSM19 sample of points as a function of the neutralino mass. The values of $m_\chi/T_{\rm fo}$ are colour-coded as indicated in the legend.

already showed an explicit dependence on the freeze-out temperature and the relic density at freeze-out.

We focus on the points in our pMSSM19 sample that have a relic density that is too large in the standard cosmological model, which leaves us almost exclusively with bino-like neutralinos. We calculated the values of κ_ϕ that give the correct relic density at $T_{RH} = T_{RH}^{\rm BBN\,lim}$, as shown in figure 9, and find a very good correlation between the relic density calculated in the standard model and $\kappa_{\phi_{\min}}$.

The points in figure 9 follow a line of slope ~ 1. Thus, the minimum value of the initial scalar field density increases with the value of the relic density in the standard model. This can be understood because the larger the relic density at freeze-out is, the stronger the dilution for a given reheating temperature must be. The small scatter of the points at low relic density is due to numerical uncertainties alone, but we note a departure from this line at large $\Omega h^2_{\rm stand}$, when $\kappa_{\phi_{\min}} \gtrsim 1$. With a scalar field density of this order of magnitude, there is also a modification of the Hubble parameter, which advances freeze-out. This mechanism tends to increase the relic density, while the entropy injection decreases it. Overall, the dilution has a stronger effect, but a larger scalar field density is required to decrease the relic density down to the measured DM density.

Next, we calculate the maximum value of η and find a clear dependence on the WIMP mass, as seen in figure 10. Indeed, the scalar field produces a fraction b of SUSY particles, which contributes as $m_\chi \times b$ to the WIMP mass density. Therefore, the larger m_χ is, the more the relic density will be increased for a given value of η, and the smaller the maximum value of η will be. At first approximation, the maximum value of η is inversely proportional to the WIMP mass. However, another mechanism is at play: for the same neutralino mass, the larger $T_{\rm fo}$ is, the larger the neutralino density at the freeze-out temperature is, and thus the smaller η must be in order to reach the correct relic density. As $T_{{\rm fo}_{\rm stand}} \approx m_\chi/20$, we can express a linear relation between $\eta_{\rm lim}$ and m_χ. However, as shown in figure 10, when $T_{\rm fo}$ departs from this approximation towards larger values, the second mechanism becomes more important, and we see a departure from the linear relation between m_χ and $\eta_{\rm lim}$. This happens for neutralino masses smaller than ~ 100 GeV in our sample of points. In any case, η must be very small, of the order of $\sim 10^{-10}$ – 10^{-9}.

IOP Conf. Series: Journal of Physics: Conf. Series **1271** (2019) 012020 doi:10.1088/1742-6596/1271/1/012020

5. Conclusion

The BBN constraints currently constitute the earliest probe of the primordial Universe. The pre-BBN period is largely unconstrained, and many phenomena could have modified the standard cosmological scenario. We studied the impact of dark matter direct and indirect detections, in conjunction with relic density constraints, on the phenomenological MSSM with neutralino dark matter, and addressed in some detail the consequences of the related uncertainties. Furthermore, we discussed the case of modified cosmological scenarios and showed that a decaying cosmological scalar field can modify the relic density by orders of magnitude, showing that the relic density is very sensitive to the properties of the early Universe. Other possible phenomena, such as quintessence, phase transitions, moduli, etc., can lead to similar conclusions.

If new particles are discovered in particle physics experiments, and the properties of the underlying scenario are determined, comparing the relic density to the measured dark matter density would allow us to probe the early Universe and, if there is a discrepancy, could lead to the discovery of new cosmological phenomena, as was illustrated here.

References

[1] Aghanim N *et al.* (Planck) 2018 (*Preprint* 1807.06209)
[2] Baro N, Boudjema F and Semenov A 2008 *Phys. Lett.* **B660** 550–560 (*Preprint* 0710.1821)
[3] Baro N, Boudjema F, Chalons G and Hao S 2010 *Phys. Rev.* **D81** 015005 (*Preprint* 0910.3293)
[4] Hindmarsh M and Philipsen O 2005 *Phys. Rev.* **D71** 087302 (*Preprint* hep-ph/0501232)
[5] Drees M, Hajkarim F and Schmitz E R 2015 *JCAP* **1506** 025 (*Preprint* 1503.03513)
[6] Kamionkowski M and Turner M S 1990 *Phys. Rev.* **D42** 3310–3320
[7] Arbey A and Mahmoudi F 2008 *Phys. Lett.* **B669** 46–51 (*Preprint* 0803.0741)
[8] Moroi T and Randall L 2000 *Nucl. Phys.* **B570** 455–472 (*Preprint* hep-ph/9906527)
[9] Arbey A and Mahmoudi F 2010 *JHEP* **05** 051 (*Preprint* 0906.0368)
[10] Aguilar M *et al.* (AMS) 2016 *Phys. Rev. Lett.* **117** 091103
[11] Cuoco A, Krämer M and Korsmeier M 2017 *Phys. Rev. Lett.* **118** 191102 (*Preprint* 1610.03071)
[12] Giesen G, Boudaud M, Génolini Y, Poulin V, Cirelli M, Salati P and Serpico P D 2015 *JCAP* **1509** 023 (*Preprint* 1504.04276)
[13] Navarro J F, Frenk C S and White S D M 1996 *Astrophys. J.* **462** 563–575 (*Preprint* astro-ph/9508025)
[14] Navarro J F, Ludlow A, Springel V, Wang J, Vogelsberger M, White S D M, Jenkins A, Frenk C S and Helmi A 2010 *Mon. Not. Roy. Astron. Soc.* **402** 21 (*Preprint* 0810.1522)
[15] Burkert A 1996 *IAU Symp.* **171** 175 [Astrophys. J.447,L25(1995)] (*Preprint* astro-ph/9504041)
[16] Delahaye T, Lineros R, Donato F, Fornengo N and Salati P 2008 *Phys. Rev.* **D77** 063527 (*Preprint* 0712.2312)
[17] Lewin J D and Smith P F 1996 *Astropart. Phys.* **6** 87–112
[18] Klos P, Menéndez J, Gazit D and Schwenk A 2013 *Phys. Rev.* **D88** 083516 [Erratum: Phys. Rev.D89,no.2,029901(2014)] (*Preprint* 1304.7684)
[19] Aprile E *et al.* (XENON) 2017 *Phys. Rev. Lett.* **119** 181301 (*Preprint* 1705.06655)
[20] Akerib D S *et al.* (LUX) 2017 *Phys. Rev. Lett.* **118** 251302 (*Preprint* 1705.03380)
[21] Read J I 2014 *J. Phys.* **G41** 063101 (*Preprint* 1404.1938)
[22] Freese K, Frieman J A and Gould A 1988 *Phys. Rev.* **D37** 3388–3405
[23] McMillan P J and Binney J J 2010 *Mon. Not. Roy. Astron. Soc.* **402** 934 (*Preprint* 0907.4685)
[24] Smith M C *et al.* 2007 *Mon. Not. Roy. Astron. Soc.* **379** 755–772 (*Preprint* astro-ph/0611671)
[25] Chaudhury S, Bhattacharjee P and Cowsik R 2010 *JCAP* **1009** 020 (*Preprint* 1006.5588)
[26] Evans N W, Carollo C M and de Zeeuw P T 2000 *Mon. Not. Roy. Astron. Soc.* **318** 1131 (*Preprint* astro-ph/0008156)
[27] Allanach B C 2002 *Comput. Phys. Commun.* **143** 305–331 (*Preprint* hep-ph/0104145)
[28] Arbey A, Boudaud M, Mahmoudi F and Robbins G 2017 *JHEP* **11** 132 (*Preprint* 1707.00426)
[29] Robbins G, Mahmoudi F, Arbey A and Boudaud M 2017 *PoS* **EPS-HEP2017** 082 (*Preprint* 1710.03658)
[30] Arbey A and Mahmoudi F 2010 *Comput. Phys. Commun.* **181** 1277–1292 (*Preprint* 0906.0369)
[31] Arbey A and Mahmoudi F 2011 *Comput. Phys. Commun.* **182** 1582–1583
[32] Arbey A, Mahmoudi F and Robbins G 2019 *Comput. Phys. Commun.* **239** 238–264 (*Preprint* 1806.11489)
[33] Arbey A, Ellis J, Mahmoudi F and Robbins G 2018 *JHEP* **10** 132 (*Preprint* 1807.00554)
[34] Arbey A, Ellis J, Mahmoudi F and Robbins G 2018 *PoS* **ICHEP2018** 035 (*Preprint* 1811.12764)
[35] Arbey A 2012 *Comput. Phys. Commun.* **183** 1822–1831 (*Preprint* 1106.1363)
[36] Arbey A, Auffinger J, Hickerson K P and Jenssen E S 2018 (*Preprint* 1806.11095)

IOP Conf. Series: Journal of Physics: Conf. Series **1271** (2019) 012021 doi:10.1088/1742-6596/1271/1/012021

Precision Determination of the Light Quark Masses from QCD Finite Energy Sum Rules

Alexes K. Mes

Department of Physics, University of Cape Town, Private Bag X3, Rondebosch 7701, South Africa

E-mail: MSXALE002@myuct.ac.za

Abstract. The light quark masses are determined from a QCD finite energy sum rule, using the pseudoscalar correlator to six-loop order in perturbative QCD, with the leading vacuum condensates and higher order quark mass corrections included. Both the fixed order perturbation theory (FOPT) method and contour improved perturbation theory (CIPT) method are explored. The results in the latter framework exhibit good convergence and stability in the window $s_0 = 3.0 \;\; 5.0$ GeV^2 for the strange quark and $s_0 = 1.5 \;\; 4.0$ GeV^2 for the up and down quarks; where s_0 is the radius of the integration contour in the complex s-plane. The results are: $\overline{m}_s(2\text{ GeV}) = 91.8 \pm 9.9$ MeV, $\overline{m}_u(2\text{ GeV}) = 2.6 \pm 0.4$ MeV, $\overline{m}_d(2\text{ GeV}) = 5.3 \pm 0.4$ MeV, and the sum $\overline{m}_{ud} \equiv (\overline{m}_u + \overline{m}_d)/2$, is $\overline{m}_{ud}(2\text{ GeV}) = 3.9 \pm 0.3$ MeV. These proceedings critically explore how the current results - computed precisely in a modern computer language, Mathematica - compare to past determinations in literature.

1. Introduction

Quantum Chromodynamics is a confining theory, meaning that quarks do not exist in nature as single, free particles; but, are rather found in hadronic bound states. Due to this quark and gluon confinement it is impossible to determine the quark masses with the same techniques as one would use to determine the mass of non-confined particles.

There are two major approaches in Quantum Chromodynamics that can be used to calculate the quark masses at some given energy scale. Broadly this divides into a numerical approach (lattice QCD [1], [2]) and an analytical approach (QCD sum rules [3,4]). Lattice QCD discretizes space and time to reduce the infinite number of field variables in QCD to a finite countable number within the path-integral formulation of Quantum Field Theory [5]. QCD sum rules separate the perturbative and nonperturbative contributions in QCD, where the latter are described by quark and gluon condensates which are present. The sum rule method relates low energy hadronic quantities (which are measurable) to high energy expressions in QCD, such that unmeasurable QCD parameters become a function of known hadronic information.

Currently, there are numerous sum rule frameworks. The finite energy sum rule method (FESR), in which QCD sum rules are placed within the complex squared energy plane, is the approach taken to determine these results.

There are two frameworks in which the contour integral in FESR can be performed: fixed order perturbation theory (FOPT) and contour improved perturbation theory (CIPT). Fixed order keeps – as its name suggests – the strong coupling, $\alpha_s(s)$, frozen on the integration contour, and performs the renormalization group improvement after integration. On the other hand, in contour improved perturbation theory the strong coupling is running and the renormalization group improvement is implemented before integration.

2. Method and Calculations

In this determination, the preferred calculation was done by considering the pseudoscalar correlator within the framework of contour improved perturbation theory. Considering the criteria of stability and convergence, this framework was favoured over fixed order perturbation theory for both the case of the correlator determining the up and down quark mass and the correlator determining the strange quark mass. Respectively, these masses were found to be $\overline{m}_u(2\,\text{GeV}) = (2.6 \pm 0.4)\,\text{MeV}$, $\overline{m}_d(2\,\text{GeV}) = (5.3 \pm 0.4)\,\text{MeV}$ and $\overline{m}_s(2\,\text{GeV}) = (91.8 \pm 9.9)\,\text{MeV}$ in the CIPT framework. The reader is referred to *Light Quark Masses from QCD Finite Energy Sum Rules* by A.K. Mes [6] where details of the method behind these calculations are provided.

The light quark masses presented here are in agreement with recent sum rule and lattice QCD determinations in this field. Figures (1) and (2) show how these results compare with the current literature.

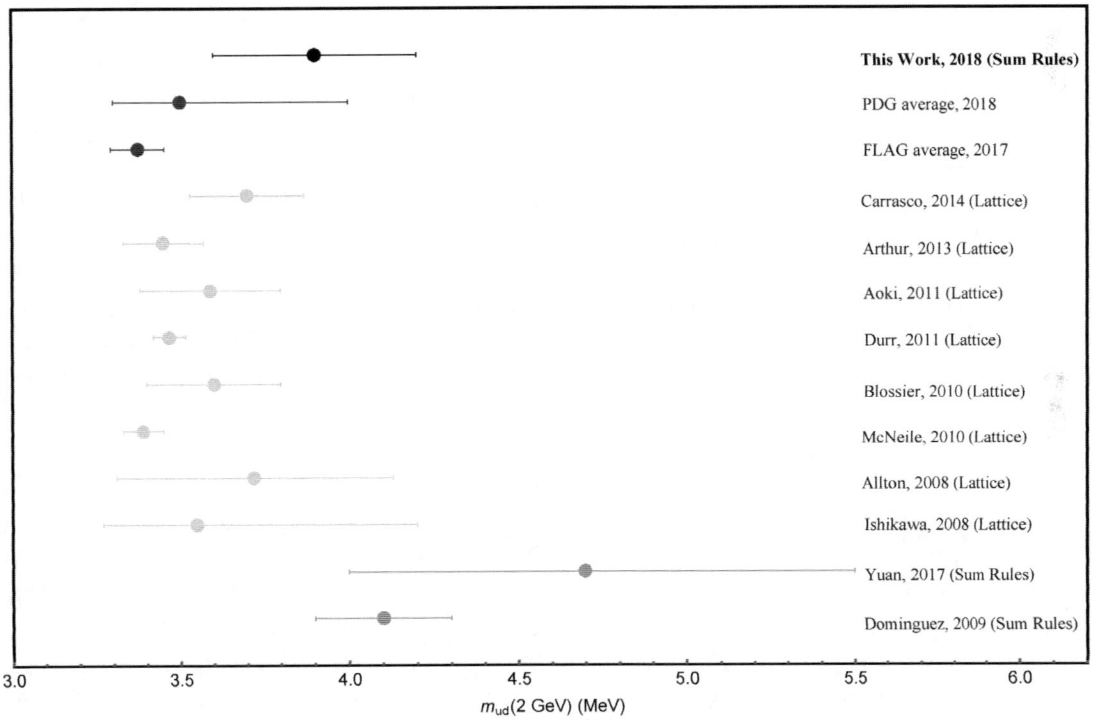

Figure 1. The up and down quark mass obtained in [6] compared to the PDG [7] and FLAG [2] averages, as well as the most recent sum rule results [8, 9] and lattice QCD results [10–17].

There is a trend for lattice QCD determinations to generally yield lower up and down quark mass values than sum rule determinations. Although this is observed, it is not apparent what the cause is - especially since the same trend is not observed in the case of the strange quark mass (figure (2)). From figure (1) we can see that this determination – through its original aspects and various improvements (examined in the next section) – is in good agreement with the recent numerical determinations; thereby reducing the existing tension between QCD sum rules and lattice QCD determinations.

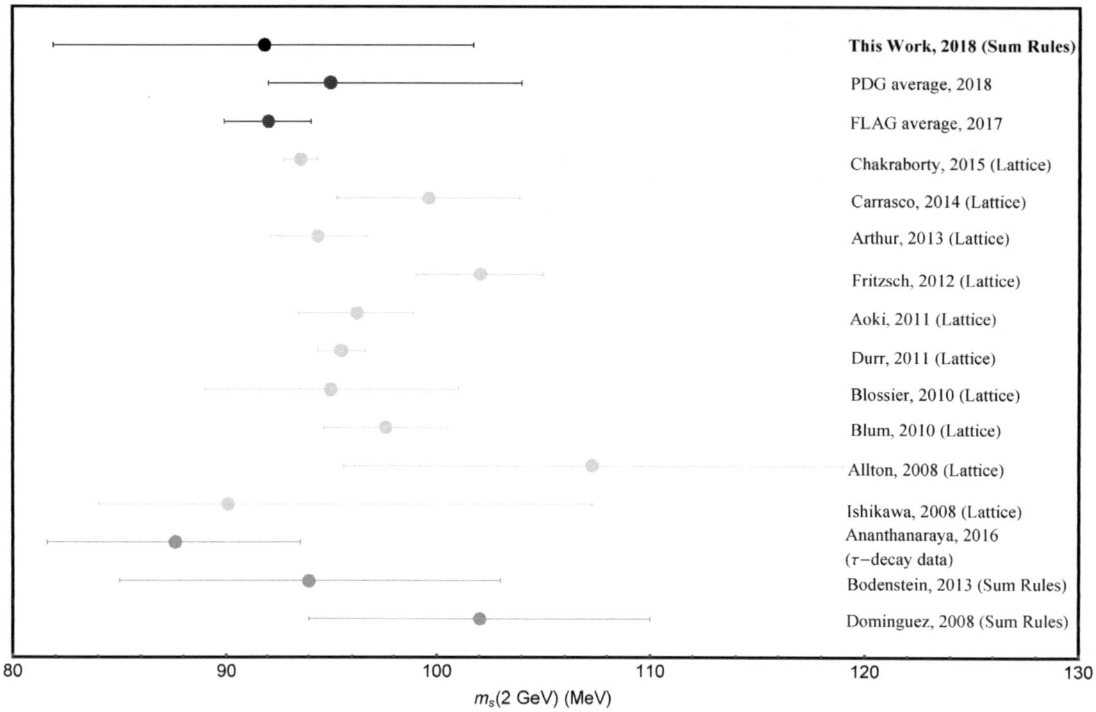

Figure 2. The strange quark mass obtained in [6] compared to the PDG [7] and FLAG [2] averages, as well as the most recent analytical results [18–20] (the Ananthanarayan et al. determination extracts the strange quark from τ-decay spectral moments using renormalization techniques, while the latter two are sum rule determinations) and lattice QCD results [10–14, 16, 17, 21–23].

The strange quark determination presented here is in good agreement with most recent results in the literature.

Examining figures (1) and (2), the question naturally emerges: *How do the precision determinations presented here differ computationally from previous sum rule calculations of the light quark masses?*, or phrased differently *What are the original aspects and improvements of the precision determinations presented here over other sum rule determinations?*. This question is addressed in the following section.

3. Comparisons

In 2018, The Particle Data Group [7] considered two phenomenological determinations by Yuan et al. [8] and Dominguez et al. [9] when determining the world average of the up and down quark masses. The latest calculation of $\overline{m}_{ud}$ by Yuan et al. [8], published in 2017, determines the up and down quark masses in the I = 0 scalar channel within the Borel transform framework. It further differs from the current determination since it includes an instanton contribution as a nonperturbative effect in the theoretical representation of the correlation function.

The previous determination by Dominguez et al. [9] – published in 2009 and also included in the Particle Data Group world average of $\overline{m}_{ud}$ – is a FESR determination of the up and down quark masses. Since this determination was also done by considering a pseudoscalar correlator, it might (on face value) seem as if [9] is very similar to the determination presented here. However, they are fundamentally different in a number of ways.

One important, noteworthy difference is how the determinations are computationally implemented. In the current precision determination the calculations have been computed in Mathematica (see [6] for the annotated notebooks), a modern 64-bit computer language with natural support for precision real and complex numbers. At a minimum, using the machine precision corresponds to 16 digits of mantissa, although higher multiples of the machine precision can be used. In comparison, the determination by Dominguez et al. [9] was performed using Fortran 90 with code edited and updated from Fortran 77; respectively these are 1991 and 1977 computer languages. Typically the early standard Fortran versions only contain single precision (numbers accurate to 7 digits of mantissa), while double precision (typically 14 digits of mantissa) can be specified for real numbers. Complex numbers present more of an issue, as the double precision must be split between the real and imaginary components. Notwithstanding this, it is the confluence of these different working numerical precisions in earlier Fortran languages that can typically result in an accumulated error. The issue of numerical precision can make a notable impact on the results. For example, in CIPT where the renormalization group equations must be solved at 10 000 points around the circular contour, with each point relying on the previous value of the strong coupling as an initial condition, the opportunity for numerical carried error is pronounced. In Mathematica the numerical underflow in such a situation can be managed with greater care.

Considering that the determination by Dominguez et al. [9] hitherto represented the most recent FESR determination of the up and down quark masses, and formed part of the world average of these quark masses as determined by the Particle Data Group [7]; it is worrying that such old computer languages were used, especially since modern alternatives are readily available. The current determination [6], where all calculations were performed in Mathematica, represents a significant advancement in achieved numerical precision of FESR determinations of the up and down quark masses.

Furthermore, the analysis of different kernels, examining the issue of the convergence of the perturbative QCD expansion, a different implementation of the running QCD coupling and a more careful error analysis are some of the considerable improvements accomplished by this present determination - the details on each of these issues can be found in [6].

Turning towards the strange quark mass, in 2018 the Particle Data Group [7] considered three phenomenological determinations Ananthanarayan et al. [18], Bodenstein et al. [19], Dominguez et al. [20] when determining the world average of the strange quark mass. The latest publication by Ananthanarayan et al. [18], published in 2016, uses a renormalization group summed perturbation theory and relates this to τ-decay spectral function data in order to extract the strange quark mass. This is not a FESR sum rule determination of $\overline{m}_s$, and as such will not be examined further here. The previous determination by Bodenstein et al. [19], published in 2013, is a FESR determination of the strange quark mass. The determination is performed in Mathematica and the convergence of the perturbative QCD expansion is examined. However, the calculations in [19] are only performed in the framework of FOPT. Here, in this current determination [6], both FOPT and CIPT are considered, leading to a different conclusion about the preferred framework in which to calculate the strange quark mass (in terms of stability and convergence) being made. The 2008 determination by Dominguez et al. [20], was calculated by the same collaboration and around the same time as the determination by Dominguez et al. [9]. Consequently it suffers from all the issues pertaining to the determination in [9] discussed previously.

4. Conclusion

The work outlined here and presented in detail in [6] gives the most accurate up, down and strange quark mass determination from a QCD finite energy sum rule to date. It is further worth noting that a publication of the up and down quark mass determinations resulting from the author's work in [6], can be found in the Journal of High Energy Physics [24].

The corresponding Mathematica notebooks to the current precision determinations in [6] can be found in the GitHub repository `https://github.com/AlexesMes/light-quark-masses`. It is still uncommon in the field of theoretical physics to make the code available with a published paper. Historically, this was not possible as it made publications unreadable and too expensive to print. However, these reasons are no longer applicable in the present day, and the author has made her code available, since she believes in the importance of modern research and open collaboration.

Acknowledgements

The author wishes to acknowledge Cesareo Dominguez and Karl Schilcher for their early supervision in the initial stages of calculating the up and down quark masses (February - July 2018).

References

[1] F. Knechtli, M. Günther, and M. Peardon. *Lattice Quantum Chromodynamics: Practical Essentials.* Springer, (2017).
[2] S. Aoki *et al.*, FLAG Coll. *Review of Lattice Results Concerning Low-energy Particle Physics. Eur. Phys. J.* ***C***, 77(2):112, (2017).
[3] C. Dominguez. *Analytical Determination of the QCD Quark Masses.* In *50 Years Of Quarks, pp.287-313.* World Scientific Publishing Co., Singapore, (2015).
[4] P. Colangelo and A. Khodjamirian. *QCD Sum Rules, a Modern Perspective.* In *At The Frontier of Particle Physics: Handbook of QCD (in 3 Volumes), pp.1495-1576.* World Scientific Publishing Co., (2001).
[5] W. Greiner, S. Schramm, and E. Stein. *Quantum Chromodynamics.* Springer, (2007).
[6] A.K. Mes. *Light Quark Masses from QCD Finite Energy Sum Rules.* `https://github.com/AlexesMes/light-quark-masses`, University of Cape Town, (2019).
[7] M. Tanabashi et al. *Particle Data Group Review. Physical Review* ***D***, 98:030001, (2018).
[8] J. Yuan, Z. Zhang, T. Steele, H. Jin, and Z. Huang. *Constraint on the Light Quark Mass* m_q *from QCD Sum Rules in the I= 0 Scalar Channel. Physical Review* ***D***, 96(1):014034, (2017).
[9] C. Dominguez, N.F Nasrallah, R. Röntsch, and K. Schilcher. *Up-and Down-quark Masses from Finite-Energy QCD Sum Rules to Five Loops. Physical Review* ***D***, 79(1):014009, (2009).
[10] N. Carrasco, A. Deuzeman, P. Dimopoulos, R. Frezzotti, V. Giménez, et al. *Up, Down, Strange and Charm Quark Masses with* $N_f = 2+1+1$ *Twisted Mass Lattice QCD. Nuclear Physics* ***B***, 887:19–68, (2014).
[11] R. Arthur, T. Blum, P.A. Boyle, N.H. Christ, N. Garron, et al. *Domain Wall QCD with Near-Physical Pions. Physical Review* ***D***, 87(9):094514, (2013).
[12] Y. Aoki, R. Arthur, T. Blum, P.A. Boyle, D. Brömmel, et al. *Continuum Limit Physics from* $2+1$ *Flavor Domain Wall QCD. Physical Review* ***D***, 83(7):074508, (2011).
[13] S. Durr, Z. Fodor, C. Hoelbling, S.D. Katz, S. Krieg, et al. *Lattice QCD at the Physical Point: Light Quark Masses. Physics Letters* ***B***, 701(2):265–268, (2011).
[14] B. Blossier, P. Dimopoulos, R. Frezzotti, V. Lubicz, M. Petschlies, et al. *Average Up/Down, Strange, and Charm Quark Masses with* $N_f = 2$ *Twisted-Mass Lattice QCD. Physical Review* ***D***, 82(11):114513, (2010).
[15] C. McNeile, C.T. Davies, E. Follana, K. Hornbostel, G.P. Lepage, et al. *High-Precision c and b Masses, and QCD Coupling from Current-Current Correlators in Lattice and Continuum QCD. Physical Review* ***D***, 82(3):034512, (2010).
[16] C. Allton, D.J. Antonio, Y. Aoki, T. Blum, P.A. Boyle, et al. *Physical Results from* $2+1$ *Flavor Domain Wall QCD and SU(2) Chiral Perturbation Theory. Physical Review* ***D***, 78(11):114509, (2008).
[17] T. Ishikawa, S. Aoki, M. Fukugita, S. Hashimoto, K.I. Ishikawa, et al. *Light-Quark Masses from Unquenched Lattice QCD. Physical Review* ***D***, 78(1):011502, (2008).
[18] B. Ananthanarayan and D. Das. *Optimal Renormalization and the Extraction of the Strange Quark Mass from Moments of the* τ*-decay Spectral Function. Physical Review* ***D***, 94(11):116014, (2016).
[19] S. Bodenstein, C. Dominguez, and K. Schilcher. *Strange Quark Mass from Sum Rules with Improved Perturbative QCD Convergence. Journal of High Energy Physics*, 2013(7):138, (2013).

IOP Conf. Series: Journal of Physics: Conf. Series **1271** (2019) 012021 doi:10.1088/1742-6596/1271/1/012021

[20] C. Dominguez, N.F Nasrallah, R. Röntsch, and K. Schilcher. *Strange Quark Mass from Finite Energy QCD Sum Rules to Five Loops. Journal of High Energy Physics*, 05:020, (2008).

[21] B. Chakraborty, C.T.H. Davies, B. Galloway, P. Knecht, J. Koponen, et al. *High-Precision Quark Masses and QCD Coupling from* $n_f = 4$ *Lattice QCD. Physical Review* ***D***, 91(5):054508, (2015).

[22] P. Fritzsch, F. Knechtli, B. Leder, M. Marinkovic, S. Schaefer, et al. *The Strange Quark Mass and Lambda Parameter of Two Flavor QCD. Nuclear Physics* ***B***, 865(3):397–429, (2012).

[23] T. Blum, R. Zhou, T. Doi, M. Hayakawa, T. Izubuchi, et al. *Electromagnetic Mass Splittings of the Low Lying Hadrons and Quark Masses from* $2 + 1$ *Flavor Lattice QCD + QED. Physical Review* **D**, 82(9):094508, (2010).

[24] C. Dominguez, A. Mes, and K. Schilcher. *Up-and Down-quark Masses from QCD Sum Rules. Journal of High Energy Physics*, 2019(2):57, (2019).

Phase transitions in finite size systems

Sylvain Mogliacci, Isobel Kolbé and W A Horowitz

Department of Physics, University of Cape Town, Rondebosch 7701, South Africa

E-mail: sylvain.mogliacci@uct.ac.za

Abstract. We follow recent developments of a finite size thermal field theoretic formalism, which is mainly relevant to the investigation of small size corrections to the thermodynamical properties of the quark-gluon plasma. In doing so, we rely on statistical mechanics and large deviation theory arguments, and investigate geometrically confined systems of infinite volume V with a finite spatial extent L. More specifically, we focus on systems in contact with a heat bath as well as on isolated ones. We find and present in detail the characteristic behaviors of a first and a second order phase transition, respectively. In the former case, we establish a connection between the phase transition and the mechanical support of the system. In the latter one, we identify a critical length $L_c \sim (V/EL)^{1/3}$, where E is the total energy of the system which is fixed. We also briefly comment on the relevance of our findings, for respectively both the Casimir experiments and the future quark-gluon plasma phenomenology.

1. Introduction

1.1. *Context and outline*

In this proceedings, we briefly report on recent advances in identifying and fully characterizing a new type of geometrically driven phase transition [1, 2]. We use the recent developments made in investigating the finite size thermodynamics of relativistic quantum fields, with especially [3, 1, 4]. These developments are mostly relevant to the need for understanding and incorporating small size corrections to the thermodynamical properties of the quark-gluon plasma, and ultimately to all its medium properties [5]. However, the procedure can and will [2] be applied following section 2 to Casimir systems such as those featuring the so called force [6, 7].

It should be noticed that we work in the thermodynamic limit ($V \to \infty$), and thus chose to implicitly rescale all thermodynamic functions by the fixed asymptotically large area parameter $V_2 \equiv V/L$, in order to be able to properly manipulate the functions. We use two simplified models of a massless single neutral scalar field, with Dirichlet spatial boundary conditions, in a geometry defined by the space in between two asymptotically large and perfectly conducting parallel plates separated by a distance L, as inspired from [1].

In section 1.2, we first present our main argument regarding the relevance and equivalence of ensembles. Our argument is based on the theory of large ensemble deviation, and fully supported by the references [8, 9]. We then describe, in section 2, a very simple model of Casimir systems in thermal equilibrium with their environments. Such a crude model is nonetheless qualitatively relevant to isothermal realistic Casimir systems in such a geometry. Furthermore, in section 3, we describe a very simple model of a non interacting gluon plasma, in the same geometry, following in a simplified manner what has been recently studied in [1, 4]. Such a simplified model is nonetheless qualitatively relevant to small size isolated systems which are created in heavy ion collisions. Finally, we conclude in section 4 with a special emphasis on the relevance of our findings for the interpretation of Casimir experimental data, as well as for the development of an improved quark-gluon plasma phenomenology.

1.2. Equivalence of ensembles

From an experimental point of view, the $\{T, L\}$ set of variables is not relevant in any of the experimental setups which are at the center of this work (see both sections 2 and 3), even if it is extremely convenient from a theoretical point of view when computing any quantity. In particular, the L variable only makes sense – experimentally speaking – for truly macroscopic systems, where one can actually use a ruler in order to measure distances and the lengths can be considered to be exactly fixed. For non-macroscopic systems[1], lengths cannot be measured exactly: They fluctuate. Therefore, in such a system, and apart from possibly the temperature T, the force or the pressure p which is applied to the system is actually what can be triggered, and thus is the pressure of the system itself. One can see that, for example, the direct measurements in Casimir systems concern forces instead of lengths [10, 7, 11, 2].

This being said, the presently investigated Phase Transitions (PT) require the promotion of $\{T, p\}$ and $\{e, p\}$ ($e \equiv E/V_2$, the scaled total energy of the system) – for the isothermal Casimir systems and those isolated which are relevant to small relativistic collisions, respectively – to the rank of experimentally relevant set of variables. Regarding the former type of system, we will, however, only consider here the $\{e, L\}$ set of variables, and refer to [1] for a complete treatment further using a refined model. Thus, we will consider the $\{T, p\}$- and $\{e, L\}$-ensembles to be the physically relevant ones for each of the system type, and analyze their respective equivalence, if any, to the $\{T, L\}$-ensemble which is used in order to perform the calculations.

Indeed, in order to make a point potentially relevant to the experiments, one must make sure that the theoretically chosen set of parameters $\{T, L\}$ gives the same answer as the aforementioned physically relevant one: One must pay attention that the two statistical ensembles are (at least locally; where applicable) equivalent, which is to say here that for example $\{T, L\} \Leftrightarrow \{T, p\}$ over part or all of the domains of variables. For the present purpose, we consider the thermodynamic equivalence of our ensembles and not the macrostate equivalence. Then, one can establish the local equivalence over some sub-domains of two ensembles – one being fully convex or concave (i.e. fully canonical or microcanonical), and the other convex in one variable and concave in the other (i.e. mixed canonical and microcanonical), which is the case here with the $\{T, p\}$- or $\{e, L\}$-ensembles and the $\{T, L\}$-ensemble, respectively – by means of the following [9]: The difference between the thermodynamic potential of, say, the $\{T, L\}$-ensemble, i.e., the Helmholtz free energy $F(T, L)$, and its single Legendre-Fenchel transform [12] back and forth $F(T, L)^{\star\star}$ must be vanishing over the said sub-domain of local equivalence [8, 9]. In other words, $F(T, L) - F(T, L)^{\star\star} = 0$ when $F(T, L)^{\star} \equiv S(e, L)$ proves the equivalence between the $\{T, L\}$-ensemble and the $\{e, L\}$ one over the related domain.

Regarding the Legendre-Fenchel transform, which is necessary in order to assess the thermodynamic equivalence of two ensembles, we use here the following procedure given that we expect our ensembles not to be globally equivalent anyway: We perform only the Legendre transform of the thermodynamic potentials over one of their variables, but restrict the domains of definition of these potentials to the sub-domains where their Hessians are not singular. On such sub-domains indeed, the Legendre-Fenchel and the Legendre Transforms are in fact equivalent, which allow us to circumvent the difficulty[2] of manipulating supporting hulls within our setups.

[1] With sizes ranging from the micro/nano-meter to the femto-meter, and involving energy scales which make them comparatively small such as the ones we consider here.

[2] Defining supporting hulls in mixed canonical and microcanonical ensembles, or simply defining them in the usual manner given that the curvature (in the sense of the second order derivative test) has the wrong sign on a whole half of the domain, are examples of such difficulties.

IOP Conf. Series: Journal of Physics: Conf. Series **1271** (2019) 012022 doi:10.1088/1742-6596/1271/1/012022

Using this procedure, the sub-domain of ensemble equivalence is defined by the overlap of both the region in which we have $F(T,L) - F(T,L)^{\star\star} = 0$, and the region in which the Legendre and the Legendre-Fenchel transforms are equivalent, i.e., in which the Hessian is non singular. If such a non empty region exists, then we can claim to have local equivalence of the two ensembles wherever on this domain [2].

2. Isothermal Casimir systems

We now present our isothermal Casimir system model, defined by the $\{T, L\}$-ensemble free energy

$$F_{\mathrm{Cs}}(T,L) = -\frac{1}{L^3} - LT^4 + \frac{\kappa}{2}(L - L_0)^2. \tag{1}$$

A few comments are in order regarding the above definition: (a) Regarding the thermal Casimir contribution to our model, we only kept both the high temperatures and short distances leading (respectively second and first above) terms, compared to the corresponding full free energy [1]. (b) For the sake of simplicity, we set the constants of these terms to one, following an unphysical but very convenient unit system $[\mathcal{U}]$. This is done, with no loss of generality, by defining specific units of temperature and length. (c) Regarding the mechanical support contribution to our model, we incorporated a spring-like term (the last above one; relevant to cantilevers in actual experimental setups [10, 7, 11, 2]). (d) This last term features a stiffness parameter κ, and a rest length parameter L_0. (e) The setting of such a new dimensionfull parameter (κ) to some arbitrary numerical value[3] is done by defining a convenient unit of energy. (f) At last, we have checked that changing the unit system, hence the numerical values of the constants and κ does not qualitatively affect the findings which we report in the following.

In figure 1 left, we present the phase diagram of the $\{T, L\}$-ensemble.

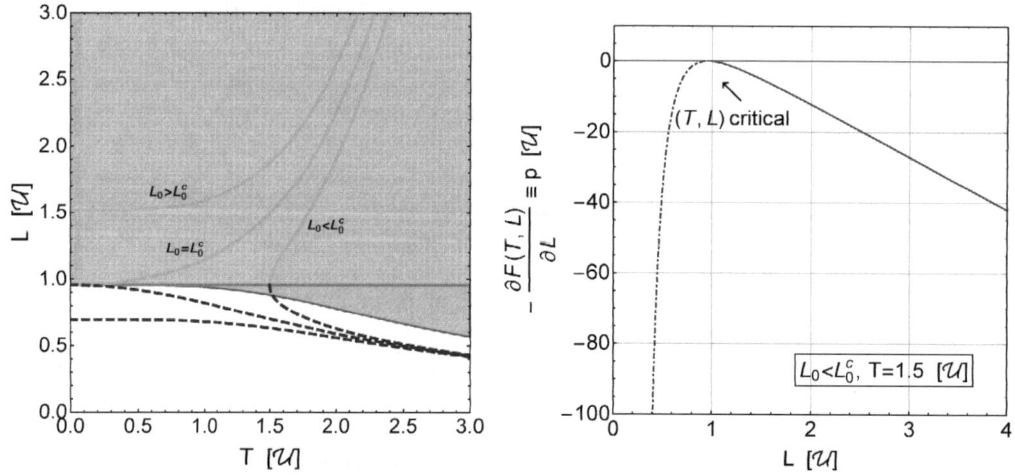

Figure 1. (Left) The $\{T, L\}$-ensemble phase diagram. The blue region marks a non singular Hessian. The pink meshed and blue sub-region marks ensemble equivalence. The red line is the line of phase transition. The Mechanical equilibrium paths, for which the total pressure vanishes, are the green and dashed black lines, each set at different values of L_0. (Right) The pressure obtained from the $\{T, L\}$-ensemble versus the length of the system, with critical point.

The blue region corresponds to the region in which the Hessian of the thermodynamic potential from equation 1 is not singular (i.e. in which the second law of thermodynamic

[3] In the following, $\kappa = 15\,\mathcal{U}$ always.

IOP Conf. Series: Journal of Physics: Conf. Series **1271** (2019) 012022 doi:10.1088/1742-6596/1271/1/012022

is not violated), and the meshed pink region corresponds to the region in which we have $F(T,L) - F(T,L)^{**} = 0$. Therefore, the overlap of the two regions is the domain of equivalence between the ensembles $\{T,L\}$ and $\{T,p\}$, as discussed in section 1.2. The red (horizontal, due to the simplicity of the model) line is the line of isothermal compressibility divergence (where the first derivative of the pressure vanishes [1]), i.e., in the $\{T,L\}$-ensemble then, the line of second order phase transition. The lines in green and dotted black represent possible paths of mechanical equilibrium (vanishing pressure), for different values of the spring rest length L_0. One can see that this parameter is compared to a specific value, namely L_0^c which happens to be the L_0 value above which no mechanical equilibrium line can meet the second order phase transition. Below that critical value, such a phase transition occurs in this ensemble.

In figure 1 right, we then show the pressure obtained from the $\{T,L\}$-ensemble versus the length of the system. We see a critical (T,L) point at a given rest length $L_0 < L_0^c$, where the pressure changes its first order derivative sign with respect to L [1]. This point lays onto the red line in the left part of the same figure, and features the said second order phase transition.

We then move to the $\{T,p\}$-ensemble, which is physically relevant for this model as we saw in section 1.2, and show in figure 2 left, the phase diagram of the $\{T,p\}$-ensemble with same notation as in figure 1. It is interesting to see that again, a phase transition features in this ensemble, in the very same manner as for the $\{T,L\}$-ensemble. However, in figure 2 right, where we display the length obtained from the $\{T,p\}$-ensemble versus the pressure of the system, that the phase transition featuring for mechanical support with rest length $L_0 < L_0^c$ is a first order one.

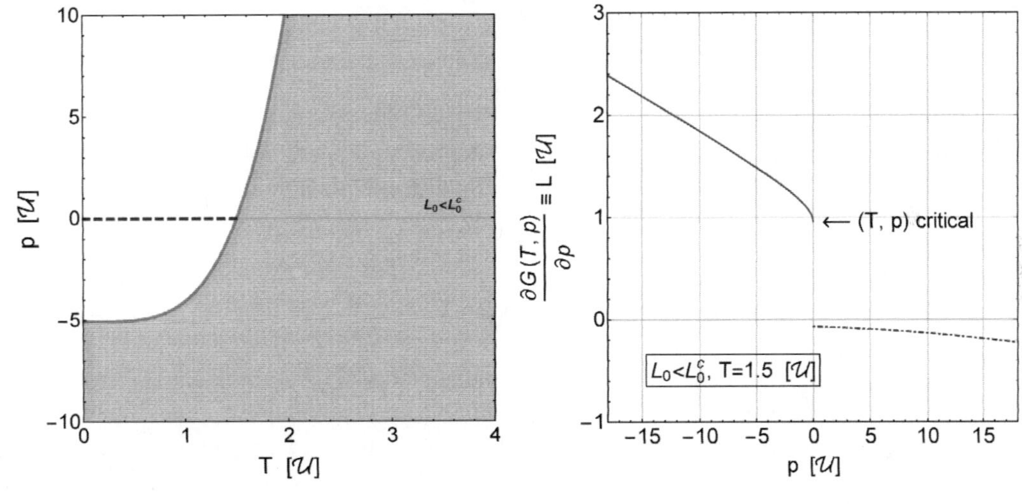

Figure 2. (Left) The phase diagram of the $\{T,p\}$-ensemble, with same notation as in figure 1. (Right) The length obtained from the $\{T,p\}$-ensemble versus the pressure of the system, with critical point.

Given the relevance of this system, it would appear that isothermal Casimir systems feature a first order phase transition for certain mechanical support setups (here reflected by the constraint $L_0 < L_0^c$).

3. Isolated systems from relativistic collisions

We now present our isolated small system model, defined by the $\{T,L\}$-ensemble free energy

$$F_{\rm Is}(T,L) = -\frac{1}{L^3} - LT^4, \tag{2}$$

where we see the exact same expressions than in equation 1, except for the mechanical support which disappeared here. Thus, the same comments than for equation 1 (a), (b), and (f) hold for the present model.

In figure 3, we show the phase diagrams of the $\{T, L\}$- and $\{e, L\}$-ensembles, respectively, with the same notation as in figure 1. The red phase transition line in figure 3 left is reported only for guidance purposes, since it is coming from the divergence of the iso-energetic compressibility which is relevant to (and computed in) the ensemble $\{e, L\}$.

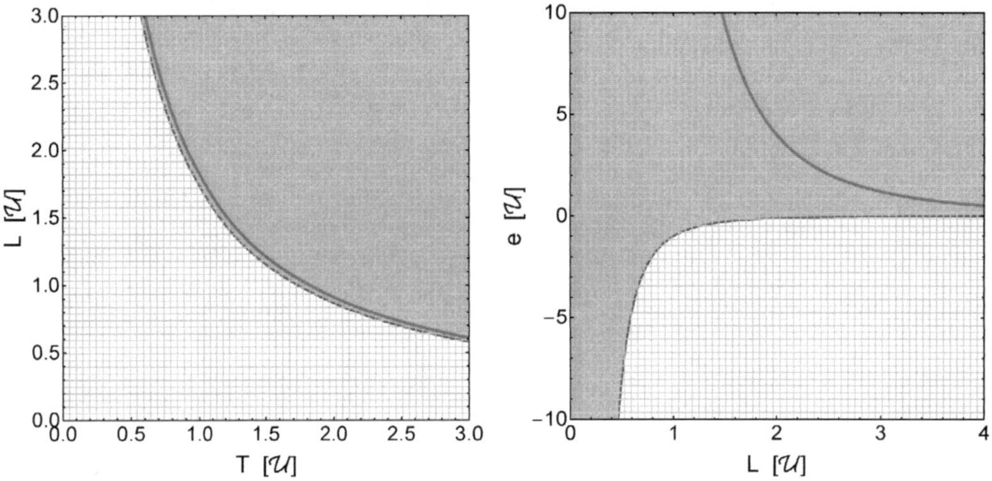

Figure 3. (Left) The phase diagram of the $\{T, L\}$-ensemble, with same notation as in figure 1. We notice the ensemble equivalence is no restricted by the necessity to keep the Hessian non singular. A red phase transition line is reported, for guidance purposes, from the divergence of the iso-energetic compressibility (pertaining then to the ensemble $\{e, L\}$). (Right) The phase diagram of the $\{e, L\}$-ensemble, with same notation as in figure 1. We also notice the ensemble equivalence is no restricted by the necessity to keep the Hessian non singular. The red line is no pertaining to the actual second order phase transition.

Finally, in figure 4, we show the divergence of the iso-energetic compressibility as computed from the (correct) $\{e, L\}$-ensemble.

The relevant ensemble for such systems is however not the $\{e, L\}$-ensemble, but the $\{e, p\}$-one, and we refer to the reference [1] for a complete treatment using a somewhat more realistic model.

4. Conclusion

We have reported some evidence about the possible existence of phase transitions in very different finite size systems: First, isothermal Casimir systems with a first order phase transition providing certain mechanical support settings. A further investigation with more realist models is currently underway [2]. Then, a possible second order phase transition into finite size systems relevant to small (quark-)gluon plasma as created in current heavy-ion collisions. A more complete such investigation is reported in reference [1].

Acknowledgments

The authors would like to thank H. Touchette for very useful discussions. S M would like to acknowledge the financial support from the Claude Leon Foundation, as well as the South Africa - CERN (SA-CERN) Collaboration for supporting his work and stay at the conference. I K and

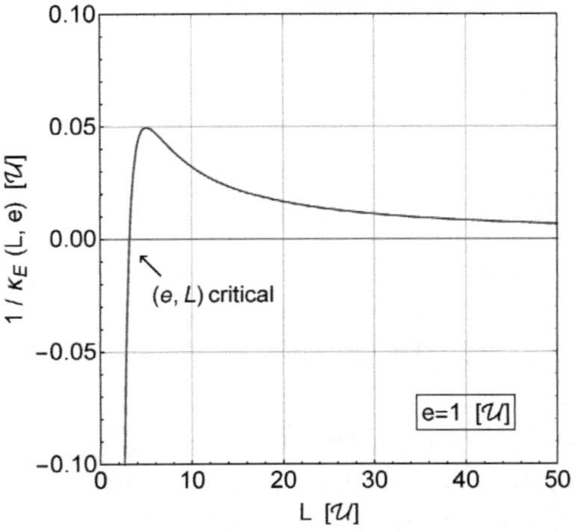

Figure 4. Iso-energetic compressibility obtained from the $\{e, L\}$-ensemble, with critical point.

W A H wish to thank the SA-CERN Collaboration, and the National Research Foundation for their support. I K wishes to further acknowledge the South African National Institute for Theoretical Physics, and the Deutscher Akademischer Austauschdienst for supporting her work.

References

[1] Mogliacci S, Kolbé I and Horowitz W A 2018 (second version) (*Preprint* 1807.07871)
[2] Mogliacci S First Order Phase Transition in Thermal Casimir Systems (to appear)
[3] Mogliacci S, Kolbé I and Horowitz W A 2018 *Universe* **4** 14 (*Preprint* 1801.08187)
[4] Kitazawa M, Mogliacci S, Kolbé I and Horowitz W A 2019 (*Preprint* 1904.00241)
[5] Kolbé I, Horowitz W A and Mogliacci S Small System Corrections to Thermal Field Theory and pQCD Energy Loss (to appear in this proceedings)
[6] Lamoreaux S K 1997 *Phys. Rev. Lett.* **78** 5–8 [Erratum: Phys. Rev. Lett.81,5475(1998)]
[7] Sushkov A O, Kim W J, Dalvit D A R and Lamoreaux S K 2011 *Nature Phys.* **7** 230–233 (*Preprint* 1011.5219)
[8] Ellis R S, Haven K and Turkington B 2000 *Journal of Statistical Physics* **101** 999–1064
[9] Touchette H 2015 *Journal of Statistical Physics* **159** 987–1016
[10] Bordag M, Klimchitskaya G L, Mohideen U and Mostepanenko V M 2009 *Advances in the Casimir effect* vol 145 (Oxford University Press)
[11] Garrett J L, Somers D A T, Sendgikoski K and Munday J N 2018 (*Preprint* 1811.07175)
[12] Touchette H 2009 *Physics Reports* **478** 1 – 69 ISSN 0370-1573 URL http://www.sciencedirect.com/science/article/pii/S0370157309001410

IOP Conf. Series: Journal of Physics: Conf. Series **1271** (2019) 012023 doi:10.1088/1742-6596/1271/1/012023

MAGIC - how MAtter's extreme phases can be revealed in Gravitational wave observations and in relativistic heavy Ion Collision experiments

Matthias Hanauske[1,2], Luke Bovard[1], Jan Steinheimer[2], Anton Motornenko[1,2], Volodymyr Vovchenko[1,2], Stefan Schramm[2], Veronica Dexheimer[3] Jens Papenfort[1], Elias R. Most[1] and Horst Stöcker[*,1,2,4]

[1] Institut für Theoretische Physik, Max-von-Laue-Straße 1, 60438 Frankfurt, Germany
[2] Frankfurt Institute for Advanced Studies, Ruth-Moufang-Straße 1, 60438 Frankfurt, Germany
[3] Department of Physics, Kent State University, Kent, OH 44243 USA
[4] GSI Helmholtzzentrum für Schwerionenforschung GmbH, 64291 Darmstadt, Germany

Abstract. Nearly one hundred years after Albert Einstein developed the field equations of general relativity and predicted the existence of gravitational waves, a gravitational wave event from a binary neutron star merger (GW170817) was detected in August 2017 by the LIGO/VIRGO collaboration. During the thereon analysis of the gravitational wave data, the equation of state of elementary matter could be constrained in the regime of high densities/temperatures. Recent simulations show, that the appearance of a hadron to quark phase transition in the interior region of a hybrid star merger remnant might change the overall properties of the merger event and could be detectable in future. On the one hand, 4D-simulations of binary neutron star mergers show that these astrophysical systems represent optimal laboratories to investigate the phase structure of quantum chromodynamics. On the other hand, accelerators like the FAIR facility at GSI Helmholtzzentrum allow one to study the properties of the quark-gluon plasma produced in relativistic collisions of heavy ions. This article combines a survey of recent advancements in two rather distinct fields, which reveal - on first sight - a surprising similarity of both, namely relativistic collisions of nuclei and of neutron star mergers.

1. Introduction

Neutron stars are, beside white dwarfs and black holes, the potential final states of the evolution of a normal star. These extremely dense astrophysical objects, which are formed in the center of supernova explosions, represent the last stable state before matter collapses to a black hole. In our galaxy, we currently know about 3000 of these exotic objects, which are mainly observed as pulsars by radio telescopes. Among the known neutron stars, there are some which are in binary systems where the companion of the neutron star is either a normal star, a planet, a white dwarf or again a neutron star.

New observational results have made it possible to study strongly interacting elementary matter in extreme astrophysical objects. The detection of a gravitational wave (GW) from a binary neutron star (BNS) merger by the LIGO/VIRGO collaboration (GW170817, [1]) marked the beginning of a new era in observational astrophysics. In August 2017, the GWs and a 1.7 seconds delayed gamma-ray burst (GRB 170817A, [2]) had been detected and, additionally, the electromagnetic counterparts of an associated kilonova [3] have been tracked by numerous observatories around the world. In contrast to electromagnetic waves, GWs travel almost unaffected with the speed of light through our universe. The little space-time oscillations (GWs) of GW170817, for example, had been traveled about 130 million years through space until they were detected in 2017. Right after the neutron stars merge, a remnant is formed, which could be stable for longer times (supramassive neutron star, SMNS) or collapse within less than one second (hypermassive neutron star, HMNS) to a rotating Kerr black hole. During this process a short gamma ray burst is emitted, releasing in less than one second the energy emitted by our Galaxy over one year [4]. Space-based gamma-ray telescopes (e.g., the Fermi's gamma-ray burst monitor or the Swift gamma-ray burst mission) detect on average approximately one gamma-ray burst per day - however, the gamma-ray burst [2], which had been associated with GW170817, is an outstanding event and in addition with the observations of the electromagnetic counterparts of the associated kilonova, provides a conclusive picture of the whole merger event.

General relativistic astrophysics of neutron stars and nuclear/elementary particle physics are strongly connected and numerical simulations in both fields strongly depend on the equation of state (EOS) of fundamental elementary matter [5]. Hot and dense matter created in high energy heavy ion collisions and mergers of a binary system of two neutron stars reach such extreme conditions, that the strongly interacting matter created is expected to undergo the deconfinement phase transition. Recently, a joint group at FIAS, the Goethe University and Kent State University discovered that the emitted gravitational waves, as predicted from general relativistic magneto- hydrodynamics from BNS merger - calculations, are extremely sensitive to the appearance of quark matter and the stiffness of the EOS of matter described by the interaction of quantum chromodynamics (QCD). The inner core of a HMNS, as well as its behavior during the gravitational collapse to a black hole depends strongly on the high density/temperature regime of the EOS [6]. This is a new observable messenger from outer space, which does provide direct signals for the phase structure of strongly interacting QCD matter at high baryon density and high temperature. Those astrophysically created extremes of thermodynamics do match, to within 20%, the values of temperatures which we find in relativistic hydrodynamics and transport theory of heavy ion collisions at the existing laboratories like LHC, SpS at Cern, RHIC at Brookhaven, and HaDes - SiS18 at GSi and at the NICA and FAIR accelerators under construction, if though at quite different rapidity windows, impact parameters and bombarding energies of the heavy nuclear systems. In this article, the simulations of BNS and binary hybrid star mergers will be presented and the different density/temperature regions within the QCD phase diagram will be visualized in order to compare it with the present and future projects of the heavy nuclear experiments.

The paper is structured as follows: After this introduction, Section 2 summarizes briefly the theoretical groundings of a BNS merger simulation. In order to motivate, that the temperature and density values reached inside a BNS merger product requires an incorporation of a hadron-quark phase transition (HQPT), Sec. 3 discusses a BNS merger in the context of the HQPT. The interior temperature and density structure of a neutron star merger product and the evolution of the hot and dense matter inside the HMNS will be analyzed and visualized in a (T-ρ) QCD phase diagram. In addition to former publications [7, 8, 9, 10, 11, 12], the different phases of a BNS merger scenario will be explained using the analogy of a variety of different dances. A brief overview about the astrophysical observables of the HQPT in a binary hybrid star merger event will be given in Section 4. Finally, a summary and outlook will be presented in Sec. 5.

2. Relativistic hydrodynamics and numerical relativity of compact star mergers

In order to calculate the properties of a BNS merger, the gravitational force needs to be implemented in the model of elementary particle physics. We used Einstein's theory of general relativity in connection with the conservation laws for energy-momentum and rest mass as the groundings of the differential equations which we solve numerically on a spatial grid using high performance supercomputers. The Einstein equation and the conservation laws are summarized in the following set of highly non-linear differential equations:

$$G_{\mu\nu} := R_{\mu\nu} - \frac{1}{2} g_{\mu\nu} R = 8\pi \, T_{\mu\nu} \, , \quad \nabla_\mu T^{\mu\nu} = 0 \, , \quad \nabla_\mu \left(\rho \, u^\mu\right) = 0 \quad , \tag{1}$$

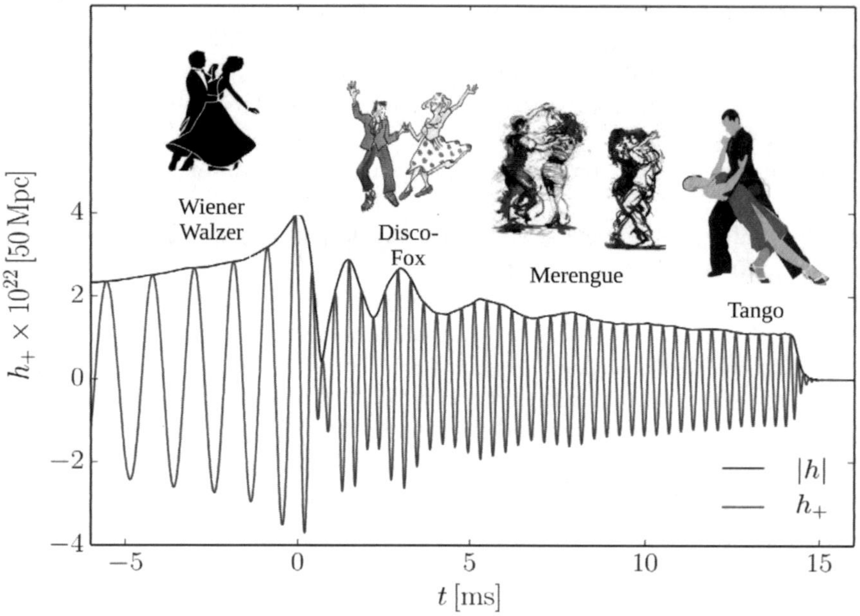

Figure 1. Illustration of the different dances within a BNS merger simulation: Gravitational-wave amplitude $|h|$ (black line) and strain amplitude in the $+$ polarization h_+ (blue line) for the `ALF2-M135` binary at a distance of 50 Mpc.

where $G_{\mu\nu}$ is the Einstein tensor, $T_{\mu\nu}$ is the energy-momentum tensor, $R_{\mu\nu}$ is the Ricci tensor, which contains first and second derivatives of the space-time metric $g_{\mu\nu}$, ∇_μ is the covariant derivative and u^μ is the four velocity of the star's fluid. The Einstein equation (first equation in (1)) describes describe on the one hand how matter moves in a curved space-time and on the other hand formulates in which way the amounts of energy-momentum curves the space-time structure. In the energy-momentum tensor $T_{\mu\nu} = (e + p)\, u_\mu u_\nu + p\, g_{\mu\nu}$ enters the energy and pressure densities of the nuclear and elementary particle physics contributions of the underlying neutron star matter and $u^\mu = dx^\mu / d\tau$ describes the four velocity of the star's fluid which is defined as the derivative of the coordinates $x^\mu = (t, x, y, z)$ by the proper time τ. In order to solve the evolution of a merging neutron star binary system numerically, equation 1 need to be reformulated, because its structure is not well posed. To reformulate equation 1, the so called $(3 + 1)$ split is used, which starts by slicing the 4-dimensional manifold $\mathcal{M}$ into 3-dimensional space-like hypersurfaces Σ_t. The space-time metric $g_{\mu\nu}$ is then sub-classified into a purely spatial metric γ_{ij}, a lapse function α and a shift vector β_i. In addition, to perform a BNS merger simulation an EOS of the compact star matter is needed.

In the following the numerical setup of the simulation of Sec. 3 will be briefly described (for details see [7, 8, 9, 10, 11, 12]). To solve the conformal traceless formulation of the Einstein equations [13, 14, 15] the `McLachlan`-thorn [16, 17] of the publicly available `Einstein Toolkit` [18] has been used. Simultaneously, the general-relativistic hydrodynamics equations are solved using the finite-volume code `Whisky` [19]. An adaptive-mesh refinement (AMR) approach had been employed using the `Carpet` mesh-refinement driver [20]. The grid hierarchy consists of six refinement levels, whose outer boundary is at $514\,M_\odot$ (i.e., $\simeq 759\,\mathrm{km}$). In contrast to [8] no π-symmetry condition across the $x = 0$ plane has been adopted. The initial configuration for the quasiequilibrium irrotational BNSs has been generated with the use of the `LORENE`-code [21] and an initial coordinate separation of the stellar centers of $45\,\mathrm{km}$ had been used. With the exception of Fig. 1, we have used the `LS220` EOS in our simulations and considered a equal-mass binary with a mass of $M_\odot$ ($1.35 M_\odot$ for each star.

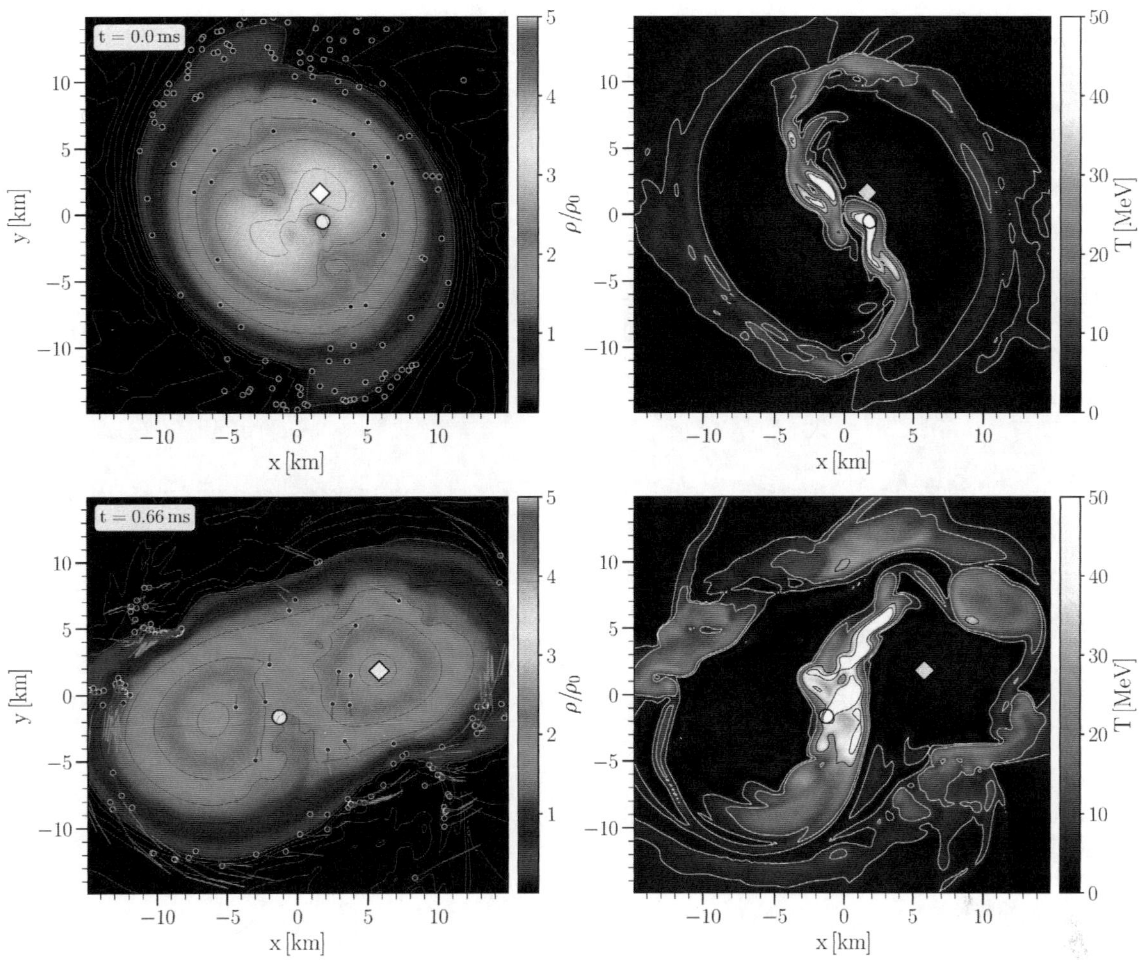

Figure 2. Distributions of the rest-mass density (left pictures) and temperature (right pictures) within the early Disco-fox phase at merger time $t = 0$ (upper panel) and at $t = 0.66$ ms (lower panel). The circles mark the maximum value of the temperature while the diamond symbols indicate the maximum of the density. Additionally, several tracer particles that remain close to the equatorial plane are visualized with small black dots.

3. The different phases of a binary neutron star merger

A BNS merger can be regarded as a great love story (see Fig. 1), ranging from the first moment of perception, the closer approach, the moment of coalescence and fusion, and can be separated in different phases. It begins with the first face-to-face encounter of the two lovers (neutron stars), which is possibly the greatest serendipity. During this inspiral phase (Viennese waltz phase) the two stars are separated by a certain distance and orbit around each other. Additionally, each neutron star rotates around its rotational axis and the whole movement looks similar to a Viennese waltz dance, where each neutron star corresponds to an individual dancer who dances with an invisible companion. Due to the emission of GWs, their separating distance decreases with time. One of the most impressive binary neutron star system is the so called *Double Pulsar: PSR J0737-3039A/B*, which has been discovered in 2003 [22, 23, 24]. The two neutron stars in this binary system, which are only separated by 800,000 km, orbit around each other with an orbital period of 147 minutes and mean velocities of one million km/h. The

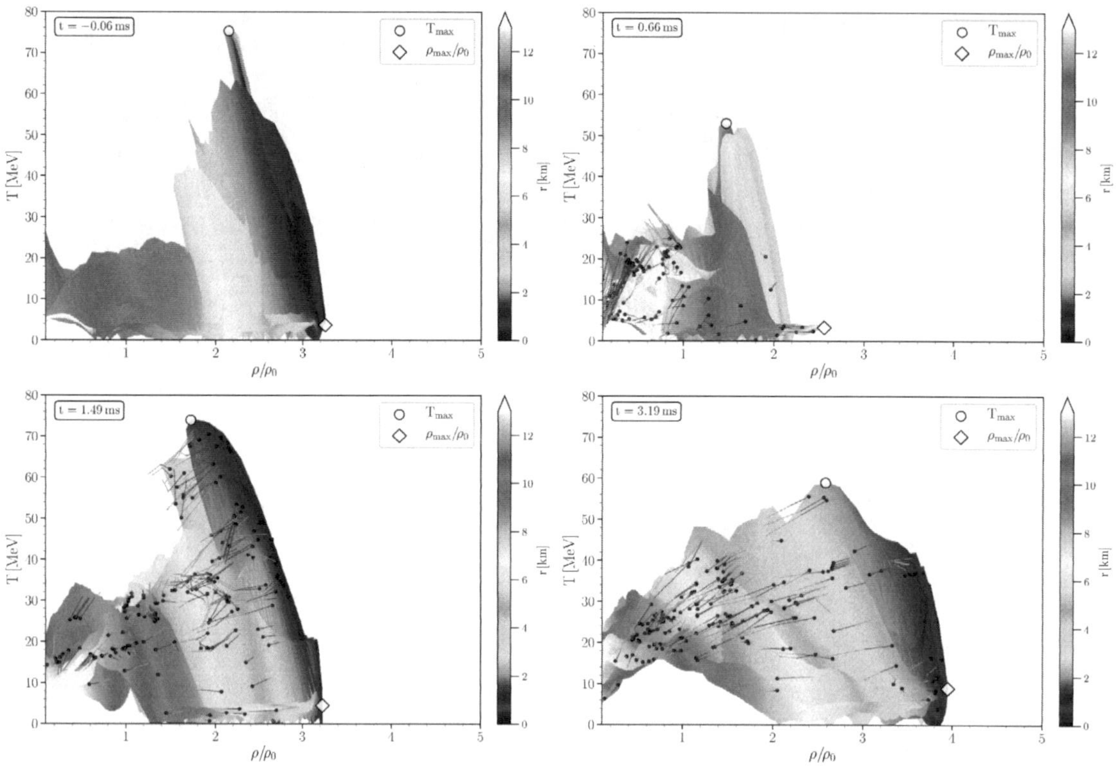

Figure 3. Density-temperature profiles (normalized to saturation density ρ_0) inside the inner region of the HMNS of the LS220-M135 simulation in the style of a (T-ρ) QCD phase diagram plot during the Disco-fox phase. The color-coding indicates the radial position r of the corresponding fluid element inside the HMNS.

distance between the two neutron stars decreases with time and, finally, the two objects need to merge and the two lovers touch each other and become a couple [8]. In this inspiral phase, the temperature effects can be neglected in good approximation and only at the stars surfaces the temperature reaches non-neglectable values. During the last orbits, the stars become tidally deformed and the temperature in the low density regime, near to the region where the two NSs touch each other, increases rapidly. During the last orbits, the stars become tidally deformed and the temperature in the low density regime, near to the region where the two NSs touch each other, increases rapidly [11, 12].

At merger time ($t = 0$ ms), where the emitted gravitational wave of the newly born remnant reaches its maximum value, the hot temperature spot of the newly born remnant reaches values up to $T \approx 75$ MeV (see white circles in the upper panel of Fig. 2). The density maximum (see white diamonds in the upper panel of Fig. 2) are almost at the center and the high temperature regions are placed between them (for details see [8, 11]). The following violent, early post-merger phase (Disco-fox phase) is characterized by a pronounced density double-core structure and hot temperature regions which are smeared out in areas between the double-core density maxima (see lower panel in Fig. 2). The movement of advanced Disco-fox dancers consist of two separate motions: A coming closer and a subsequently removing of the two bodies and a shared rotation with respect to the static dance floor. The two dancers correspond now to the two, still separately visible double-core density maxima of the HMNS (see left picture in the lower panel of Fig. 2) and the movement of the double-core structure, within the first ≈ 2 ms after the merger, looks therefore quite similar to a Disco-fox dance.

IOP Conf. Series: Journal of Physics: Conf. Series **1271** (2019) 012023 doi:10.1088/1742-6596/1271/1/012023

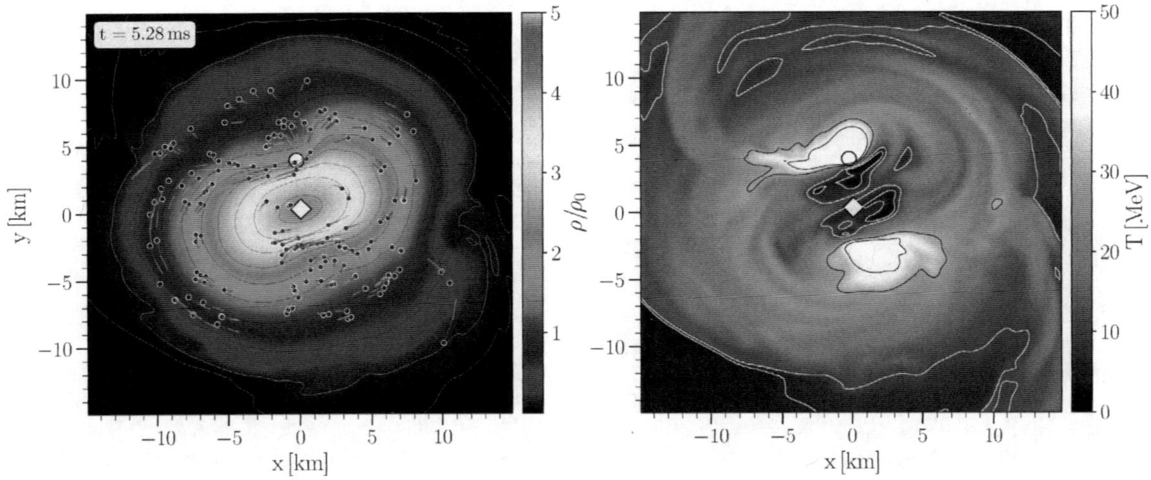

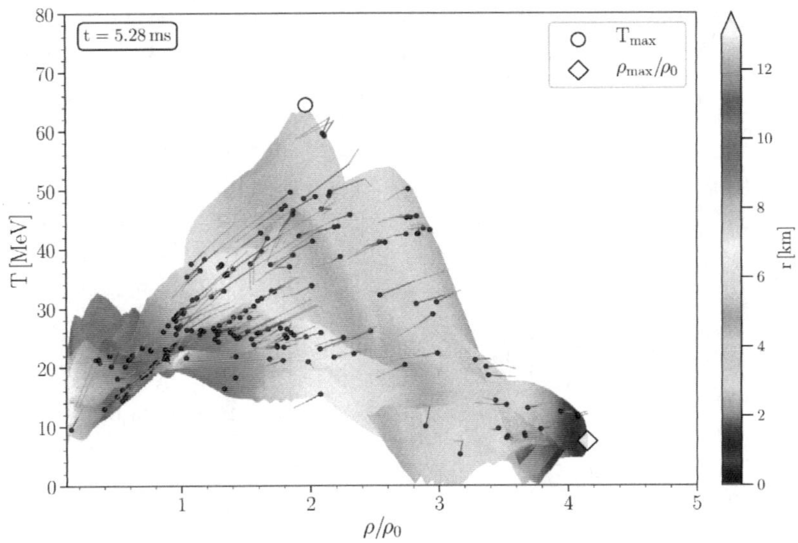

Figure 4. Same as Fig. 2 and Fig. 3 but at $t = 5.28$ ms within the early Merengue phase of a BNS mmerger. At this post-merger time, the double-core structure has disappeared and the density distribution has a 'peanut' shape. The HMNS center has become ultra dense ($\rho/\rho_0 \geq 4$) at moderate temperatures ($T \approx 10$ MeV) and two hot temperature spots ($T \approx 60$ MeV) appear at densities around $\rho/\rho_0 \approx 2$.

Fig. 3 illustrates the spatial allocation of the density and temperature values reached in a (T-ρ/ρ_0) plot during the early post-merger phase (Disco-fox phase). The color of a density/temperature point ((T-ρ)-fluid element) indicates its radial position r measured from the origin of the simulation $(x, y) = (0, 0)$ on the equatorial plane ($z = 0$). In order to track the motion of individual fluid cells, tracer particles have been used within the `LS220-M135` simulation (for details see [25, 26]). Initially placed at $t = 0$ near to the surface of the newly born remnant (see left picture in the upper panel of Fig. 2) the tracer particles diffuse both spatially and in the (T-ρ)-plane, however, in the following, only the tracers that stay in the inner region of the HMNS will be shown. The flowlines of these tracer particles can be visualized using the method of a "corotating frame". In a corotating frame, each grid point is rotating at a frequency that is half the angular frequency of the instantaneous emitted gravitational waves, $\Omega_{\rm GW}$ and it corresponds to the collective rotation of the whole HMNS. In a "corotating frame" the observer is transformed within the rotating frame of the dancing couple and the movement of the two density double-cores (dancers) takes place on a straight line. In a corotating frame the Disko-fox looks more like a West Coast Swing dance and the merger snapshot (see upper panel of Fig. 2) can be regarded as the first "sugar push" of the West Coast Swing couple.

IOP Conf. Series: Journal of Physics: Conf. Series **1271** (2019) 012023 doi:10.1088/1742-6596/1271/1/012023

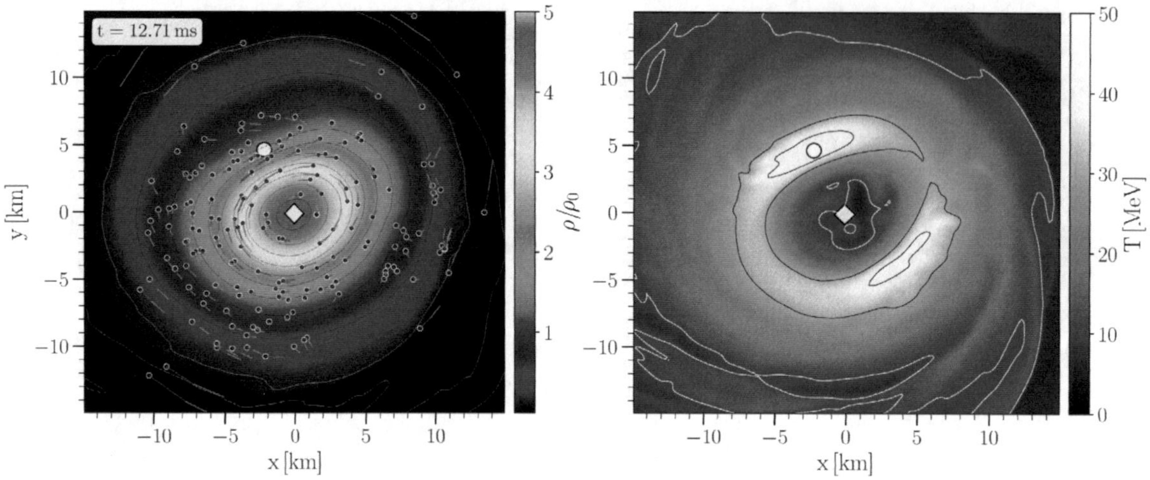

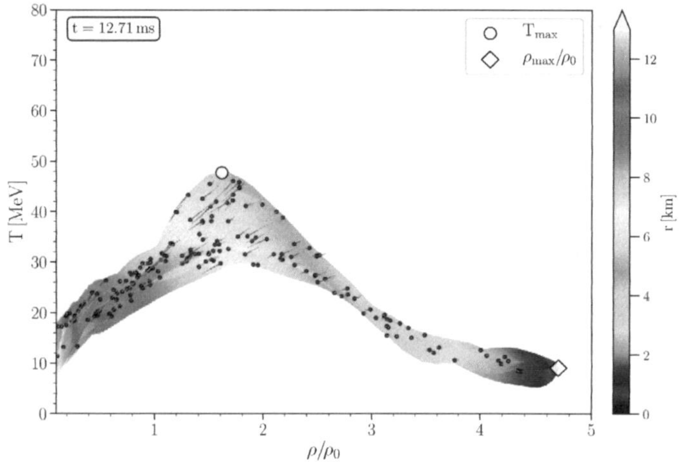

Figure 5. Same as Fig. 4 but at $t = 12.71$ ms within the Merengue phase of a BNS merger. Around this post-merger time, two hot temperature spots smeared out to form a ring-like structure. The neutron star matter populates a smaller area $\mathcal{A}$ in the (T-ρ/ρ_0) plane compared to the early part of the Merengue phase (see Fig. 4). The motions of the tracer particles are restricted within the surface of $\mathcal{A}$ and at its outer borders they are reflected.

After the violent and transforming, early post-merger phase, a new phase begins (the Merengue phase). In the time interval 2 ms $< t <$ 4 ms, the double-core structure disappears and the maximum density value shifts to the central region of the HMNS. The high temperature regions transform into two hot temperature spots and they move further out. The density distribution at this post-merger time has been named 'peanut' shape but the highest value of the density ρ is located in the center of the HMNS. In a Merengue dance, the dancers are so close together that distinguishing the individuals is difficult (no double-core structure). In close embrace, the united couple quickly rotates around each other and this movement describes in a clear way the motion of the HMNS at $t > 4$ ms. The high temperature values ($T > 40$ MeV) are reached now in regions where the density is in a range of $1 - 2.5\rho_0$, while the maximum density values are always at moderate temperatures $T < 20$ MeV. The hot temperature spots have moved further out and the interior of the HMNS, where the maximum of the density is located. The tracer particles have diffused over the entire inner region of the HMNS and populate almost the whole area of the (T-ρ/ρ_0) plane. Some of these tracers circulate around the high temperature hot spots, others populate the low temperature high dense inner region, and some are moving in the outer surface of the HMNS within the low density regime (see Fig. 4).

Fig. 5 shows a snapshot at later times in the Merengue phase ($t = 12.71$ ms). The two pronounces temperature maxima at $t = 5.94$ ms have almost merged to one maximum at this post-merger time (see lower left picture in Fig. 5). At these post-merger times, the temperature hot spots have smeared out to become a ring like structure, the 'peanut' shape has been dissolved and the area populated in the (T-ρ)-plane has been constricted to a small quasi stable region. The central region of the HMNS consists of high dense matter ($\rho/\rho_0 \approx 5$) at moderate temperature values $T \approx 10$ MeV, while the maximum of the temperature is reached at the highest point of the temperature ring like structure at $r \approx 6$ km at moderate density values ($\rho/\rho_0 \approx 2$). The unusual temperature ring-like structure (see right picture in the upper panel of Fig. 5) is closely related with the rotation profile Ω of the differentially rotating HMNS ($\Omega = \alpha\, v^\phi \; - \; \beta^\phi$, where v^ϕ and β^ϕ describes the ϕ-component of the three-velocity and shift vector, see [7] for details).

The results of the `LS220-M135` simulation presented so far show that within the first 20 ms the matter inside the HMNS populate areas in the QCD phase diagram where an inclusion of the quark degrees of freedom in the EOS is necessary. Especially in the interior region of the HMNS for $t > 10$ ms, the density reaches values where a non neglectable amount of deconfined quark matter is expected to be present. If the total mass of the BNS merger system is above a certain value or the EOS behaves differently in its ultra high density regime, the HMNS will collapse to a black hole and this last dance of the two lovers can be understood by a Tango dance, where the motion of the dancers is suddenly freezing and the emission of gravitational wave stops. Fig. 1 summarizes the different dances of a BNS merger simulation and connects them with the emitted gravitational-wave amplitude of a BNS merger simulation (`ALF2-M135` run, for details see [7]). During the collapse of the HMNS to a black hole, the density in the inner region increases rapidly, the quarks get free and the color charge of the deconfiened pure quark phase becomes visible - however, the formation of the event horizon prevents that this new degree of color can be observed from the outside and the whole deconfiened quark phase gets macroscopically confined by general relativity (see p.186-188 in [27]). The event horizon of a black hole marks a certain threshold and the collapse of a neutron star to a black hole looks for an outside observer looks like a frozen picture of the collapsing star - however, the color of this frozen picture will be infinitely gravitationally red-shifted quite rapidly.

4. Binary hybrid star mergers

The thermodynamical origin of the transition from hadronic matter to quark matter is still not known. The appearance of deconfined quarks with increasing chemical potential and/or temperature is determined by the type of the transition. In case of the first order phase transition the quarks appear jump-like, so with increasing chemical potential the number of quarks increases discontinuously. The construction of the phase transition can be done in different ways, using a Gibbs construction between the hadron and quark models or, assuming a large surface tension of the first deconfined quark bubbles, in a sharp discontinuous way, using a Maxwell construction between the two separate models. Within this work we concentrate on the assumption of first order phase transition associated with appearance of the quarks. The crossover type of deconfinement transition is also possible, then quarks appear smoothly without separation into confined/deconfined phases. This mechanism is developed in [28, 29] where the approach that also reproduces the properties of QCD at high temperatures is applied to model the neutron stars where the QCD matter is cold and dense.

The astrophysical observables of the QGP, which are connected with compact stars can be grouped in the following way [27]: Evidence of the HQPT based on mass and radius properties, rotational behaviors, twin star properties and by means of a future gravitational wave detection. The effects of a strong HQPT have been already investigated in the context of static [30, 31, 32] and uniformly rotating hybrid stars [33, 34] and the results show that tremendous changes in the star properties might occur including the existence of a third family of compact stars - the so called "twin stars" [35]. In general, the GW-frequency at the moment of collision in a neutron star merger ($f_{\rm peak}$) is lower than in a hybrid or quark star merger [27, 36].

With the use of the observed tidal deformations of the two neutron stars from the late inspiral phase and other properties of GW170817, the EOS of dense matter could be severely constrained [37, 38, 39, 40, 41, 42, 43, 44, 45, 46, 47, 48]. Unfortunately the ultra dense and high temperature regime of the EOS, which is only reached within the post-merger phase, has not been observed in GW170817. However, in prospecting merger events within the next observing run, these extreme conditions will be possibly detected [49]. During the post-merger time of a BNS merger, the value of central rest-mass density can increase to several times of normal nuclear matter. For such high densities, the EOS is still poorly constrained by observations from heavy-ion collisions. By analyzing the power spectral density profile of the post-merger emission of a future event within the next observing run of the LIGO/VIRGO collaboration, the GW signal can set tight constraints on the high density regime of the EOS of elementary matter [50]. Numerical simulations which includes a density/temperature and composition dependent EOS with a HQPT, the so called hybrid star merger simulations, have only been performed recently.

In [51] a temperature-dependent EOS [52] with a strong HQPT has been used for the first time in a binary hybrid star merger simulation. Within this model it was shown that the phase transition, which happens during the post-merger phase, leads to a hot and dense quark core that, when it collapses to a black hole, produces a ringdown signal different from the hadronic one. The formation of a hot and dense quark core for later post-merger times changes the evolution of the hypermassive hybrid star. In contrast to the numerical results performed within a purely hadronic model, the temperature in the inner core of the merger remnant increases, resulting in a different evolution of the temperature profile. In [36] a different temperature-dependent EOS for a hybrid star model was used in a merger simulation and it was shown that the dominant post-merger GW frequency f_{peak} exhibit a significant deviation from the empirical relation between f_{peak} and the tidal deformability Λ, if a strong first-order phase transition leads to the formation of a gravitationally stable quark matter core in the hypermassive hybrid star. In [53] the numerical results of binary compact star merger simulations have been presented within a two-family scenario assuming the Bodmer-Witten "strange matter" hypothesis'. Especially, within an EOS that includes the possibility of a twin star behavior, the astrophysical observables of a HQPT might be detectable by future detection of compact star merger events [9, 11, 45, 46, 12].

5. Summary and outlook

The possible appearance of a transition from confined hadronic to deconfined quark matter (hadron-quark phase transition, HQPT), and the formation of regions of deconfined quark matter in the interior of a compact star merger product have been discussed within this article. The temperature and density structure of a neutron star merger product and the evolution of hot and dense matter inside the produced HMNS/SMNS advises an incorporation of a HQPT in the EOS. The occurrence of hot temperature regions and their spatial location is closely connected with the rotational properties of the HMNS/SMNS [7]. Additionally, the possibility of a viscousless superfluid quark phase might change the overall properties as viscous dissipation and energy transport can play a significant role in the survival time of the post-merger object [10]. Binary hybrid star mergers represent therefore optimal astrophysical laboratories to investigate the phase structure of QCD and in addition with the observations from heavy-ion collisions will possibly provide a conclusive picture on the QCD phase structure at high density and temperature [8]. We demonstrated how the gravitational wave signals from future advanced LIGO-Virgo - to radiowave signals from SKA-events can be combined with the high multiplicity fluctuation - and flow measurements in heavy ion detectors in the lab to pin down the EOS and the phase structure of dense matter.

Acknowledgments

We would like to thank Luciano Rezzolla. Without his profound knowledge and his comprehensive expertise in the field of numerical relativity and general relativistic hydrodynamics the presented simulations and the whole article would not have been possible. MH gratefully acknowledges support from the Frankfurt Institute for Advanced Studies (FIAS) and the Goethe University Frankfurt, VD acknowledges support from the National Science Foundation under grant PHY-1748621, while HS acknowledges the Judah M. Eisenberg laureatus Professur endowment.

References

[1] Abbott B P *et al.* (LIGO Scientific Collaboration and Virgo Collaboration) 2017 *Phys. Rev. Lett.* **119**(16) 161101 URL `https://link.aps.org/doi/10.1103/PhysRevLett.119.161101`

[2] LIGO Scientific Collaboration, Virgo Collaboration, Gamma-Ray Burst Monitor F and INTEGRAL 2017 *Astrophys. J. Lett.* **848** L13 (*Preprint* `1710.05834`) URL `http://stacks.iop.org/2041-8205/848/i=2/a=L13`

[3] The LIGO Scientific Collaboration, the Virgo Collaboration, Abbott B P, Abbott R, Abbott T D, Acernese F, Ackley K, aAdams C, Adams T, Addesso P and et al (LIGO Scientific Collaboration and Virgo Collaboration) 2017 *Astrophys. J. Lett.* **848** L12 URL `http://stacks.iop.org/2041-8205/848/i=2/a=L12`

[4] Rezzolla L, Giacomazzo B, Baiotti L, Granot J, Kouveliotou C and Aloy M A 2011 *Astrophys. J. Letters* **732** L6 (*Preprint* `1101.4298`)

[5] Hanauske M, Steinheimer J, Bovard L, Mukherjee A, Schramm S, Takami K, Papenfort J, Wechselberger N, Rezzolla L and Stöcker H 2017 *Journal of Physics: Conference Series* vol 878 (IOP Publishing) p 012031

[6] Most E R, Papenfort L J, Dexheimer V, Hanauske M, Schramm S, Stöcker H and Rezzolla L 2019 *Phys. Rev. Lett.* **122** 061101 (*Preprint* `1807.03684`)

[7] Hanauske M, Takami K, Bovard L, Rezzolla L, Font J A, Galeazzi F and Stöcker H 2017 *Phys. Rev. D* **96** 043004 (*Preprint* `1611.07152`)

[8] Hanauske M, Steinheimer J, Bovard L, Mukherjee A, Schramm S, Takami K, Papenfort J, Wechselberger N, Rezzolla L and Stöcker H 2017 *Journal of Physics: Conference Series, Volume 878, Issue 1, article id. 012031 (2017).* vol 878 p 012031

[9] Hanauske M, Yilmaz Z S, Mitropoulos C, Rezzolla L and Stöcker H 2018 *17th International Conference on Strangeness in Quark Matter (SQM 2017), Utrecht, The Netherlands, Edited by Mischke, A.; Kuijer, P.; EPJ Web of Conferences, Volume 171, id.20004* vol 171 p 20004

[10] Alford M G, Bovard L, Hanauske M, Rezzolla L and Schwenzer K 2018 *Phys. Rev. Lett.* **120** 041101

[11] Hanauske M and Bovard L 2018 *Journal of Astrophysics and Astronomy* **39** 45

[12] Hanauske M, Steinheimer J, Motornenko A, Vovchenko V, Bovard L, Most E, Papenfort L, Schramm S and Stöcker H 2019 *Particles* **2** 44–56

[13] Nakamura T, Oohara K and Kojima Y 1987 *Progress of Theoretical Physics Supplement* **90** 1–218

[14] Shibata M and Nakamura T 1995 *Phys. Rev. D* **52** 5428–5444

[15] Baumgarte T W and Shapiro S L 1999 *Phys. Rev. D* **59** 024007 (*Preprint* `gr-qc/9810065`)

[16] Brown D, Diener P, Sarbach O, Schnetter E and Tiglio M 2009 *Phys. Rev. D* **79** 044023 (*Preprint* `0809.3533`) URL `http://dx.doi.org/10.1103/PhysRevD.79.044023`

[17] Löffler F, Faber J, Bentivegna E, Bode T, Diener P, Haas R, Hinder I, Mundim B C, Ott C D, Schnetter E, Allen G, Campanelli M and Laguna P 2012 *Class. Quantum Grav.* **29** 115001 (*Preprint* `arXiv:1111.3344[gr-qc]`)

[18] The Einstein Toolkit Consortium: `einsteintoolkit.org`

[19] Baiotti L, Hawke I, Montero P J, Löffler F, Rezzolla L, Stergioulas N, Font J A and Seidel E 2005 *Phys. Rev. D* **71** 024035 (*Preprint* `gr-qc/0403029`)

[20] Schnetter E, Hawley S H and Hawke I 2004 *Class. Quantum Grav.* **21** 1465–1488 (*Preprint* `gr-qc/0310042`)

[21] Gourgoulhon E, Grandclément P, Taniguchi K, Marck J A and Bonazzola S 2001 *Phys. Rev. D* **63** 064029 (*Preprint* `gr-qc/0007028`)

[22] Lyne A, Burgay M, Kramer M, Possenti A, Manchester R, Camilo F, McLaughlin M, Lorimer D, D'Amico N, Joshi B *et al.* 2004 *Science* **303** 1153–1157

[23] Kramer M and Stairs I 2008 *Annu. Rev. Astron. Astrophys.* **46** 541–572

[24] Burgay M, Kramer M and McLaughlin M 2014 *Bulletin of the Astronomical Society of India* **42** 101–119

[25] Bovard L and Rezzolla L 2017 *Classical and Quantum Gravity* **34** 215005 (*Preprint* `1705.07882`) URL `http://stacks.iop.org/0264-9381/34/i=21/a=215005`

[26] Bovard L, Martin D, Guercilena F, Arcones A, Rezzolla L and Korobkin O 2017 *Phys. Rev. D* **96** 124005 (*Preprint* `1709.09630`)

[27] Hanauske M 2004 *Properties of Compact Stars within QCD-motivated Models* Ph.D. thesis J.W. Goethe-University Frankfurt, Germany

[28] Steinheimer J, Schramm S and Stocker H 2011 *J. Phys.* **G38** 035001 (*Preprint* `1009.5239`)

[29] Motornenko A, Vovchenko V, Steinheimer J, Schramm S and Stoecker H 2018 *27th International Conference on Ultrarelativistic Nucleus-Nucleus Collisions (Quark Matter 2018) Venice, Italy, May 14-19, 2018* (*Preprint* `1809.02000`)

[30] Hanauske M and Greiner W 2001 *General Relativity and Gravitation* **33** 739–755

[31] Mishustin I N, Hanauske M, Bhattacharyya A, Satarov L M, Stöcker H and Greiner W 2003 *Physics Letters B* **552** 1–8 (*Preprint* `hep-ph/0210422`)

[32] Shovkovy I, Hanauske M and Huang M 2003 *Phys. Rev. D* **67** 103004

[33] Banik S, Hanauske M, Bandyopadhyay D and Greiner W 2004 *Phys. Rev. D* **70** 123004

[34] Bhattacharyya A, Ghosh S K, Hanauske M and Raha S 2005 *Phys. Rev. C* **71** 048801

[35] Glendenning N K and Kettner C 2000 *Astron. Astrophys.* **353** L9 (*Preprint* astro-ph/9807155)

[36] Bauswein A, Bastian N U F, Blaschke D B, Chatziioannou K, Clark J A, Fischer T and Oertel M 2018 *ArXiv e-prints* arXiv:1809.01116 (*Preprint* 1809.01116)

[37] Bauswein A, Just O, Janka H T and Stergioulas N 2017 *Astrophys. J. Lett.* **850** L34 (*Preprint* 1710.06843)

[38] Rezzolla L, Most E R and Weih L R 2018 *Astrophys. J. Lett.* **852** L25 (*Preprint* 1711.00314)

[39] Margalit B and Metzger B D 2017 *ApJ* **850** L19 (*Preprint* 1710.05938)

[40] Paschalidis V, Yagi K, Alvarez-Castillo D, Blaschke D B and Sedrakian A 2018 *Phys. Rev. D* **97** 084038 (*Preprint* 1712.00451)

[41] Most E R, Weih L R, Rezzolla L and Schaffner-Bielich J 2018 *Phys. Rev. Lett.* **120** 261103 (*Preprint* 1803.00549)

[42] De S, Finstad D, Lattimer J M, Brown D A, Berger E and Biwer C M 2018 *Phys. Rev. Lett.* **121** 091102 (*Preprint* 1804.08583)

[43] The LIGO Scientific Collaboration and the Virgo Collaboration 2018 *Phys. Rev. Lett.* **121** 161101 (*Preprint* 1805.11581)

[44] Abbott B, Abbott R, Abbott T, Acernese F, Ackley K, Adams C, Adams T, Addesso P, Adhikari R, Adya V *et al.* 2018 *arXiv preprint arXiv:1805.11579*

[45] Christian J E, Zacchi A and Schaffner-Bielich J 2018 *ArXiv e-prints* arXiv:1809.03333 (*Preprint* 1809.03333)

[46] Montana G, Tolos L, Hanauske M and Rezzolla L 2018 *ArXiv e-prints* arXiv:1811.10929 (*Preprint* 1811.10929)

[47] Han S and Steiner A W 2018 *arXiv preprint arXiv:1810.10967*

[48] Kiuchi K, Kyutoku K, Shibata M and Taniguchi K 2019 *arXiv preprint arXiv:1903.01466*

[49] The LIGO Scientific Collaboration, the Virgo Collaboration, Abbott B P, Abbott R, Abbott T D, Acernese F, Ackley K, Adams C, Adams T, Addesso P and et al 2017 *ArXiv e-prints* (*Preprint* 1710.09320)

[50] Rezzolla L and Takami K 2016 *Phys. Rev. D* **93** 124051 (*Preprint* 1604.00246)

[51] Most E R, Papenfort L J, Dexheimer V, Hanauske M, Schramm S, Stöcker H and Rezzolla L 2018 *ArXiv e-prints* arXiv:1807.03684 (*Preprint* 1807.03684)

[52] Dexheimer V A and Schramm S 2010 *Phys. Rev. C* **81** 045201 (*Preprint* 0901.1748)

[53] De Pietri R, Pagliara G, Wiktorowicz G, Traversi S, Drago A, Feo A and Pasquali M 2019 Merger of compact stars in the two-families scenario Tech. rep.

Institute of Physics
Dirac House, Temple Back
Bristol BS1 6BE UK

ISSN: 1742-6588
ISBN 978-1-5108-9188-3

9 781510 891883

CURRAN ASSOCIATES INC.
proceedings.com